Praise for Embedded Android

"This is the definitive book for anyone wanting to create a system based on Android. If you don't work for Google and you are working with the low-level Android interfaces, you need this book."

—*Greg Kroah-Hartman, Core Linux Kernel Developer*

"If you or your team works on creating custom Android images, devices, or ROM mods, you want this book! Other than the source code itself, this is the only place where you'll find an explanation of how Android works, how the Android build system works, and an overall view of how Android is put together. I especially like the chapters on the build system and frameworks (4, 6, and 7), where there are many nuggets of information from the AOSP source that are hard to reverse-engineer. This book will save you and your team a lot of time. I wish we had it back when our teams were starting on the Frozen Yogurt version of Android two years ago. This book is likely to become required reading for new team members working on Intel Android stacks for the Intel reference phones."

—*Mark Gross, Android/Linux Kernel Architect, Platform System Integration/Mobile Communications Group/Intel Corporation*

"Karim methodically knocks out the many mysteries Android poses to embedded system developers. This book is a practical treatment of working with the open source software project on all classes of devices, beyond just consumer phones and tablets. I'm personally pleased to see so many examples provided on affordable hardware, namely BeagleBone, not just on emulators."

—*Jason Kridner, Sitara Software Architecture Manager at Texas Instruments and cofounder of BeagleBoard.org*

"This book contains information that previously took hundreds of hours for my engineers to discover. It is required reading for any new person that is working with Android on my team."

—*Dr. Mark Micire, Researcher in Space and Mobile Field Robotics, Carnegie Mellon University*

"Thanks to this book, for the first time embedded system developers have access to an open and vertically integrated stack that contains everything they need to build robust and high-performing Linux-based products. Android's revolutionary execution model transcends phones and tablets, and its application developer platform is unmatched in the industry for features and development speed. This book will give developers a valuable resource for understanding everything between the application layer and the kernel, and how to extend and change things to create an infinite variety of Androids."

—*Zach Pfeffer*, Tech Lead for Linaro's Android team

"Finally, a book on the Android platform from a systems perspective! There are plenty of books on creating Android applications, but for too long no single, comprehensive source for information on Android's internals. In Embedded Android, Karim has collected a vast quantity of material that is essential and helpful for Android systems programmers and integrators (although, to be sure, application developers would benefit from a reading as well). Karim's copious examples, references, and explanations are gleaned from his extensive experience with and analysis of Android. It's the book I wish I had had when I walked my own trail of tears learning Android for work at Sony. With this book, I could have saved myself months learning the ins and outs of Android. No doubt this will be the canonical reference book for Android system developers for years to come."

—*Tim Bird*, Senior Staff Engineer, Sony Network Entertainment, and Architecture Group Chair, CE Workgroup of the Linux Foundation

"Karim Yaghmour's book is an excellent guide for those wishing to get into the burgeoning field of Android-based embedded projects and products. The book covers the full range from kernel support through licensing and trademark issues, including information on running Android systems in "headless" mode as well. This book deserves a place on every serious embedded Android developer's bookshelf."

—*Paul E. McKenney*, IBM Distinguished Engineer and Linux Kernel RCU Maintainer

"Although Android is officially designed for mobile and tablet segments, it's unquestionably getting considered for many other product segments, like automotive, UI panels like HMI, wearable gadgets, and so on. This book is highly recommended, as it covers all the essential fundamentals and concepts that help developers port and develop Android-based solutions for both mobile and nonmobile product segments."

—*Khasim Syed Mohammed*, Lead Engineer, Texas Instruments

"A great resource not only for embedded Android developers, but also for Android app developers to learn the wiring below the Java surface."

—*Lars Vogel*, *CEO, vogella GmbH*

"Once again, Karim has hit the nail on the head. If you're interested in porting Android to a new device or just interested in the guts of how Android runs on a piece of hardware, this is the book you've been searching for. This book leads you through all of the facets of build-environment setup, getting the AOSP sources, adding your hardware to the Android sources and deploying a new Android build to the hardware. It discusses the underpinnings of Android including the HAL and how to give your custom hardware support within the Android framework. In short, of all the books on Android, this is the one book that targets the Android device builder rather than Android application developer or end user. I just wish this book would have been available when I first got into Android porting. It could have saved me months of trial and error efforts."

—*Mike Anderson*, *Chief Scientist, The PTR Group, Inc.*

"Embedded Android has been a great resource for our company. It is a must-have when porting Android to new hardware or integrating new features at a low level. Karim is a great instructor, and his writing captures his style well."

—*Jim Steele*, *VP of Engineering, Sensor Platforms*

"Embedded Android is a must-read for anyone who wants to seriously work the Android internals and bring up Android on new platforms. It helps in navigating the extensive AOSP codebase, and understanding the overall architecture and design of the system."

—*Balwinder Kaur*, *Senior Member, Technical Staff, Aptina Imaging*

"So you thought you knew about Android internals? Well, think again! Chapter after chapter, you'll discover what's behind the scenes and why Android is not just another embedded Linux distribution. Get yourself ready for stepping into a whirlpool, 'cause Embedded Android is a gold mine for anyone looking to do serious hacking on Google's OS."

—*Benjamin Zores*, *Android Platform Architect, Alcatel-Lucent*

"Definitely one of the most valuable and complete resources about the Android system stack. A must-have for every Android system engineer."

—*Maxime Ripard*, Android Lead, Free Electrons

"When I was handed a development board running Linux, and was told to 'get Android running on it,' it was difficult to find much information about how to bring Android up on a new device. Luckily for me, Embedded Android became available about the same time that I was beginning development. What a lifesaver! Embedded Android gave me the kick-start I needed to understand the underpinnings of Android and what I would need to do to bring Android up on a new piece of hardware. I loved all the details and background, from the boot sequence to the build system. After having read Embedded Android, I felt I had a much better grasp of Android and how it interacted with the Linux kernel."

—*Casey Anderson*, Embedded Systems Architect, Trendril

Embedded Android

Karim Yaghmour

Beijing · Cambridge · Farnham · Köln · Sebastopol · Tokyo

Embedded Android

by Karim Yaghmour

Copyright © 2013 Karim Yaghmour. All rights reserved.

Printed in the United States of America.

Published by O'Reilly Media, Inc., 1005 Gravenstein Highway North, Sebastopol, CA 95472.

O'Reilly books may be purchased for educational, business, or sales promotional use. Online editions are also available for most titles (*http://my.safaribooksonline.com*). For more information, contact our corporate/institutional sales department: 800-998-9938 or *corporate@oreilly.com*.

Editors: Andy Oram and Mike Hendrickson
Production Editor: Kara Ebrahim
Copyeditor: Rebecca Freed
Proofreader: Julie Van Keuren

Indexer: Bob Pfahler
Cover Designer: Randy Comer
Interior Designer: David Futato
Illustrator: Rebecca Demarest

March 2013:   First Edition

Revision History for the First Edition:

2013-03-11:   First release

See *http://oreilly.com/catalog/errata.csp?isbn=9781449308292* for release details.

Nutshell Handbook, the Nutshell Handbook logo, and the O'Reilly logo are registered trademarks of O'Reilly Media, Inc. *Embedded Android*, the image of a Moorish wall gecko, and related trade dress are trademarks of O'Reilly Media, Inc.

Many of the designations used by manufacturers and sellers to distinguish their products are claimed as trademarks. Where those designations appear in this book, and O'Reilly Media, Inc., was aware of a trademark claim, the designations have been printed in caps or initial caps.

While every precaution has been taken in the preparation of this book, the publisher and author assume no responsibility for errors or omissions, or for damages resulting from the use of the information contained herein.

ISBN: 978-1-449-30829-2

[LSI]

To Anaïs, Thomas, and Vincent.

May your journeys be filled with the joys of sharing and discovery.

Table of Contents

Preface. xi

1. Introduction. 1
   History   1
   Features and Characteristics   2
   Development Model   5
      Differences From "Classic" Open Source Projects   5
      Feature Inclusion, Roadmaps, and New Releases   7
   Ecosystem   7
      A Word on the Open Handset Alliance   8
   Getting "Android"   9
   Legal Framework   10
      Code Licenses   10
      Branding Use   13
      Google's Own Android Apps   15
      Alternative App Markets   15
      Oracle versus Google   15
      Mobile Patent Warfare   16
   Hardware and Compliance Requirements   17
      Compliance Definition Document   18
      Compliance Test Suite   21
   Development Setup and Tools   22

2. Internals Primer. 25
   App Developer's View   25
      Android Concepts   26
      Framework Intro   30
      App Development Tools   31
      Native Development   32

Overall Architecture	33
Linux Kernel	34
Wakelocks	36
Low-Memory Killer	37
Binder	39
Anonymous Shared Memory (ashmem)	40
Alarm	41
Logger	42
Other Notable Androidisms	45
Hardware Support	46
The Linux Approach	46
Android's General Approach	47
Loading and Interfacing Methods	49
Device Support Details	51
Native User-Space	52
Filesystem Layout	53
Libraries	54
Init	57
Toolbox	58
Daemons	59
Command-Line Utilities	60
Dalvik and Android's Java	60
Java Native Interface (JNI)	63
System Services	63
Service Manager and Binder Interaction	68
Calling on Services	70
A Service Example: the Activity Manager	70
Stock AOSP Packages	71
System Startup	73

3. AOSP Jump-Start 79

Development Host Setup	79
Getting the AOSP	80
Inside the AOSP	86
Build Basics	91
Build System Setup	91
Building Android	94
Running Android	99
Using the Android Debug Bridge (ADB)	101
Mastering the Emulator	105

4. The Build System 111

Comparison with Other Build Systems	111
Architecture	113
Configuration	115
envsetup.sh	118
Function Definitions	124
Main Make Recipes	125
Cleaning	127
Module Build Templates	128
Output	132
Build Recipes	134
The Default droid Build	134
Seeing the Build Commands	134
Building the SDK for Linux and Mac OS	135
Building the SDK for Windows	136
Building the CTS	136
Building the NDK	137
Updating the API	138
Building a Single Module	139
Building Out of Tree	140
Building Recursively, In-Tree	142
Basic AOSP Hacks	143
Adding a Device	143
Adding an App	148
Adding an App Overlay	149
Adding a Native Tool or Daemon	150
Adding a Native Library	151

5. Hardware Primer 155

Typical System Architecture	155
The Baseband Processor	157
Core Components	158
Real-World Interaction	159
Connectivity	160
Expansion, Development, and Debugging	160
What's in a System-on-Chip (SoC)?	161
Memory Layout and Mapping	165
Development Setup	169
Evaluation Boards	171

6. Native User-Space 175

Filesystem	175
The Root Directory	179

/system	180
/data	182
SD Card	185
The Build System and the Filesystem	185
adb	191
Theory of Operation	191
Main Flags, Parameters, and Environment Variables	193
Basic Local Commands	194
Device Connection and Status	195
Basic Remote Commands	197
Filesystem Commands	202
State-Altering Commands	204
Tunneling PPP	207
Android's Command Line	208
The Shell Up to 2.3/Gingerbread	209
The Shell Since 4.0/Ice-Cream Sandwich	210
Toolbox	211
Core Native Utilities and Daemons	220
Extra Native Utilities and Daemons	227
Framework Utilities and Daemons	228
Init	228
Theory of Operation	228
Configuration Files	230
Global Properties	238
ueventd	243
Boot Logo	245

7. Android Framework. 249

Kick-Starting the Framework	250
Core Building Blocks	250
System Services	254
Boot Animation	257
Dex Optimization	260
Apps Startup	262
Utilities and Commands	266
General-Purpose Utilities	266
Service-Specific Utilities	278
Dalvik Utilities	292
Support Daemons	297
installd	298
vold	299
netd	301

	rild	302
	keystore	303
	Other Support Daemons	304
	Hardware Abstraction Layer	304

A. Legacy User-Space.. 307

B. Adding Support for New Hardware.. 323

C. Customizing the Default Lists of Packages.................................... 337

D. Default init.rc Files.. 341

E. Resources.. 367

Index.. 373

Preface

Android's growth is phenomenal. In a very short time span, it has succeeded in becoming one of the top mobile platforms in the market. Clearly, the unique combination of open source licensing, aggressive go-to-market, and trendy interface is bearing fruit for Google's Android team. Needless to say, the massive user uptake generated by Android has not gone unnoticed by handset manufacturers, mobile network operators, silicon manufacturers, and app developers. Products, apps, and devices "for," "compatible with," or "based on" Android seem to be coming out ever so fast.

Beyond its mobile success, however, Android is also attracting the attention of yet another, unintended crowd: embedded systems developers. While a large number of embedded devices have little to no human interface, a substantial number of devices that would traditionally be considered "embedded" do have user interfaces. For a goodly number of modern machines, in addition to pure technical functionality, developers creating user-facing devices must also contend with human-computer interaction (HCI) factors. Therefore, designers must either present users with an experience they are already familiar with or risk alienating users by requiring them to learn a lesser-known or entirely new user experience. Before Android, the user interface choices available to the developers of such devices were fairly limited and limiting.

Clearly, embedded developers prefer to offer users an interface they are already familiar with. Although that interface might have been window-based in the past—and hence a lot of embedded devices were based on classic window-centric, desktop-like, or desktop-based interfaces—Apple's iOS and Google's Android have forever democratized the use of touch-based, iPhone-like graphical interfaces. This shift in user paradigms and expectations, combined with Android's open source licensing, have created a groundswell of interest about Android within the embedded world.

Unlike Android app developers, however, developers wanting to do any sort of platform work in Android, including porting or adapting Android to an embedded device, rapidly run into quite a significant problem: the almost total lack of documentation on how to do that. So, while Google provides app developers with a considerable amount of online documentation, and while there are a number of books on the topic, such as O'Reilly's *Learning Android*, embedded developers have to contend with the minimalistic set of documents provided by Google at *http://source.android.com*. In sum, embedded developers seriously entertaining the use of Android in their systems were essentially reduced to starting with Android's source code.

The purpose of this book is to remedy that situation and to enable you to embed Android in any device. You will, therefore, learn about Android's architecture, how to navigate its source code, how to modify its various components, and how to create your own version for your particular device. In addition, you will learn how Android integrates into the Linux kernel and understand the commonalities and differences it has with its Linux roots. For instance, we will discuss how Android leverages Linux's driver model to create its very own hardware layer and how to take "legacy" Linux components such as glibc and BusyBox and package them as part of Android. Along the way, you will learn day-to-day tips and tricks, such as how to use Android's *repo* tool and how to integrate with or modify Android's build system.

Learning How to Embed Android

I've been involved with open source software since the mid-'90s. I was fortunate enough to join in before it became recognized as the powerful software movement that it is today and, therefore, witness its rise firsthand in the early 2000s. I've also made my share of open source contributions and, yes, participated in a couple of, shall we say, colorful flame wars here and there. Among other things, I also wrote the first edition of O'Reilly's *Building Embedded Linux Systems*.

So when Android—which I knew was Linux-based—started becoming popular, I knew enough about Linux's history and embedded Linux to know that it was worth investigating. Then, I was naively thinking: "I know Linux fairly well and Android is based on Linux; how hard could it be?" That is, until I actually started to seriously look into and, most importantly, inside Android. That's when I realized that Android was very foreign. Little of what I knew about Linux and the packages it's commonly used with in embedded systems applied to Android. Not only that, but the abstractions built in Android were even weirder still.

So began a very long (and ongoing) quest to figure things out. How does Android work? How is it different from regular Linux? How can I customize it? How can I use it in an embedded system? How do I build it? How does its app development API translate into what I know about Linux's user-space? etc. And the more I dug into Android, the more alien it felt and the more questions I had.

The first thing I did was to actually go to *http://developer.android.com* and *http://source.android.com* and print out everything I could get my hands on, save for the actual developer API reference. I ended up with a stack of about 8 to 10 inches of paper. I read through most of it, underlined a lot of the key passages I found, added plenty of notes in the margins, and created a whole list of questions I couldn't find answers for. In parallel, I started exploring the sources made available by Google through the Android Open Source Project (AOSP). In all honesty, it took me about 6 to 12 months before I actually started feeling confident enough to navigate within the AOSP.

The book you presently hold is a result of the work I've done on Android since starting to explore it—including the various projects I've been involved in, such as helping different development teams customizing Android for use in their embedded designs. And I've learned enough about Android to say this: By no means is this book exhaustive. There are a lot of things about Android and its internals that this book doesn't and can't cover. This book should, nevertheless, allow you to jump-start your efforts in molding Android to fit your needs.

Audience for This Book

This book is primarily geared toward developers who intend to create embedded systems based on Android or who would like to take Android and customize it for specific uses. It's assumed you know about embedded systems development and have at least a good handle on how Linux works and how to interact with its command line.

I don't assume you have any knowledge of Java, and you can get away without knowing Java for quite a few of the tasks required to customize Android. However, as your work within Android progresses, you'll find it necessary to start becoming familiar with Java to a certain degree. Indeed, many of Android's key parts are written in Java, and you'll therefore need to learn the language in order to properly integrate most additions to specific parts of the stack.

This book isn't, however, about either app development or Java programming in any way. If these are the topics you are interested in, I recommend you look elsewhere. There are quite a few books on each of these topics already available. This book isn't about embedded systems, either, and there are books on that topic, too. Finally, this book isn't about embedded Linux, which also has its own books. Still, being familiar with Linux's use in embedded systems is something of a plus when it comes to Android. Indeed, though Android is a departure from all things traditionally known as "embedded Linux," many of the techniques typically used for creating embedded Linux systems can guide and help in the creation of embedded Android systems.

This book will also be helpful to you if you're interested in understanding Android's internals. Indeed, customizing Android for use in embedded systems requires knowing at least some basics about its internals. So while the discussion isn't geared toward a thorough exploration of Android's sources, the explanations do show how to interact with the various parts of the Android stack at a fairly intimate level.

Organization of the Material

Like many other titles, this book gradually builds in complexity as it goes, with the early chapters serving as background material for later chapters. If you're a manager and just want to grab the essentials, or if you're wondering which set of chapters you have to read through before you can start skipping chapters and read material selectively, I recommend you at least read through the first three chapters. That doesn't mean that the rest isn't relevant, but the content is much more modular after that.

Chapter 1, *Introduction*, covers the general things you should know about Android's use in embedded systems, such as where it comes from, how its development model and licensing differ from conventional open source projects, and the type of hardware required to run Android.

Chapter 2, *Internals Primer*, digs into Android's internals and exposes you to the main abstractions it comprises. We start by introducing the app development model that app developers are accustomed to. Then we dig into the Android-specific kernel modifications, how hardware support is added in Android, the Android native user-space, Dalvik, the system server, and the overall system startup.

Chapter 3, *AOSP Jump-Start*, explains how to get the Android sources from Google, how to compile them into a functional emulator image, and how to run that image and shell into it. Using the emulator is an easy way to explore Android's underpinnings without requiring actual hardware.

Chapter 4, *The Build System*, provides a detailed explanation of Android's build system. Indeed, unlike most open source projects out there, Android's build system is nonrecursive. This chapter explains the architecture of Android's build system, how it's typically used within the AOSP, and how to add your own modifications to the AOSP.

Chapter 5, *Hardware Primer*, introduces you to the types of hardware for which Android is designed. This includes covering the System-on-Chips (SoCs) typically used with Android, the memory layout of typical Android systems, the typical development setup to use with Android, and a couple of evaluation boards you can easily use for prototyping embedded Android systems.

Chapter 6, *Native User-Space*, covers the root filesystem layout, the *adb* tool, Android's command line, and its custom *init*.

Chapter 7, *Android Framework*, discusses how the Android Framework is kick-started, the utilities and commands used to interact with it, and the support daemons required for it to operate properly.

Appendix A, *Legacy User-Space*, explains how to get a legacy stack of "embedded Linux" software to coexist with Android's user-space.

Appendix B, *Adding Support for New Hardware*, shows you how to extend the Android stack to add support for new hardware. This includes showing you how to add a new system service and how to extend Android's Hardware Abstraction Layer (HAL).

Appendix C, *Customizing the Default Lists of Packages*, provides you with pointers to help you customize what's included by default in AOSP-generated images.

Appendix D, *Default init.rc Files*, contains a commented set of the default *init.rc* files used in version 2.3/Gingerbread and version 4.2/Jelly Bean.

Appendix E, *Resources*, lists a number of resources you may find useful, such as websites, mailing lists, books, and events.

Software Versions

If you hadn't already guessed it when you picked up this book, the versions we cover here are likely way behind the current Android version. And that is likely to be the case forever forward. In fact, I don't ever expect any version of this book to be able to apply to the latest release of Android. The reason is very simple: Android releases occur every six months. It took almost two years to write this book and, from past experience, it takes anywhere from six months to a year, if not more, to update an existing title to the latest version of the software it covers.

So either you stop reading right now and return this book right away, or you read on for a cogent explanation on how to best use this book despite its almost guaranteed obsolescence.

Despite its very rapid release cycle, Android's internal architecture and the procedures for building it have remained almost unchanged since its introduction about five years ago. So while this book was first written with 2.3/Gingerbread in mind, it's been relatively straightforward to update it to also cover 4.2/Jelly Bean with references included to other versions, including 4.0/Ice-Cream Sandwich and 4.1/Jelly Bean where relevant. Hence, while new versions add new features, and many of the software components we discuss here will be enriched with every new version, the underlying procedures and mechanisms are likely to remain applicable for quite some time still.

Therefore, while you can be assured that I am committed to continuing to monitor Android's development and updating this title as often as I humanly can, you should still be able to benefit from the explanations contained in this book for quite a few more versions than the ones covered.

Some actually expect 2.3/Gingerbread to be around for a very long time given that its hardware requirements are much more modest than later versions. At the AnDevCon IV conference in December 2012, for instance, the keynote speaker from Facebook explained that it expected to have to support its app on devices running 2.3/Gingerbread for a very long time, given that that version runs on cheaper hardware than more recent versions.

Conventions Used in This Book

The following typographical conventions are used in this book:

Italic
: Indicates new terms, URLs, email addresses, filenames, and file extensions.

`Constant width`
: Used for program listings, as well as within paragraphs to refer to program elements such as variable or function names, databases, data types, environment variables, statements, and keywords.

`Constant width bold`
: Shows commands or other text that should be typed literally by the user.

`Constant width italic`
: Shows text that should be replaced with user-supplied values or by values determined by context.

 This icon signifies a tip, suggestion, or general note.

 This icon indicates a warning or caution.

Using Code Examples

This book is here to help you get your job done. In general, you may use the code in this book in your programs and documentation. You do not need to contact us for permission unless you're reproducing a significant portion of the code. For example, writing a program that uses several chunks of code from this book does not require permission. Selling or distributing a CD-ROM of examples from O'Reilly books does require permission. Answering a question by citing this book and quoting example code

does not require permission. Incorporating a significant amount of example code from
this book into your product's documentation does require permission.

We appreciate, but do not require, attribution. An attribution usually includes the title,
author, publisher, and ISBN. For example: "*Embedded Android* by Karim Yaghmour
(O'Reilly). Copyright 2013 Karim Yaghmour, 978-1-449-30829-2."

If you feel your use of code examples falls outside fair use or the permission given above,
feel free to contact us at *permissions@oreilly.com*.

Safari® Books Online

Safari Books Online is an on-demand digital library that lets you
easily search over 7,500 technology and creative reference books and
videos to find the answers you need quickly.

With a subscription, you can read any page and watch any video from our library online.
Read books on your cell phone and mobile devices. Access new titles before they are
available for print, and get exclusive access to manuscripts in development and post
feedback for the authors. Copy and paste code samples, organize your favorites, download chapters, bookmark key sections, create notes, print out pages, and benefit from
tons of other time-saving features.

O'Reilly Media has uploaded this book to the Safari Books Online service. To have full
digital access to this book and others on similar topics from O'Reilly and other publishers, sign up for free at *http://my.safaribooksonline.com*.

How to Contact Us

Please address comments and questions concerning this book to the publisher:

> O'Reilly Media, Inc.
> 1005 Gravenstein Highway North
> Sebastopol, CA 95472
> 800-998-9938 (in the United States or Canada)
> 707-829-0515 (international or local)
> 707-829-0104 (fax)

We have a web page for this book, where we list errata, examples, and any additional
information. You can access this page at *http://oreil.ly/embedded-android*.

To comment or ask technical questions about this book, send email to:

> *bookquestions@oreilly.com*

For more information about our books, courses, conferences, and news, see our website at *http://www.oreilly.com*.

Find us on Facebook: *http://facebook.com/oreilly*

Follow us on Twitter: *http://twitter.com/oreillymedia*

Watch us on YouTube: *http://www.youtube.com/oreillymedia*

Acknowledgments

This is my second book ever and my first in 10 years. I'm somewhat skeptical about self-diagnosis, especially when I'm doing it myself—as I'm doing right here, but I clearly seem to have a tendency to be naively drawn to exploring uncharted territory. When I set out to write my first book, *Building Embedded Linux Systems*, in 2001, there wasn't any book describing in full what embedded Linux was about. It took me two years to write down what was in fact half the material I originally thought would take me one year to write. In the same way, there was practically no information about embedded Android when I set out to write the present book in 2011. Somewhat coincidentally, it's taken me two years to finish the manuscript you're presently holding in your hands (or, these days, looking at on your screen, tablet, phone, or whichever device hadn't yet been conceived as I'm writing these lines.)

Overall, I've found that writing books feels like attrition warfare. Maybe that's because of the topics I choose, or maybe it's just my own quirks. Still, akin to attrition warfare, writing books on ambitious topics isn't something that can be done alone. Indeed, when I set out writing this book, I knew but a fraction of what you'll find in these pages. While you can bet that I've done a tremendous amount of research, I should also highlight that what you have here is the result of a very large number of interactions I've had with many talented developers, each of whom taught me a little bit more than what I knew then. Therefore, if you've ever asked me a question at a conference or during a class, or if you've ever explained to me what you're doing with Android or what problems you're encountering with it or, better yet, have sent me in the right direction when I was lost with Android, know that part of you is somewhere in here.

It also takes a special breed of publisher to make this type of book possible. As with my first book, everyone at O'Reilly has simply been exceptional. I would like to first thank Mike Hendrickson for believing in this project and in my ability to deliver it. It's also been a tremendous privilege to once more have the chance to work with Andy Oram as an editor. He's again done a fantastic job at vetting the text you're reading and, oftentimes, pointing out technical issues. In addition to Andy, I'd also like to thank Rachel Roumeliotis and Maria Stallone for gently reminding me to continue pushing this book forward.

Another aspect of writing this type of book is that utmost caution has to be exercised in order to ensure technical accuracy. It was therefore crucial for me to have a strong technical review team. As such, I would like to start by thanking Magnus Bäck, Mark Gross, and Amit Pundir for agreeing very early on in this project to review the book and for having provided generous feedback over the long period when it was written. This initial group was joined along the way by many other talented individuals. Hardware guru David Anders provided key feedback on the hardware chapter. Robert PJ Day did a great job of making sure it made sense for those who've never been exposed to Android. Benjamin Zores ironed out several aspects of the stack's internals. Finally, some readers of the book's early versions, such as Andrew Van Uitert and Maxime Ripard, gracefully shared with me some issues they found along the way.

I would like to most especially thank Linaro's Android team and Bernhard Rosenkränzer specifically for almost single-handedly pointing out the vast majority of discrepancies between the earlier version of this book, which was very 2.3/Gingerbread-centric, and 4.2/Jelly Bean. If you're happy to hold a book that covers two major Android versions, one of which is the latest one at the time of this writing, thank Bernhard. Not only did he force my hand in updating the book, but his input was by far the most extensive— and often the most detailed. I would therefore like to warmly thank Zach Pfeffer for offering his team's help and making it possible for Bernhard to contribute, along with Vishal Bhoj, Fahad Kunnathadi, and YongQin Liu.

As I said earlier, people I've met along the way at conferences have been instrumental in this writing. I would therefore like to single out two organizations that have gone out of their way to make it possible for me to participate in their conferences. First, I'd like to thank the BZ Media team, who've been organizing the AnDevCon conferences since early 2011 and who trusted me early on to talk about Android's internals and have continued inviting me since. Special thanks to Alan Zeichick, Ted Bahr, Stacy Burris, and Katie Serignese. I'd also like to thank the Linux Foundation for giving me the chance to keynote, speak, and participate in a number of events they've been organizing over the years, including the Android Builders Summit, the Embedded Linux Conference, and the Embedded Linux Conference Europe. Special thanks to Mike Woster, Amanda McPherson, Angela Brown, Craig Ross, Maresa Fowler, Rudolf Streif, Dominic Duval, Ibrahim Haddad, and Jerry Cooperstein.

A special thanks also to the team at RevolutionLinux, especially Benoit des Ligneris, Bruno Lambert, and Patrick Turcotte, for agreeing to be my guinea pigs early on. Your trust has borne fruit.

Finally, a very special thanks to Google's Android team for having created one of the best brain-teasers I've run into in a while. I say this sincerely: Exploring this operating system has been one of the funnest things I've done in some time. Kudos to the entire team for creating an amazing piece of software and making it available so generously under such a permissive license. And while I understand this is an unconventional open-source project where transparency isn't (for good reason) on the agenda, I'd like to thank those Android developers who've helped (or in some cases at least tried) in various ways. Thanks to Brian Swetland for filling in the blanks every so often on LWN and to Chet Haase.

These acknowledgments would be incomplete without closing with those who are closest to my heart. Thank you Sonia, Anaïs, Thomas, and Vincent for your loving patience throughout. *Les mains invisibles qui ont écrit les espaces entre les lignes sont les leurs et je leur en suis profondémment reconnaissant.*[1]

---

1. The invisible hands that wrote the spaces between the lines are theirs, and for this I am profoundly grateful to them.

CHAPTER 1
Introduction

Putting Android on an embedded device is a complex task involving an intricate understanding of its internals and a clever mix of modifications to the Android Open Source Project (AOSP) and the kernel on which it runs, Linux. Before we get into the details of embedding Android, however, let's start by covering some essential background that embedded developers should factor in when dealing with Android, such as Android's hardware requirements, as well as the legal framework surrounding Android and its implications within an embedded setting. First, let's look at where Android comes from and how it was developed.

History

The story goes[1] that back in early 2002, Google's Larry Page and Sergey Brin attended a talk at Stanford about the development of the then-new Sidekick phone by Danger Inc. The speaker was Andy Rubin, Danger's CEO at the time, and the Sidekick was one of the first multifunction, Internet-enabled devices. After the talk, Larry went up to look at the device and was happy to see that Google was the default search engine. Soon after, both Larry and Sergey became Sidekick users.

Despite its novelty and enthusiastic users, however, the Sidekick didn't achieve commercial success. By 2003, Rubin and Danger's board agreed it was time for him to leave. After trying out a few things, Rubin decided he wanted to get back into the phone OS business. Using a domain name he owned, *android.com*, he set out to create an open OS for phone manufacturers. After investing most of his savings in the project and

---

1. Coinciding with Android's initial announcement in November 2007, *The New York Times* ran an article entitled "I, Robot: The Man Behind the Google Phone" (*http://www.nytimes.com/2007/11/04/technology/04google.html*) by John Markoff, which gave an insightful background portrait of Andy Rubin and his career. By extension, it provided a lot of insight on the story behind Android. This section is partly based on that article.

having received some additional seed money, he set out to get the company funded. Soon after, in August 2005, Google acquired Android Inc. with little fanfare.

Between its acquisition and its announcement to the world in November 2007, Google released little to no information about Android. Instead, the development team worked furiously on the OS while deals and prototypes were being worked on behind the scenes. The initial announcement was made by the Open Handset Alliance (OHA), a group of companies unveiled for the occasion with its stated mission being the development of open standards for mobile devices and Android being its first product. A year later, in September 2008, the first open source version of Android, 1.0, was made available.

Several Android versions have been released since then, and the OS's progression and development is obviously more public. As we will see later, though, much of the work on Android continues to be done behind closed doors. Table 1-1 provides a summary of the various Android releases and the most notable features found in the corresponding AOSP.

Table 1-1. Android versions

Version	Release date	Codename	Most notable feature(s)	Open source
1.0	September 2008	Unknown		Yes
1.1	February 2009	Unknown[a]		Yes
1.5	April 2009	Cupcake	Onscreen soft keyboard	Yes
1.6	September 2009	Donut	Battery usage screen and VPN support	Yes
2.0, 2.0.1, 2.1	October 2009	Eclair	Exchange support	Yes
2.2	May 2010	Froyo	Just-in-Time (JIT) compile	Yes
2.3	December 2010	Gingerbread	SIP and NFC support	Yes
3.0	January 2011	Honeycomb	Tablet form-factor support	No
3.1	May 2011	Honeycomb	USB host support and APIs	No
4.0	November 2011	Ice-Cream Sandwich	Merged phone and tablet form-factor support	Yes
4.1	June 2012	Jelly Bean	Lots of performance optimizations	Yes
4.2	November 2012	Jelly Bean	Multiuser support	Yes

[a] This version is rumored to have been called "Petit Four." Have a look at this Google+ post (*https://plus.google.com/107797272029781254158/posts/CABJ1RdxH8G*) for more information.

Features and Characteristics

Around the time 2.3.x/Gingerbread was released, Google used to advertise the following features about Android on its developer site:

Application framework
: The application framework used by app developers to create what is commonly referred to as Android apps. The use of this framework is documented online (*http://developer.android.com*) and in books like O'Reilly's *Learning Android*.

Dalvik virtual machine
: The clean-room byte-code interpreter implementation used in Android as a replacement for the Sun Java virtual machine (VM). While the latter interprets *.class* files, Dalvik interprets *.dex* files. These files are generated by the *dx* utility using the *.class* files generated by the Java compiler part of the JDK.

Integrated browser
: Android includes a WebKit-based browser as part of its standard list of applications. App developers can use the `WebView` class to use the WebKit engine within their own apps.

Optimized graphics
: Android provides its own 2D graphics library but relies on OpenGL ES[2] for its 3D capabilities.

SQLite
: This is the standard SQLite database found here (*http://www.sqlite.org*) and made available to app developers through the application framework.

Media support
: Android provides support for a wide range of media formats through StageFright, its custom media framework. Prior to 2.2, Android used to rely on PacketVideo's OpenCore framework.

GSM telephony support[3]
: The telephony support is hardware dependent, and device manufacturers must provide a HAL module to enable Android to interface with their hardware. HAL modules will be discussed in the next chapter.

Bluetooth, EDGE, 3G, and WiFi
: Android includes support for most wireless connection technologies. While some are implemented in Android-specific fashion, such as EDGE and 3G, others are provided in the same way as in plain Linux, as in the case of Bluetooth and WiFi.

---

2. OpenGL *ES* is a version of the OpenGL standard aimed at embedded systems.
3. Android obviously supports more than just GSM telephony. Nevertheless, this is the feature's name as it was officially advertised.

Camera, GPS, compass, and accelerometer
: Interfacing with the user's environment is key to Android. APIs are made available in the application framework to access these devices, and some HAL modules are required to enable their support.

Rich development environment
: This is likely one of Android's greatest assets. The development environment available to developers makes it very easy to get started with Android. A full SDK is freely available to download, along with an emulator, an Eclipse plug-in, and a number of debugging and profiling tools.

There are of course a lot more features that could be listed for Android, such as USB support, multitasking, multitouch, SIP, tethering, voice-activated commands, etc., but the previous list should give you a good idea of what you'll find in Android. Also note that every new Android release brings in its own new set of features. Check the Platform Highlights published with every version for more information on features and enhancements.

In addition to its basic feature set, the Android platform has a few characteristics that make it an especially interesting platform for embedded development. Here's a quick summary:

Broad app ecosystem
: At the time of this writing, there were 700,000 apps in Google Play, previously known as the Android Market. This compares quite favorably to the Apple App Store's 700,000 apps and ensures that you have a large pool to choose from should you want to prepackage applications with your embedded device. Bear in mind that you likely need to enter into some kind of agreement with an app's publisher before you can package that app. The app's availability in Google Play doesn't imply the right for you as a third party to redistribute it.

Consistent app APIs
: All APIs provided in the application framework are meant to be forward-compatible. Hence, custom apps that you develop for inclusion in your embedded system should continue working in future Android versions. In contrast, modifications you make to Android's source code are not guaranteed to continue applying or even working in the next Android release.

Replaceable components
: Because Android is open source, and as a benefit of its architecture, a lot of its components can be replaced outright. For instance, if you don't like the default Launcher app (home screen), you can write your own. More fundamental changes

can also be made to Android. The GStreamer[4] developers, for example, were able to replace StageFright, the default media framework in Android, with GStreamer without modifying the app API.

Extendable
Another benefit of Android's openness and its architecture is that adding support for additional features and hardware is relatively straightforward. You just need to emulate what the platform is doing for other hardware or features of the same type. For instance, you can add support for custom hardware to the HAL by adding a handful of files, as is explained in Appendix B.

Customizable
If you'd rather use existing components, such as the existing Launcher app, you can still customize them to your liking. Whether it be tuning their behavior or changing their look and feel, you are again free to modify the AOSP as needed.

Development Model

When considering whether to use Android, it's crucial that you understand the ramifications its development process may have on any modifications you make to it or to any dependencies you may have on its internals.

Differences From "Classic" Open Source Projects

Android's open source nature is one of its most trumpeted features. Indeed, as we've just seen, many of the software engineering benefits that derive from being open source apply to Android.

Despite its licensing, however, Android is unlike most open source projects in that its development is done mostly behind closed doors. The vast majority of open source projects, for example, have public mailing lists and forums where the main developers can be found interacting with one another, and public source repositories providing access to the main development branch's tip. No such thing can be found for Android.

This is best summarized by Andy Rubin himself: "Open source is different than a community-driven project. Android is light on community-driven, somewhat heavy on open source."

Whether we like it or not, Android is mainly developed within Google by the Android development team, and the public is not privy to either internal discussions nor the tip of the development branch. Instead, Google makes code-drops every time a new version of Android ships on a new device, which is usually every six months. For instance, a

---

4. GStreamer is the default media framework used in most desktop Linux environments, including Gnome, KDE, and XFCE.

few days after the Samsung Nexus S was released in December 2010, the code for the new version of the Android it was running, 2.3/Gingerbread, was made publicly available at *http://android.googlesource.com/*.

Obviously there is a certain amount of discomfort in the open source community with the continued use of the term "open source" in the context of a project whose development model contradicts the standard modus operandi of open source projects, especially given Android's popularity. The open source community has not historically been well served by projects that have adopted a similar development model. Others fear this development model also makes them vulnerable to potential changes in Google's business objectives.

Political issues aside, though, Android's development model means that as a developer, your ability to make contributions to Android is limited. Indeed, unless you become part of the Android development team at Google, you will not be able to make contributions to the tip of the development branch. Also, save for a handful of exceptions, it's unlikely you will be able to discuss your enhancements one-on-one with the core development team members. However, you are still free to submit enhancements and fixes to the AOSP code dumps made available at *http://android.googlesource.com/*.

The worst side effect of Google's approach is that you have absolutely no way to get inside information about the platform decisions being made by the Android development team. If new features are added within the AOSP, for example, or if modifications are made to core components, you will find out how such changes are made and how they impact changes you might have made to a previous version only by analyzing the next code dump. Furthermore, you will have no way to learn about the underlying requirement, restriction, or issue that justified the modification or inclusion. Had this been a true open source project, a public mailing list archive would exist where all this information, or pointers to it, would be available.

That being said, it's important to remember how significant a contribution Google is making by distributing Android under an open source license. Despite its awkward development model from an open source community perspective, it remains that Google's work on Android is a godsend for a large number of developers. Plus, it has accomplished one thing no other open source project was ever able to: created a massively successful Linux distribution. It would, therefore, be hard to fault Android's development team for its work.

Furthermore, it can easily be argued that from a business and go-to-market perspective that a community-driven process would definitely knock the wind out of any product announcements Google would attempt to release, making it impossible to create "buzz" around press announcements and the like, since every new feature would be developed in the open. That is to say nothing of the nondeterministic nature of community-driven processes that can see a group of people take years to agree on the best way to implement a given feature set. And, simply based on track record, Android's success has definitely benefited from Google's ability to rapidly move it forward and to generate press interest based on releases of cool new products.

Feature Inclusion, Roadmaps, and New Releases

In brief, there is no publicly available roadmap for features and capabilities in future Android releases. At best, Google will announce ahead of time the name and approximate release date of the next version. Usually you can expect a new Android release to be made in time for the Google I/O conference, which is typically held in May, and another release by year-end. What will be in that release, though, is anyone's guess.

Typically, however, Google will choose a single manufacturer to work with on the next Android release. During that period, Google will work very closely with that single manufacturer's engineers to ready the next Android version to work on a targeted upcoming lead (or flagship) device. During that period, the manufacturer's team is reported to have access to the tip of the development branch. Once the device is put on the market, the corresponding source code dump is made to the public repositories. For the next release, it chooses another manufacturer and starts over.

There is one notable exception to that cycle: Android 3.x/Honeycomb. In that specific case, Google didn't release the source code to the corresponding lead device, the Motorola Xoom. The rationale seems to have been that the Android development team essentially forked the Android codebase at some point in time to start getting a tablet-ready version of Android out ASAP, in response to market timing prerogatives. Hence, in that version, very little regard was given to preserving backward compatibility with the phone form factor. And given that, Google did not wish to make the code available to avoid fragmentation of its platform. Instead, both phone and tablet form factor support were merged into the subsequent Android 4.0/Ice-Cream Sandwich release.

Ecosystem

As of January 2013:

- 1.3 million Android phones are activated each day, up from 400,000 in June 2011 and 200,000 in August 2010.
- Google Play contains around 700,000 apps. In comparison, the Apple App Store has about the same number of apps.[5]
- Android holds 72% of the global smartphone market.

---

5. At the time of this writing, it's the first time ever that Google Play catches up to the number of apps in the App Store.

Android is clearly on the upswing. In fact, Gartner predicted (*http://reut.rs/UDAEr9*) in October 2012 that Android would be the dominant OS, besting the venerable Windows, by 2016. Much as Linux disrupted the embedded market about a decade ago, Android is poised to make its mark. Not only will it flip the mobile market on its head, eliminating or sidelining even some of the strongest players, but in the embedded space it is likely going to become the de facto standard UI for a vast majority of user-centric embedded devices. There are even signs that it might displace classic "embedded Linux" in headless (non-user-centric) devices.

An entire ecosystem is therefore rapidly building around Android. Silicon and System-on-Chip (SoC) manufacturers such as ARM, TI, Qualcomm, Freescale, and Nvidia have added Android support for their products, and handset and tablet manufacturers such as Motorola, Samsung, HTC, Sony, LG, Archos, Dell, and ASUS ship an ever-increasing number of Android-equipped devices. This ecosystem also includes a growing number of diverse players, such as Amazon, Verizon, Sprint, and Barnes & Noble, creating their own application markets.

Grassroots communities and projects are also starting to sprout around Android, even though it is developed behind closed doors. Many of those efforts are done using public mailing lists and forums, like classic open source projects. Such community efforts typically start by forking the official Android source releases to create their own Android distributions with custom features and enhancements. Such is the case, for instance, with the CyanogenMod (*http://cyanogenmod.com*) project, which provides aftermarket images for power users. There are also efforts by various silicon vendors to provide Android versions enabled or enhanced for their platforms. For example, Linaro—a nonprofit organization created by ARM SoC vendors to consolidate their platform-enablement work—provides its own optimized Android tree. Other efforts follow in the footsteps of phone modders, which essentially rely on hacking the binaries provided by the manufacturers to create their own modifications or variants. Have a look at Appendix E for a full list of AOSP forks and the communities developing them.

A Word on the Open Handset Alliance

As I mentioned earlier, the OHA was the initial vehicle through which Android was first announced. It describes itself on its website as "a group of 82 technology and mobile companies who have come together to accelerate innovation in mobile and offer consumers a richer, less expensive, and better mobile experience. Together we have developed Android, the first complete, open, and free mobile platform."

Beyond the initial announcement, however, it is unclear what role the OHA plays. For example, an attendee at the "Fireside Chat with the Android Team" at Google I/O 2010 asked the panel what privileges were conferred to him as a developer for belonging to a company that is part of the OHA. After asking around the panel, the speaker essentially answered that the panel didn't know because they aren't the OHA. Hence, it would appear that OHA membership benefits are not clear to the Android development team itself.

The role of the OHA is further blurred by the fact that it does not seem to be a full-time organization with board members and permanent staff. Instead, it's just an "alliance." In addition, there is no mention of the OHA within any of Google's Android announcements, nor do any new Android announcements emanate from the OHA. In sum, one would be tempted to speculate that Google likely put the OHA together mainly as a marketing front to show the industry's support for Android, but that in practice it has little to no bearing on Android's development.

Getting "Android"

There are two main pieces required to get Android working on your embedded system: an Android-compatible Linux kernel and the Android Platform.

For a very long time, getting an Android-compatible Linux kernel was a difficult task; it continues to be in some cases at the time of this writing. Instead of using a "vanilla" kernel from *http://kernel.org* to run the Platform, you needed either to use one of the kernels available within the AOSP or to patch a vanilla kernel to make it Android-compatible. The underlying issue was that many additions were made to the kernel by the Android developers in order to allow their custom Platform to work. In turn, these additions' inclusion in the official mainline kernel were historically met with a lot of resistance.

While we'll discuss kernel issues in greater detail in the next chapter, know that starting from the Kernel Summit of 2011 in Prague, the kernel developers decided to proactively seek to mainline the features required to run the Android Platform on top of the official Linux kernel releases. As such, many of the required features have since been merged, while others have been (or, at the time of this writing, are currently being) replaced or superseded by other mechanisms. At the time of this writing, the easiest way to get yourself an Android-ready kernel was to ask your SoC vendor. Indeed, given Android's popularity, most major SoC vendors provide active support for all Android-required components for their products.

The Android Platform is essentially a custom Linux distribution containing the user-space packages that make up what is typically called "Android." The releases listed in Table 1-1 are actually Platform releases. We will discuss the content and architecture of the Platform in the next chapter. For the time being, keep in mind that a Platform release has a role similar to that of standard Linux distributions such as Ubuntu or Fedora. It's a self-coherent set of software packages that, once built, provides a specific user experience with specific tools, interfaces, and developer APIs.

While the proper term to identify the source code corresponding to the Android distribution running on top of an Android-compatible kernel is "Android Platform," it is commonly referred to as "the AOSP"—as is the case in fact throughout this book—even though the Android Open Source Project proper, which is hosted on this site (*http://android.google source.com/*), contains a few more components in addition to the Platform, such as sample Linux kernel trees and additional packages that would not typically be downloaded when the Platform is fetched using the usual *repo* command.

Hacking Binaries

Lack of access to Android sources hasn't discouraged passionate modders from actually hacking and customizing Android to their liking. For example, the fact that Android 3.x/Honeycomb wasn't available didn't preclude modders from getting it to run on the Barnes & Noble Nook. They achieved this by retrieving the executable binaries found in the emulator image provided as part of the Honeycomb SDK and used those as is on the Nook, albeit forfeiting hardware acceleration. The same type of hack has been used to "root" or update versions of various Android components on actual devices for which the manufacturer provides no source code.

Legal Framework

Like any other piece of software, Android's use and distribution is limited by a set of licenses, intellectual property restrictions, and market realities. Let's look at a few of these.

 Obviously I'm not a lawyer and this isn't legal advice. You should talk to competent legal counsel to see how any of the applicable terms or licenses apply to your specific case. Still, I've been around open source software long enough that you could consider what follows as an engineer's educated point of view.

Code Licenses

As we discussed earlier, there are two parts to "Android": an Android-compatible Linux kernel and an AOSP release. Even though it's modified to run the AOSP, the Linux kernel continues to be subject to the same GNU GPLv2 license that it has always been under. As such, remember that you are not allowed to distribute any modifications you make to the kernel under any other license than the GPL. Hence, if you take a kernel version from *http://android.googlesource.com* or your SoC vendor and modify it to make it run on your system, you are allowed to distribute the resulting kernel image in your product only so long as you abide by the GPL. This means you must make the sources used to create the image, including your modifications, available to recipients under the terms of the GPL.

The *COPYING* file in the kernel's sources includes a notice by Linus Torvalds that clearly identifies that only the kernel is subject to the GPL, and that applications running on top of it are **not** considered "derived works." Hence, you are free to create applications that run on top of the Linux kernel and distribute them under the license of your choice.

These rules and their applicability are generally well understood and accepted within open source circles and by most companies that opt to support the Linux kernel or to use it as the basis for their products. In addition to the kernel, a large number of key components of Linux-based distributions are typically licensed under one form or another of the GPL. The GNU C library (glibc) and the GNU compiler (GCC), for example, are licensed under the LGPL and the GPL respectively. Important packages commonly used in embedded Linux systems such as uClibc and BusyBox are also licensed under the LGPL and the GPL.

Not everyone is comfortable with the GNU GPL, however. Indeed, the restrictions it imposes on the licensing of derived works can pose a serious challenge to large organizations, especially given geographic distribution, cultural differences among the various locations of development subunits, and the reliance on external subcontractors. A manufacturer selling a product in North America, for example, might have to deal with dozens, if not hundreds, of suppliers to get that product to the market. Each of these suppliers might deliver a piece that may or may not contain GPL'ed code. Yet the manufacturer whose name appears on the item sold to the customer will be bound to provide the sources to the GPL components regardless of which supplier originated them. In addition, processes must be put in place to ensure that engineers who work on GPL-based projects are abiding by the licenses.

When Google set out to work with manufacturers on its "open" phone OS, therefore, it appears that very rapidly it became clear that the GPL had to be avoided as much as possible. In fact, other kernels than Linux were apparently considered, but Linux was chosen because it already had strong industry support, particularly from ARM silicon manufacturers, and because it was fairly well isolated from the rest of the system, so that its GPL licensing would have little impact.[6]

It was decided, though, that every effort would be made to make sure that the vast majority of user-space components would be based on licenses that did not pose the same logistical issues as the GPL. That is why many of the common GPL- and LGPL-

---

6. See this LWN post by Brian Swetland (*http://lwn.net/Articles/446371/*), a member of Android's kernel development team, for more information on the rationale behind these choices.

licensed components typically found in embedded Linux systems, such as glibc, uClibc, and BusyBox, aren't included in the AOSP. Instead, the bulk of the components created by Google for the AOSP are published under the Apache License 2.0 (a.k.a. ASL) with some key components, such as the Bionic library (a replacement for glibc and uClibc) and the Toolbox utility (a replacement for BusyBox), licensed under the BSD license. Some classic open source projects are also incorporated, mostly as is in source form under their original licensing, into the AOSP within the *external/* directory. This means that parts of the AOSP are made of software that is neither ASL nor BSD. The AOSP does, in fact, still contain GPL and LGPL components. The distribution of the binaries resulting from the compiling of such components, however, should not pose any problems since they aren't meant to be typically customized by the OEM (i.e., no derived works are expected to be created) and the original sources of those components as used in the AOSP are readily available for all to download at *http://android.googlesource.com*, thereby complying, where necessary, with the GPL's requirement that redistribution of derivative works continue being made under the GPL.

Unlike the GPL, the ASL does not require that derivative works be published under a specific license. In fact, you can choose whatever license best suits your needs for the modifications you make. Here are the relevant portions from the ASL (the full license is available from the Apache Software Foundation (*http://www.apache.org/licenses/*)):

- "Subject to the terms and conditions of this License, each Contributor hereby grants to You a perpetual, worldwide, non-exclusive, no-charge, royalty-free, irrevocable copyright license to reproduce, prepare Derivative Works of, publicly display, publicly perform, sublicense, and distribute the Work and such Derivative Works in Source or Object form."
- "You may add Your own copyright statement to Your modifications and may provide additional or different license terms and conditions for use, reproduction, or distribution of Your modifications, or for any such Derivative Works as a whole, provided Your use, reproduction, and distribution of the Work otherwise complies with the conditions stated in this License."

Furthermore, the ASL explicitly provides a patent license grant, meaning that you do not require any patent license from Google for using the ASL-licensed Android code. It also imposes a few "administrative" requirements—such as the need to clearly mark modified files, to provide recipients with a copy of the ASL license, and to preserve *NOTICE* files as is. Essentially, though, you are free to license your modifications under the terms that fit your purpose. The BSD license that covers Bionic and Toolbox allows similar binary-only distribution.

Hence, manufacturers can take the AOSP and customize it to their needs while keeping those modifications proprietary if they wish, so long as they continue abiding by the rest of the provisions of the ASL. If nothing else, this diminishes the burden of having to implement a process to track all modifications in order to provide those modifications back to recipients who would be entitled to request them had the GPL been used instead.

> ## Adding GPL-Licensed Components
>
> Although every effort has been made to keep the GPL out of Android's user-space as much as possible, there are cases where you may want to explicitly add GPL-licensed components to your Android distribution. For example, you want to include either glibc or uClibc, which are POSIX-compliant C libraries—in contrast to Android's Bionic, which is not—because you would like to run preexisting Linux applications on Android without having to port them over to Bionic. Or you may want to use BusyBox in addition to Toolbox, since the latter is much more limited in functionality than the former.
>
> These additions may be specific to your development environment and may be removed in the final product, or they may be permanent fixtures of your own customized Android. No matter which avenue you decide on, whether it be plain Android or Android with some additional GPL packages, remember that you must follow the licenses' requirements.

Branding Use

While being very generous with Android's source code, Google controls most Android-related branding elements more strictly. Let's take a look at some of those elements and their associated terms of use. For the official list, along with the official terms, have a look at this site (*http://bit.ly/Zu5HCV*).

Android robot
:   This is the familiar green robot seen everywhere around all things Android. Its role is similar to the Linux penguin, and the permissions for its use are similarly permissive. In fact, Google states that it "can be used, reproduced, and modified freely in marketing communications." The only requirement is that proper attribution be made according to the terms of the Creative Commons Attribution license.

Android logo
:   This is the set of letters in custom typeface that spell out A-N-D-R-O-I-D and that appear during the device and emulator bootup, and on the Android website (*http://android.com*). You are not authorized to use that logo under any circumstance. Chapter 7 shows you how to replace the bootup logo.

Android custom typeface
:   This is the custom typeface used to render the Android logo, and its use is as restricted as the logo.

"Android" in official names and messaging
   As Google states, " 'Android' by itself cannot be used in the name of an application name or accessory product. Instead use 'for Android.' " Therefore, you can't say "Android MediaPlayer," but you can say "MediaPlayer for Android." Google also states that "Android may be used as a descriptor, as long as it is followed by a proper generic term" such as "Android™ application" for example. Of course, proper trademark attribution must always be made. In sum, you can't name your product "Android Foo" without Google's permission, though "Foo for Android" is fine.

"Android"-branded devices
   As the FAQ for the Android Compatibility Program (ACP) states (*http://source.android.com/faqs.html#compatibility*): "[I]f a manufacturer wishes to use the Android name with their product...they must first demonstrate that the device is compatible." Branding your device as being "Android" is therefore a privilege that Google intends to police. In essence, you will have to make sure your device is compliant and then talk to Google and enter into some kind of agreement with it before you can advertise your device as being "Foo Android." We will cover the Android Compatibility Program later in this chapter.

"Droid" in official names
   You may not use "Droid" alone in a name, such as "Foo Droid," for example. For some reason the I haven't yet entirely figured out, "Droid" is a trademark of Lucasfilm. Achieve a Jedi rank, you likely must, before you can use it.

Word (and Brand) Play

While Google holds strict control over the use of the Android brand, the ASL used for licensing the bulk of the AOSP states the following: "This License does not grant permission to use the trade names, trademarks, service marks, or product names of the Licensor, except as required for reasonable and customary use in describing the origin of the Work and reproducing the content of the *NOTICE* file."

While this clearly says you have no right to use the associated trademark, the "reasonable and customary use in describing the origin" exception is seen by many as allowing you to state that your device is "AOSP based." Some push this further and simply state that their product is "based on Android" or "Android based." You'll even find some clever marketing material sporting the Android robot to advertise a product without mentioning the word "Android."

Probably one of the sneakiest wordplays I've seen is when a product lists the following as part of one of its features: "Runs Android applications." You can bet yourself a couple of green robots that if it runs Android applications, it's almost guaranteed to contain the AOSP in some way, shape, or form.

Google's Own Android Apps

While the AOSP contains a core set of applications that are available under the ASL, "Android"-branded phones usually contain an additional set of "Google" applications that are not part of the AOSP, such as Play Store (the "app market" app), YouTube, "Maps and Navigation," Gmail, etc. Obviously, users expect to have these apps as part of Android, and you might therefore want to make them available on your device. If that is the case, you will need to abide by the ACP and enter into an agreement with Google, very much in line with what you need to do to be allowed to use "Android" in your product's name. We will cover the ACP shortly.

Alternative App Markets

Though the main app market (i.e., Google Play) is the one hosted by Google and made available to users through the Play Store app installed on "Android"-branded devices, other players are leveraging Android's open APIs and open source licensing to offer alternative app markets. Such is the case with online merchants such as Amazon and Barnes & Noble, as well as mobile network operators such as Verizon and Sprint. In fact, I know of nothing that would preclude you from creating your own app store. There is even at least one open source project, the Affero-licensed F-Droid Repository (*http://f-droid.org/repository/*), that provides both an app market application and a corresponding server backend under the GPL.

Oracle versus Google

As part of acquiring Sun Microsystems, Oracle also acquired Sun's intellectual property (IP) rights to the Java language and, according to Java creator James Gosling,[7] it was clear during the acquisition process that Oracle intended from the outset to go after Google with Sun's Java IP portfolio. And in August 2010 it did just that, filing suit against Google, claiming that it infringed on several patents and committed copyright violations.

Without going into the merits of the case, it's obvious that Android does indeed heavily rely on Java. And clearly Sun created Java and owned a lot of intellectual property around the language it created. In what appears to have been an effort to anticipate any claims Sun may put forward against Android, the Android development team went out of its way to use as little of Sun's Java in the Android OS as possible. Java is in fact composed mainly of three things: the language and its semantics, the virtual machine that runs the Java byte-code generated by the Java compiler, and the class library that contains the packages used by Java applications at runtime.

---

7. See Gosling's blog postings on the topic at *http://nighthacks.com/roller/jag/entry/the_shit_finally_hits_the* and *http://nighthacks.com/roller/jag/entry/quite_the_firestorm* for more details.

The official versions of the Java components are provided by Oracle as part of the Java Development Kit (JDK) and the Java Runtime Environment (JRE). Android, on the other hand, relies only on the Java compiler found in the JDK for building parts of the AOSP; that compiler isn't included as part of the images generated by the AOSP. Also, instead of using Oracle's Java VM, Android relies on Dalvik, a VM custom built for Android, and instead of using the official class library, Android relies on Apache Harmony, a clean-room reimplementation of the class library. Hence, it would seem that Google made every reasonable effort to at least avoid any copyright and/or distribution issues.

Still, it remains to be seen where these legal proceedings will go. Although by May 2012 Google had prevailed on both the copyright and patent fronts of the initial trial, Oracle appealed the verdict in October of that same year. There is of course a lot at stake, and it will likely take many years for this saga to play itself out. If you want to follow the latest round of these proceedings or read up on past episodes, I suggest you have a look at the Groklaw website (*http://groklaw.net*) and consult the relevant Wikipedia entry (*http://en.wikipedia.org/wiki/Oracle_v._Google*).

Another indirectly related, yet very relevant, development is that IBM joined Oracle's OpenJDK efforts in October 2010. IBM had been the driving force behind the Apache Harmony project, which is the class library used in Android, and its departure pretty much ensures that the project will become orphaned. How this development impacts Android is unknown at the time of this writing.

Incidentally, though he later left, James Gosling joined Google in March 2011.

Mobile Patent Warfare

The previous section is to some extent but the tip of the iceberg with regard to litigation and legal wranglings ongoing in the mobile world at the time of this writing. Sales of mobile phones have overtaken the sales of traditional PCs, and the mobile market's growth has resulted in the majority of players being somehow involved in legal maneuvers against and/or because of its competitors. There's even a Wikipedia entry entitled Smartphone wars (*http://en.wikipedia.org/wiki/Smartphone_wars*) dedicated to listing the ongoing battles.

It's hard to say where any of this will go. There seems to be no end to the variety of strategies companies will employ or the lengths to which they'll go to ensure they prevail. Apple and Samsung, for instance, are at the time of this writing involved in court cases against each other (*http://bit.ly/XKRssa*) in quite a few countries. Microsoft is also rumored to be contacting various manufacturers to request royalties for the use of Android; as evidenced by some of the filings made by Barnes & Noble with the courts after it was sued by Microsoft for refusing to pay.

How any of this might affect your own product is difficult to say. As always, consult with competent legal counsel as needed. Usually it's a question of volume. So if your product is for a niche market, you're probably too small a fish to matter. If you're creating a mass-market product, on the other hand, you'll likely want to make sure you've covered all your bases.

Hardware and Compliance Requirements

In principle, Android should run on any hardware that runs Linux. Android has in fact been made to run on ARM, x86, MIPS, SuperH, and PowerPC—all architectures supported by Linux. A corollary to this is that if you want to port Android to your hardware, you must first port Linux to it. Beyond being able to run Linux, though, there are few other hardware requirements for running the AOSP, apart from the logical requirement of having some kind of display and pointer mechanism to allow users to interact with the interface. Obviously, you might have to modify the AOSP to make it work on your hardware configuration, if you don't support a peripheral it expects. For instance, if you don't have a GPS unit in your product, you might want to provide a mock GPS HAL module, as the Android emulator does, to the AOSP. You will also need to make sure you have enough memory to store the Android images and a sufficiently powerful CPU to give the user a decent experience.

In sum, therefore, there are few restrictions if you just want to get the AOSP up and running on your hardware. If, however, you are working on a device that must carry "Android" branding or must include the standard Google-owned applications found in typical consumer Android devices—such as the Maps or Play Store applications—you need to go through the Android Compatibility Program (ACP) mentioned earlier. There are two separate yet complementary parts to the ACP: the Compliance Definition Document (CDD) and the Compliance Test Suite (CTS). Even if you don't intend to participate in the ACP, you might still want to take a look at the CDD and the CTS, as they give a very good idea about the general mind-set that went into the design goals of the Android version you intend to use.

Every Android release has its own CDD and CTS. You must therefore use the CDD and CTS that match the version you intend to use for your final product. If you switch Android releases midway through your project—because, for instance, a new Android release comes out with cool new features you'd like to have—you will need to make sure you comply with that release's CDD and CTS. Keep in mind also that you need to interact with Google to confirm compliance. Hence, switching may involve jumping through a few hoops and potential product delivery delays.

The overarching goal of the ACP, and therefore the CDD and the CTS, is to ensure a uniform ecosystem for users and application developers. Hence, before you are allowed to ship an "Android"-branded device, Google wants to make sure you aren't fragmenting the Android ecosystem by introducing incompatible or crippled products. This, in turn, makes sense for manufacturers since they are benefiting from the compliance of others when they use the "Android" branding. Look at this site (*http://source.android.com/compatibility/*) for more details about the ACP.

Note that Google reserves the right to decline your participation in the Android ecosystem, and therefore prevent your ability to ship the Play Store app with your device and use the "Android" branding. As stated on their site: "Unfortunately, for a variety of legal and business reasons, we aren't able to automatically license Google Play to all compatible devices."

Compliance Definition Document

The CDD is the policy part of the ACP and is available at the ACP URL above. It specifies the requirements that must be met for a device to be considered compatible. The language in the CDD is based on RFC2119, with a heavy use of "MUST," "SHOULD," "MAY," etc. to describe the different attributes. Around 25 pages in length, it covers all aspects of the device's hardware and software capabilities. Essentially, it goes over every aspect that cannot simply be automatically tested using the CTS. Let's go over some of what the CDD requires.

This discussion is based on the Android 2.3/Gingerbread CDD. The specific version you use will likely have slightly different requirements.

Software

This section lists the Java and native APIs along with the web, virtual machine, and user interface compatibility requirements. Essentially, if you are using the AOSP, you should readily conform to this section of the CDD.

Application packaging compatibility

This section specifies that your device must be able to install and run *.apk* files. All Android apps developed using the Android SDK are compiled into *.apk* files, and these are the files that are distributed through Google Play and installed on users' devices.

Multimedia compatibility

Here the CDD describes the media codecs (decoders and encoders), audio recording, and audio latency requirements for the device. The AOSP includes the StageFright multimedia framework, and you can therefore conform to the CDD by using the AOSP. However, you should read the audio recording and latency sections, as they contain specific technical information that may impact the type of hardware or hardware configuration your device must be equipped with.

Developer tool compatibility

This section lists the Android-specific tools that must be supported on your device. Basically, these are the common tools used during app development and testing: *adb*, *ddms*, and *monkey*. Typically, developers don't interact with these tools directly. Instead, they usually develop within the Eclipse development environment and use the Android Development Tool (ADT) plug-in, which takes care of interacting with the lower-level tools.

Hardware compatibility

This is probably the most important section for embedded developers, as it likely has profound ramifications on the design decisions made for the targeted device. Here's a summary of what each subsection spells out.

Display and graphics
- Your device's screen must be at least 2.5 inches in physical diagonal size.
- Its density must be at least 100dpi.
- Its aspect ratio must be between 4:3 and 16:9.
- It must support dynamic screen orientation from portrait to landscape and vice versa. If orientation can't be changed, then it must support letterboxing, since apps may force orientation changes.
- It must support OpenGL ES 1.0, though it may omit 2.0 support.

Input devices
- Your device must support the Input Method Framework, which allows developers to create custom onscreen, soft keyboards.
- It must provide at least one soft keyboard.
- It can't include a hardware keyboard that doesn't conform to the API.
- It must provide Home, Menu, and Back buttons.
- It must have a touch screen, whether it be capacitive or resistive.
- It should support independent tracked points (multitouch) if possible.

Sensors
> While all sensors are qualified using "SHOULD," meaning that they aren't compulsory, your device must accurately report the presence or absence of sensors and must return an accurate list of supported sensors.

Data connectivity
> The most important item here is an explicit statement that Android may be used on devices that don't have telephony hardware. This was added to allow for Android-based tablet devices. Furthermore, your device should have hardware support for 802.11x, Bluetooth, and near field communication (NFC). Ultimately, your device must support some form of networking that permits a bandwidth of 200Kbps.

Cameras
> Your device should include a rear-facing camera and may include a front-facing one as well.

Memory and storage
> - Your device must have at least 128MB for storing the kernel and user-space.
> - It must have at least 150MB for storing user data.
> - It must have at least 1GB of "shared storage." This is typically, though not always, the removable SD card.
> - It must also provide a mechanism to access shared data from a PC. In other words, when the device is connected through USB, the content of the SD card must be accessible on the PC.

USB
> This requirement is likely the one that most heavily demonstrates how user-centric "Android"-branded devices must be, since it essentially assumes that the user owns the device and therefore requires you to allow users to fully control the device when it's connected to a computer. In some cases this might be a showstopper for you, as you may not actually want or may not be able to have users connect your embedded device to a computer. Nevertheless, the CDD requires the following:
>
> - Your device must implement a USB client, connectable through USB-A.
> - It must implement the Android Debug Bridge (ADB) protocol as provided in the *adb* command over USB.
> - It must implement USB mass storage, thereby allowing the device's SD card to be accessed on the host.

Newer CDDs obviously have evolved from this list. There's no longer a need to have physical Home, Menu, and Back buttons since 3.0, since those can be displayed onscreen. OpenGL ES 2.0 support is also now mandatory. In addition to USB mass storage support, the device can also now provide Media Transfer Protocol (MTP) instead.

Performance compatibility

Although the CDD doesn't specify CPU speed requirements, it does specify app-related time limitations that will impact your choice of CPU speed. For instance:

- The Browser app must launch in less than 1300ms.
- The MMS/SMS app must launch in less than 700ms.
- The AlarmClock app must launch in less than 650ms.
- Relaunching an already-running app must take less time than the original launch.

Security model compatibility

Your device must conform to the security environment enforced by the Android application framework, Dalvik, and the Linux kernel. Specifically, apps must have access and be submitted to the permission model described as part of the SDK's documentation. Apps must also be constrained by the same sandboxing limitations they have by running as separate processes with distinct user IDs (UIDs) in Linux. The filesystem access rights must also conform to those described in the developer documentation. Finally, if you aren't using Dalvik, whatever VM you use should impose the same security behavior as Dalvik.

Software compatibility testing

Your device must pass the CTS, including the human-operated CTS Verifier part. In addition, your device must be able to run specific reference applications from Google Play.

Updatable software

There has to be a mechanism for your device to be updated. This may be done over the air (OTA) with an offline update via reboot. It also may be done using a "tethered" update via a USB connection to a PC, or be done "offline" using removable storage.

Compliance Test Suite

The CTS comes as part of the AOSP, and we will discuss how to build and use it in Chapter 4. The AOSP includes a special build target that generates the *cts* command-line tool, the main interface for controlling the test suite. The CTS relies on *adb* to push and run tests on the USB-connected target. The tests are based on the JUnit Java unit testing framework, and they exercise different parts of the framework, such as the APIs,

Dalvik, Intents, Permissions, etc. Once the tests are done, they will generate a ZIP file containing XML files and screenshots that you need to submit to *cts@android.com*.

Development Setup and Tools

There are two separate sets of tools for Android development: those used for application development and those used for platform development. If you want to set up an application development environment, have a look at *Learning Android* or at Google's online documentation (*http://developer.android.com/*). If you want to do platform development, as we will do here, your tool needs will vary, as you will see later in this book.

At the most basic level, though, you need to have a Linux-based workstation to build the AOSP. In fact, at the time of this writing, Google's only supported build environment is 64-bit Ubuntu 10.04. That doesn't mean that another Ubuntu version or even another Linux distribution won't work or, in the case of Android versions up to Gingerbread, that you won't be able to build the AOSP on a 32-bit system,[8] but essentially that configuration reflects Google's own Android compile farms configuration. An easy way to get your hands dirty with AOSP work without changing your workstation OS is to create an Ubuntu virtual machine using your favorite virtualization tool. I typically use VirtualBox (*http://www.virtualbox.org/*), since I've found that it makes it easy to access the host's serial ports in the guest OS.

In some cases, even though 32-bit build support wasn't available for a given Android version, patches were created to make such compiling possible. This is especially true for Gingerbread. So even though the official tree may not support 32-bit builds, you may be able to find another tree that does or a mailing list posting that explains how to do it. Still, it remains that newer AOSP versions require more and more powerful machines to build in a reasonable amount of time, and most of these systems end up being 64 bit. Hence, the impetus for supporting builds on 32-bit systems diminishes with every new version of Android.

No matter what your setup is, keep in mind that the AOSP is several gigabytes in size before building, and its final size is much larger. Gingerbread, for example, is about 3GB in size uncompiled and grows to about 10GB once compiled, while 4.2/Jelly Bean is 6GB uncompiled and grows to about 24GB once compiled.[9] When you factor in that you are likely going to operate on a few separate versions—for testing purposes if for no other

---

8. More recent versions such as JellyBean 4.1 and 4.2 can be built only on 64-bit systems.
9. These uncompiled numbers don't count the space taken by the *.git* and *.repo* directories in the tree. The uncompiled size of 2.3.7/Gingerbread with those directories is 5.5GB and that of 4.2/Jelly Bean is 18GB.

reason—you rapidly realize that you'll need tens if not hundreds of gigabytes for serious AOSP work. Also note that during the period this book was written (2011 to 2013), build times for the latest AOSP on the highest-end machines have always hovered between 30 minutes to an hour. Even minor modifications may result in a five-minute run to complete the build or regenerate output images. You will therefore also likely want to make sure you have a fairly powerful machine when developing Android-based embedded systems. We'll discuss the AOSP build and its requirements in greater detail in Chapter 4.

CHAPTER 2
Internals Primer

As we've just seen, Android's sources are freely available for you to download, modify, and install for any device you choose. In fact, it is fairly trivial to just grab the code, build it, and run it in the Android emulator. To customize the AOSP to your device and its hardware, however, you'll need to first understand Android's internals to a certain extent. So you'll get a high-level view of Android internals in this chapter, and have the opportunity in later chapters to dig into parts of internals in greater detail, including tying said internals to the actual AOSP sources.

As mentioned in the Preface, this book is mainly based on 2.3.x/Gingerbread. That said, Android's internals had remained fairly stable over its lifetime up to that version of Android, and they've changed very little from that version to the current 4.2/Jelly Bean. Still, while the bulk of the internals remains relatively unchanged, critical changes can come unannounced thanks to Android's closed development process. For instance, in 2.2/Froyo and previous versions, the Status Bar was an integral part of the System Server. In 2.3/Gingerbread, the Status Bar was moved out of the System Server and now runs independently from it.[1]

App Developer's View

Given that Android's development API is unlike any other existing API, including anything found in the Linux world, it's important to spend some time understanding what "Android" looks like from the app developers' perspective, even though it's very different from what Android looks like for anyone hacking the AOSP. As an embedded developer

---

1. Some speculate that this change was triggered because some app developers were doing too many fancy tricks with notification that were having negative impacts on the System Server, and that the Android team hence decided to make the Status Bar a separate process from the System Server.

working on embedding Android on a device, you might not have to actually deal directly with the idiosyncrasies of Android's app development API, but some of your colleagues might. If nothing else, you might as well share a common lingo with app developers. Of course, this section is merely a summary, and I recommend you read up on Android app development for more in-depth coverage.

Android Concepts

Application developers must take a few key concepts into account when developing Android apps. These concepts shape the architecture of all Android apps and dictate what developers can and cannot do. Overall, they make users' lives better, but they can sometimes be challenging to deal with.

Components

Android applications consist of loosely tied *components*. Components of one app can invoke or use components of other apps. Most importantly, there is no single entry point to an Android app: no main() function or any equivalent. Instead, there are predefined events called *intents* that developers can tie their components to, thereby enabling their components to be activated on the occurrence of the corresponding events. A simple example is the component that handles the user's contacts database, which is invoked when the user presses a Contacts button in the Dialer or another app. An app, therefore, can have as many entry points as it has components.

There are four main types of components:

Activities
    Just as the "window" is the main building block of all visual interaction in window-based GUI systems, activities are the main building block in an Android app. Unlike a window, however, activities cannot be "maximized," "minimized," or "resized." Instead, activities always take the entirety of the visual area and are made to be stacked on top of one another in the same way as a browser remembers web pages in the sequence they were accessed, allowing the user to go back to where he was previously. In fact, as described in the previous chapter, all Android devices have a Back button, whether it be a physical button on the device or a soft button displayed onscreen, to make this behavior available to the user. In contrast to web browsing, though, there is no button corresponding to the "forward" browsing action; only "back" is possible.

    One globally defined Android intent allows an activity to be displayed as an icon on the app launcher (the main app list on the device). Because the vast majority of apps want to appear on the main app list, they provide at least one activity that is defined as capable of responding to that intent. Typically, the user will start from a particular activity and move through several others and end up creating a stack of activities all related to the original one they launched; this stack of activities is called a *task*. The user can then switch to another task by clicking the Home button and starting another activity stack from the app launcher.

Services
  Android services are akin to background processes or daemons in the Unix world. Essentially, a service is activated when another component requires its services and typically remains active for the duration required by its caller. Most importantly, though, services can be made available to components outside an app, thereby exposing some of that app's core functionality to other apps. There is usually no visual sign of a service being active.

Broadcast receivers
  Broadcast receivers are akin to interrupt handlers. When a key event occurs, a broadcast receiver is triggered to handle that event on the app's behalf. For instance, an app might want to be notified when the battery level is low or when "airplane mode" (to shut down the wireless connections) has been activated. When not handling a specific event for which they are registered, broadcast receivers are otherwise inactive.

Content providers
  Content providers are essentially databases. Usually, an app will include a content provider if it needs to make its data accessible to other apps. If you're building a Twitter client app, for instance, you could give other apps on the device access to the tweet feed you're presenting to the user through a content provider. All content providers present the same API to apps, regardless of how they are actually implemented internally. Most content providers rely on the SQLite functionality included in Android, but they can also use files or other types of storage.

Intents

Intents are one of the most important concepts in Android. They are the late-binding mechanisms that allow components to interact. An app developer could send an intent for an activity to "view" a web page or "view" a PDF, hence making it possible for the user to view a designated HTML or PDF document even if the requesting app itself doesn't include the capabilities to do so. More fancy use of intents is also possible. An app developer could, for instance, send a specific intent to trigger a phone call.

Think of intents as polymorphic Unix signals that don't necessarily have to be predefined or require a specific designated target component or app. If you are familiar with Qt, you can think of an intent as similar to, though not entirely the same as, a Qt signal. The intent itself is a passive object. The effects of its dispatching will depend on its content, the mechanism used to dispatch it, the system's built-in rules, and the set of installed apps. One of the system's rules, for instance, is that intents are tied to the type of component they are sent to. An intent sent to a service, for example, can be received only by a service, not by an activity or a broadcast receiver.

Components can be declared as capable of dealing with given intent types using filters in the *manifest* file. The system will thereafter match intents to that filter and trigger the corresponding component at runtime. This is typically called an "implicit" intent. An intent can also be sent to a specific component in an "explicit" fashion, bypassing the need to declare that intent within the receiving component's filter. The explicit invocation, though, requires the app to know about the designated component ahead of time, which typically applies only when intents are sent within components of the same app.

Component lifecycle

Another central tenet of Android is that the user shouldn't have to manage task switching. While there are a number of ways to switch among tasks, including a built-in mechanism that's typically accessed with a long press on the Home button, as well as a number of task manager apps available for Android, the user experience doesn't rely on those. Instead, the user is expected to start as many apps as he wants and "switch" among them by clicking Home to go to the home screen and clicking any other app. The app he clicks may be an entirely new one, or one that he previously started and for which an activity stack (a.k.a. a "task") already exists.

The corollary to, or consequence of, this design decision is that apps gradually use up more and more system resources as they are started, a process that can't go on forever. At some point, the system will have to start reclaiming the resources of the least recently used or nonpriority components in order to make way for newly activated components. Still, this resource recycling should be entirely transparent to the user. In other words, when a component is taken down to make way for a new one, and then the user returns to the original component, it should start up at the point where it was taken down and act as if it had been waiting in memory all along.

To make this behavior possible, Android defines a standard *lifecycle* for each component type. An app developer must manage her components' lifecycle by implementing a series of callbacks for each component. These callbacks are then triggered by events related to the component lifecycle. For instance, when an activity is no longer in the foreground (and therefore more likely to be destroyed than if it's in the foreground), its on Pause() callback is triggered. Google uses a state diagram (*https://developer.android.com/images/training/basics/basic-lifecycle.png*) to explain the activity's lifecycle to app developers.

Managing component lifecycles is one of the greatest challenges faced by app developers, because they must carefully save and restore component states on key transitional events. The desired end result is that the user never needs to "task switch" between apps or be aware that components from previously used apps were destroyed to make way for new ones he started.

Manifest file

If there has to be a "main" entry point to an app, the manifest file is likely it. Basically, it informs the system of the app's components, the capabilities required to run the app, the minimum level of the API required, any hardware requirements, etc. The manifest is formatted as an XML file and resides at the topmost directory of the app's sources as *AndroidManifest.xml*. The apps' components are typically all described statically in the manifest file. In fact, apart from broadcast receivers, which can be registered at runtime, all other components must be declared at build time in the manifest file.

Processes and threads

Whenever an app's component is activated, whether it be by the system or by another app, a process will be started to house that app's components. And unless the app developer does anything to override the system defaults, all other components of that app that start after the initial component is activated will run within the same process as that component. In other words, all components of an app are contained within a single Linux process. Hence, developers should avoid making long or blocking operations in standard components and use threads instead.

And because the user is essentially allowed to activate as many components as he wants, several Linux processes are typically active at any time to serve the many apps containing the user's components. When there are too many processes running to allow for new ones to start, the Linux kernel's out-of-memory (OOM) killing mechanisms will kick in. At that point, Android's in-kernel OOM handler will get called, and it will determine which processes must be killed to make space.

Put simply, the entirety of Android's behavior is predicated on low-memory conditions.

If the developer of the app whose process is killed by Android's OOM handler has implemented his components' lifecycles properly, the user shouldn't see any adverse behavior. For all practical purposes, in fact, the user shouldn't even notice that the process housing the app's components went away and got re-created "automagically" later.

Remote procedure calls (RPCs)

Much like many other components of the system, Android defines its own RPC/IPC (remote procedure call/inter-process communication) mechanism: *Binder*. So communication across components is not typically done using the usual socket or System V IPC. Instead, components use the in-kernel Binder mechanism, accessible through */dev/binder*, which will be covered later in this chapter.

App developers, however, do not use the Binder mechanism directly. Instead, they must define and interact with interfaces using Android's Interface Definition Language (IDL). Interface definitions are usually stored in an *.aidl* file and are processed by the *aidl* tool to generate the proper stubs and marshaling/unmarshaling code required to transfer objects and data back and forth using the Binder mechanism.

Framework Intro

In addition to the concepts we just discussed, Android also defines its own development framework, which allows developers to access functionality typically found in other development frameworks. Let's take a brief look at this framework and its capabilities.

User interface

UI elements in Android include traditional widgets such as buttons, text boxes, dialogs, menus, and event handlers. This part of the API is relatively straightforward, and developers usually find their way around it fairly easily if they've already coded for any other UI framework.

All UI objects in Android are built as descendants of the View class and are organized within a hierarchy of ViewGroups. An activity's UI can actually be specified either statically in XML (which is the usual way) or declared dynamically in Java. The UI can also be modified at runtime in Java if need be. An activity's UI is displayed when its content is set as the root of a ViewGroup hierarchy.

Data storage

Android presents developers with several storage options. For simple storage needs, Android provides *shared preferences*, which allow developers to store key-value pairs either in a data set shared by all components of the app or within a specific separate file. Developers can also manipulate files directly. These files may be stored privately by the app, so they are inaccessible to other apps, or they can be made readable and/or writable by other apps. App developers can also use the SQLite functionality included in Android to manage their own private databases. Such a database can then be made available to other apps by hosting it within a content provider component.

Security and permissions

Security in Android is enforced at the process level. In other words, Android relies on Linux's existing process isolation mechanisms to implement its own policies. To that end, every app installed gets its own UID and group identifier (GID). Essentially, it's as if every app is a separate "user" in the system. And as in any multiuser Unix system, these "users" cannot access one another's resources unless permissions are explicitly granted to do so. In effect, each app lives in its own separate sandbox.

To exit the sandbox and access key system functionality or resources, apps must use Android's permission mechanisms, which require developers to statically declare the permissions needed by an app in its manifest file. Some permissions, such as the right to access the Internet (i.e., use sockets), dial the phone, or use the camera, are predefined by Android. Other permissions can be declared by app developers and then be required for other apps to interact with a given app's components. When an app is installed, the user is prompted to approve the permissions required to run an app.

Access enforcement is based on per-process operations and requests to access a specific URI (universal resource identifier), and the decision to grant access to a specific functionality or resource is based on certificates and user prompts. The certificates are the ones used by app developers to sign the apps they make available through Google Play. Hence, developers can restrict access to their apps' functionality to other apps they themselves created in the past.

The Android development framework provides a lot more functionality, of course, than can be covered here. I invite you to read up on Android app development elsewhere or visit *http://developer.android.com* for more information on 2D and 3D graphics, multimedia, location and maps, Bluetooth, NFC, etc.

App Development Tools

The typical way to develop Android applications is to use the freely available Android Software Development Kit (SDK) (*http://developer.android.com/sdk/index.html*). This SDK—along with Eclipse, its corresponding Android Development Tools (ADT) plug-in, and the QEMU-based emulator in the SDK—allows developers to do the vast majority of development work straight from their workstations. Developers will also usually want to test their apps on real devices prior to making them available through Google Play, as there are usually runtime behavior differences between the emulator and actual devices. Some software publishers take this to the extreme and test their apps on several dozen devices before shipping a new release.

Testing on Several Hundred Devices

Obviously, app developers can't be expected to have every possible device at their disposal for testing. A few companies have therefore sprung up to allow app developers to test their apps on several hundred devices by simply uploading their apps to these companies' websites.

These companies typically have a web interface allowing developers to submit their app for execution on their device farm. Developers are then given detailed reports about failures and sometimes fairly explicit output from the failed devices' logs. Have a look at Apkudo (*http://www.apkudo.com/*), Bitbar's Testdroid products (*http://testdroid.com/*), and LessPainful (*http://www.lesspainful.com/*) if you need such functionality.

> Interestingly, Apkudo also provides a service to allow you to test devices prior to their release by running several hundred popular apps on the device to ensure that the AOSP it runs performs correctly.

Even if you don't plan to develop any apps for your embedded system, I highly suggest you set up the development environment on your workstation. If nothing else, this will allow you to validate the effects of modifications you make to the AOSP using basic test applications. It will also be essential if you plan to extend the AOSP's API and create and distribute your own custom SDK.

To set up an app development environment, follow the instructions provided by Google for the SDK, or have a look at the book *Learning Android* by Marko Gargenta (O'Reilly).

Native Development

While the majority of apps are developed exclusively in Java using the development environment we just discussed, certain developers need to run natively compiled code. To this end, Google has made the Native Development Kit (NDK) (*http://developer.android.com/tools/sdk/ndk/index.html*) available. As advertised, this is mostly aimed at game developers needing to squeeze every last bit of performance out of the device their game is running on. As such, the APIs made available in the NDK are mostly geared toward graphics rendering and sensor input retrieval. The infamous Angry Birds game, for example, relies heavily on code running natively.

Another possible use of the NDK is obviously to port over an existing codebase to Android. If you've developed a lot of legacy C code over several years (a common situation for development houses that have created applications for other mobile devices), you won't necessarily want to rewrite it in Java. Instead, you can use the NDK to compile it for Android and package it with some Java code to use some of the more Android-specific functionality made available by the SDK. The Firefox browser, for instance, relies heavily on the NDK to run some of its legacy code on Android.

As I just hinted, the nice thing about the NDK is that you can combine it with the SDK and therefore have parts of your app in Java and parts of your app in C. That said, it's crucial to understand that the NDK gives you access only to a very limited subset of the Android API. There is, for instance, presently no API allowing you to send an intent from within C code compiled with the NDK; the SDK must be used to do it in Java instead. Again, the APIs made available through the NDK are mostly geared toward game development.

Sometimes embedded and system developers coming to Android expect to be able to use the NDK to do platform-level work. The word "native" in the NDK can be misleading in that regard, because the use of the NDK still involves all the limitations and requirements that apply to Java app developers. So, as an embedded developer, remember that the NDK is useful for app developers to run native code that they can call from their Java code. Apart from that, the NDK will be of little to no use for the type of work you are likely to undertake.

Overall Architecture

Figure 2-1 is probably one of the most important diagrams presented in this book, and I suggest you find a way to bookmark its location, as I will often refer back to it, if not explicitly then implicitly. Although it's a simplified view—and we will get the chance to enrich it as we go—it gives a pretty good idea of Android's architecture and how the various bits and pieces fit together.

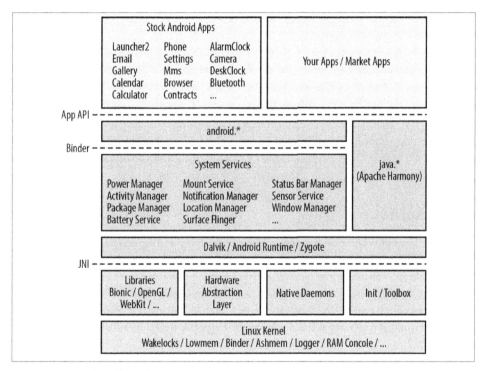

Figure 2-1. Android's architecture

If you are familiar with some form of Linux development, then the first thing that should strike you is that beyond the Linux kernel itself, there is little in that stack that resembles anything typically seen in the Linux or Unix world. There is no glibc, no X Window System, no GTK, no BusyBox, no bash shell, and so on. Many veteran Linux and embedded Linux practitioners have indeed noted that Android feels very alien. Though the Android stack starts from a clean slate with regard to user-space, we will discuss how to get "legacy" or "classic" Linux applications and utilities to coexist side by side with the Android stack in Appendix A.

 The Google developer documentation presents a different architectural diagram (*http://bit.ly/15HLHBZ*) from that shown in Figure 2-1. The former is likely well suited for app developers, but it omits key information that must be understood by embedded developers. For instance, Google's diagram and developer documentation offer little to no reference at the time of this writing to the System Server. Yet, as an embedded developer, you need to know what that component is, because it's one of the most important parts of Android, and you might need to extend or interact with it directly.

This is especially important to understand because you'll see Google's diagram presented and copied in several documents and presentations. If nothing else, remember that the internals and significance of the System Server are rarely if at all explained to app developers, and that the bulk of information out there is aimed at app developers, not developers doing platform work.

Let's take a deeper look into each part of Android's architecture, starting from the bottom of Figure 2-1 and going up. Once we are done covering the various components, we'll end this chapter by going over the system's startup process.

Linux Kernel

The Linux kernel is the centerpiece of all distributions traditionally labeled as "Linux," including mainstream distributions such as Ubuntu, Fedora, and Debian. And while it's available in "vanilla" form from the Linux Kernel Archives (*http://kernel.org*), most distributions apply their own patches to it to fix bugs and enhance the performance or customize the behavior of certain aspects before distributing it to their users. Android, as such, is no different in that the Android developers patch the "vanilla" kernel to meet their needs.

Historically, Android differed from standard practice, however, in relying on several custom functionalities that were significantly different from what was found in the "vanilla" kernel. In fact, whereas the kernel shipped by a Linux distribution can easily be replaced by a kernel from *kernel.org* with little to no impact on the rest of the distribution's components, Android's user-space components would simply not work unless they were running on an "Androidized" kernel. As I mentioned in the previous chapter, Android kernels were, up until recently, major forks from the mainline kernel. As I also mentioned, the situation has since progressed a lot, and many of the features required to run Android are finding their way into the mainline kernel.

Hopefully things will have progressed enough by the time you read this that you can just grab a kernel straight from *http://kernel.org* and run the AOSP on top of it. However, if past is prelude and the history of embedded Linux is an indication of what's to come, then your best source for getting a proper, Android-compatible kernel to run on your hardware is likely going to be the vendor of the SoC you're using.

Although it's beyond the scope of this book to discuss the Linux kernel's internals, let's go over the main "Androidisms" added to the kernel. You can get information about the kernel's internals by having a look at Robert Love's *Linux Kernel Development, 3rd ed.* (Addison-Wesley Professional, 2010) and starting to follow the Linux Weekly News (LWN) (*http://lwn.net*) site. LWN contains several seminal articles on the kernel's internals and provides the latest news regarding the Linux kernel's development.

Note that the following subsections cover only the most important Androidisms. Androidized kernels typically contain several hundred patches over the standard kernel, often to provide device-specific functionality, fixes, and enhancements. You can use *git*[2] to do an exhaustive analysis of the commit deltas between one of the kernels at *http://android.googlesource.com* and the mainline kernel it was forked from. Also, note that some of the functionality in some Androidized kernels, such as the PMEM driver, is device-specific and isn't necessarily used in all Android devices.

Creating Your Own Androidized Kernel

If you'd like to know how to create Androidized kernels from scratch or if you're tasked with this, say because you work for an SoC vendor, have a look at the Androidization of linux kernel (*http://bit.ly/16e1k5l*) blog post by Linaro engineer Vishal Bhoj, published in March 2012. In this post, Vishal explains how to create an Androidized kernel using the *git rebase* command. For more information about that specific command, have a look at the corresponding online git documentation (*http://git-scm.com/book/en/Git-Branching-Rebasing*).

Incidentally, Linaro, whose role is to assist its members with platform enablement, maintains an Androidized kernel that closely follows Linus's HEAD. For more information on this work, have a look at this thread (*http://bit.ly/Z5yG1m*).

---

2. Git is a distributed source code management tool created by Linus Torvalds to manage the kernel sources. You can find more information about it at *http://git-scm.com/*.

Wakelocks

Of all the Androidisms, this is likely the one that was most contentious. The discussion threads covering its inclusion in the mainline kernel generated close to 2,000 emails, and even then there was no clear path for merging the wakelock functionality. It was only after the 2011 Kernel Summit, where kernel developers agreed to merge most Androidisms into the mainline, that efforts were made to try to rehabilitate the wakelock mechanism or, as was ultimately decided, to create an equivalent that was more palatable to the rest of the kernel development community.

As of the end of May 2012, equivalents to the wakelocks and their correlated *early suspend* mechanisms have been merged into the mainline kernel. The early suspend replacement is called *autosleep*, and the wakelock mechanism has been replaced by a new `epoll()` flag called `EPOLLWAKEUP`. The API is also therefore different from the original functionality added by the Android team, but the resulting functionality is effectively the same. At the time of this writing, it's expected that the new versions of the AOSP would start using the new mechanisms instead of the old ones.

To understand what wakelocks are and do, we must first discuss how power management is typically used in Linux. The most common use case of Linux's power management is a laptop computer. When the lid is closed on a laptop running Linux, it will usually go into "suspend" or "sleep" mode. In that mode, the system's state is preserved in RAM, but all other parts of the hardware are shut down. Hence, the computer uses as little battery power as possible. When the lid is raised, the laptop "wakes up," and the user can resume using it almost instantaneously.

That modus operandi works fine for laptops and desktop-like devices, but it doesn't fit mobile devices such as handsets as well. Hence, Android's development team devised a mechanism that changes the rules slightly to make them more palatable for such use cases. Instead of letting the system be put to sleep at the user's behest, an Androidized kernel is made to go to sleep as soon and as often as possible. And to keep the system from going to sleep while important processing is being done or while an app is waiting for the user's input, wakelocks are provided to keep the system awake.

The wakelocks and early suspend functionality are actually built on top of Linux's existing power management functionality. However, they introduce a different development model, since application and driver developers must explicitly grab wakelocks whenever they conduct critical operations or must wait for user input. Usually, app developers don't need to deal with wakelocks directly, because the abstractions they use automatically take care of the required locking. They can, nonetheless, communicate with the Power Manager Service if they require explicit wakelocks. Driver developers, on the other hand, can call on the added in-kernel wakelock primitives to grab and release wakelocks. The downside of using wakelocks in a driver, however, used to be that it became impossible to push that driver into the mainline kernel, because the mainline didn't include wakelock support. Given the recent inclusion of equivalent functionality into the mainline, this is no longer an issue.

The following LWN articles describe wakelocks in more detail and explain the various issues surrounding their inclusion in the mainline kernel:

- Wakelocks and the embedded problem (*http://lwn.net/Articles/318611/*)
- From wakelocks to a real solution (*http://lwn.net/Articles/319860/*)
- Suspend block (*http://lwn.net/Articles/385103/*)
- Blocking suspend blockers (*http://lwn.net/Articles/388131/*)
- What comes after suspend blockers (*http://lwn.net/Articles/390369/*)
- An alternative to suspend blockers (*http://lwn.net/Articles/416690/*)
- KS2011: Patch review (*http://lwn.net/Articles/464298/*)
- Bringing Android closer to the mainline (*http://lwn.net/Articles/472984/*)
- Autosleep and wake locks (*http://lwn.net/Articles/479841/*)
- 3.5 merge window part 2 (*http://lwn.net/Articles/498693/*)

Low-Memory Killer

As mentioned earlier, Android's behavior is very much predicated on low-memory conditions. Hence, out-of-memory behavior is crucial. For this reason, the Android development team has added an additional low-memory killer to the kernel that kicks in before the default kernel OOM killer. Android's low-memory killer applies the policies described in the app development documentation, weeding out processes hosting components that haven't been used in a long time and are not high priority.

Android's low-memory killer is based on the OOM adjustments mechanism available in Linux that enables the enforcement of different OOM kill priorities for different processes. Basically, the OOM adjustments allow the user-space to control part of the kernel's OOM killing policies. The OOM adjustments range from −17 to 15, with a higher number meaning the associated process is a better candidate for being killed if the system is out of memory.

Android therefore attributes different OOM adjustment levels to different types of processes according to the components they are running and configures its own low-memory killer to apply different thresholds for each category of process. This effectively allows it to preempt the activation of the kernel's own OOM killer—which kicks in only when the system has no memory left—by kicking in when the given thresholds are reached, not when the system runs out of memory.

The user-space policies are themselves applied by the init process at startup (see "Init" on page 57), and readjusted and partly enforced at runtime by the Activity Manager Service, which is part of the System Server. The Activity Manager is one of the most important services in the System Server and is responsible for, among many other things, carrying out the component lifecycle presented earlier.

 Have a look at the Taming the OOM killer (*http://lwn.net/Articles/317814/*) LWN article if you'd like to get more information regarding the kernel's OOM killer and how Android traditionally builds on it.

At the time of this writing, Android's low-memory killer is found in the kernel's staging tree along with many of the other Android-specific drivers. Work is currently under way to rewrite this functionality within a more general framework for low-memory conditions. Have a look at the Userspace low memory killer daemon (*http://lwn.net/Articles/511731/*) post to the Linux Kernel Mailing List (LKML) and the linux-vmevent (*http://git.infradead.org/users/cbou/linux-vmevent.git*) patch for a glimpse of what's currently being worked on. Essentially, the goal is to move the decision process about what to do in low-memory conditions to a daemon in user-space.

> ## Android and the Linux Staging Tree
>
> At the time of this writing, many of the drivers required to run Android have been merged into the *staging* tree. While this means they are still found in mainline kernels available at *http://kernel.org*, it also means that kernel developers believe those drivers require work before being considered mature enough to be merged alongside the "clean" set of drivers found in the rest of the kernel tree.
>
> Specifically, many Android drivers are currently found in the *drivers/staging/android* directory of the kernel. They should remain there until they have been refactored or rewritten to suit the criteria for them to be admitted as official Linux drivers into the relevant location within the *drivers/* directory.
>
> If you aren't familiar with the staging tree, have a look at Greg Kroah-Hartman's[3] The Linux Staging Tree, what it is and is not (*http://www.kroah.com/log/linux/linux-staging-update.html*) blog post from March 2009: "The Linux Staging tree (or just 'staging' from now on) is used to hold standalone drivers and filesystems that are not ready to be

---

3. Greg is one of the top kernel developers and maintainers.

merged into the main portion of the Linux kernel tree at this point in time for various technical reasons. It is contained within the main Linux kernel tree so that users can get access to the drivers much easier than before, and to provide a common place for the development to happen, resolving the 'hundreds of different download sites' problem that most out-of-tree drivers have had in the past."

Binder

Binder is an RPC/IPC mechanism akin to COM under Windows. Its roots actually date back to work done within BeOS prior to Be's assets being bought by Palm. It continued life within Palm, and the fruits of that work were eventually released as the *OpenBinder* project. Though OpenBinder never survived as a standalone project, a few key developers who had worked on it, such as Dianne Hackborn and Arve Hjønnevåg, eventually ended up working on the Android development team.

Android's Binder mechanism is therefore inspired by that previous work, but Android's implementation does not derive from the OpenBinder code. Instead, it's a clean-room rewrite of a subset of the OpenBinder functionality. The OpenBinder Documentation (*http://www.angryredplanet.com/~hackbod/openbinder/*) remains a must-read if you want to understand the mechanism's underpinnings and its design philosophy, and so is Dianne Hackborn's explanation (*http://lkml.org/lkml/2009/6/25/3*) on the LKML of how the Binder is used in Android.

In essence, Binder attempts to provide remote object invocation capabilities on top of a classic OS. In other words, instead of reengineering traditional OS concepts, Binder "attempts to embrace and transcend them." Hence, developers get the benefits of dealing with remote services as objects without having to deal with a new OS. It therefore becomes very easy to extend a system's functionality by adding remotely invocable objects instead of implementing new daemons for providing new services, as would usually be the case in the Unix philosophy. The remote object can therefore be implemented in any desired language and may share the same process space as other remote services or have its own separate process. All that is needed to invoke its methods is its interface definition and a reference to it.

And as you can see in Figure 2-1, Binder is a cornerstone of Android's architecture. It's what allows apps to talk the System Server, and it's what apps use to talk to each others' service components, although, as I mentioned earlier, app developers don't actually talk to the Binder directly. Instead, they use the interfaces and stubs generated by the *aidl* tool. Even when apps interface with the System Server, the `android.*` APIs abstract its services, and the developer never actually sees that Binder is being used.

Though they sound semantically similar, there is a very big difference between services running within the System Server and services exposed to other apps through the "service" component model I introduced in "Components" on page 26 as being one of the components available to app developers. Most importantly, service components are subject to the same system mechanics as any other component. Hence, they are lifecycle-managed and run within the same privilege sandbox associated with the app they are part of. Services running within the System Server, on the other hand, typically run with system privileges and live from boot to reboot. The only things these two types of services share are: a) their name, and b) the use of Binder to interact with them.

The in-kernel driver part of the Binder mechanism is a character driver accessible through */dev/binder*. It's used to transmit parcels of data between the communicating parties using calls to ioctl(). It also allows one process to designate itself as the "Context Manager." The importance of the Context Manager, along with the actual user-space use of the Binder driver, will be discussed in more detail later in this chapter.

Since the 3.3 release of the Linux kernel, the Binder driver has been merged into the staging tree. There is currently no project under way to clean this driver up or to rewrite it to make it applicable and/or useful for more general-purpose use in standard Linux desktop and server systems. It's therefore likely to remain in *drivers/staging/android/* for the foreseeable future.

Anonymous Shared Memory (ashmem)

Another IPC mechanism available in most OSes is shared memory. In Linux, this is usually provided by the POSIX SHM functionality, part of the System V IPC mechanisms. If you look at the *bionic/libc/docs/SYSV-IPC.TXT* file included in the AOSP, however, you'll discover that the Android development team seems to have a dislike for SysV IPC. Indeed, the argument is made in that file that the use of SysV IPC mechanisms in Linux can lead to resource leakage within the kernel, opening the door for malicious or misbehaving software to cripple the system.

Though it isn't stated as such by Android developers or any of the documentation within the ashmem code or surrounding its use, ashmem very likely owes part of its existence to SysV IPC's shortcomings as seen by the Android development team. Ashmem is therefore described as being similar to POSIX SHM "but with different behavior." For instance, it uses reference counting to destroy memory regions when all processes referring to them have exited, and will shrink mapped regions if the system is in need of memory. "Unpinning" a region allows it to be shrunk, whereas "pinning" a region disallows the shrinking.

Typically, a first process creates a shared memory region using ashmem, and uses Binder to share the corresponding file descriptor with other processes with which it wishes to share the region. Dalvik's JIT code cache, for instance, is provided to Dalvik instances through ashmem. A lot of System Server components, such as the Surface Flinger and the Audio Flinger, rely on ashmem—through the IMemory interface, rather than directly.

IMemory is an internal interface available only within the AOSP, not to app developers. The closest class exposed to app developers is Memory File.

At the time of this writing, the ashmem driver is included in the mainline's *drivers/staging/android/* directory and is slated for rewriting.

Alarm

The alarm driver added to the kernel is another case where the default kernel functionality wasn't sufficient for Android's requirements. Android's alarm driver is actually layered on top of the kernel's existing Real-Time Clock (RTC) and High-Resolution Timers (HRT) functionalities. The kernel's RTC functionality provides a framework for driver developers to create board-specific RTC functions, while the kernel exposes a single hardware-independent interface through the main RTC driver. The kernel HRT functionality, on the other hand, allows callers to get woken up at very specific points in time.

In "vanilla" Linux, application developers typically call the setitimer() system call to get a signal when a given time value expires; for more information, see the setitimer()'s man page. The system call allows for a handful of types of timers, one of which, ITIMER_REAL, uses the kernel's HRT. This functionality, however, doesn't work when the system is suspended. In other words, if an application uses setitimer() to request being woken up at a given time and then in the interim the device is suspended, that application will get its signal only when the device is woken up again.

Separately from the setitimer() system call, the kernel's RTC driver is accessible through */dev/rtc* and enables its users to use an ioctl() to, among other things, set an alarm that will be activated by the RTC hardware device in the system. That alarm will fire off whether the system is suspended or not, since it's predicated on the behavior of the RTC device, which remains active even when the rest of the system is suspended.

Android's alarm driver cleverly combines the best of both worlds. By default, the driver uses the kernel's HRT functionality to provide alarms to its users, much like the kernel's own built-in timer functionality. However, if the system is about to suspend itself, it programs the RTC so that the system gets woken up at the appropriate time. Hence, whenever an application from user-space needs a specific alarm, it just needs to use

Android's alarm driver to be woken up at the appropriate time, regardless of whether the system is suspended in the interim.

From user-space, the alarm driver appears as the */dev/alarm* character device and allows its users to set up alarms and adjust the system's time (wall time) through `ioctl()` calls. There are a few key AOSP components that rely on */dev/alarm*. For instance, Toolbox and the `SystemClock` class, available through the app development API, rely on it to set/get the system's time. Most importantly, though, the Alarm Manager service part of the System Server uses it to provide alarm services to apps that are exposed to app developers through the `AlarmManager` class.

Both the driver and Alarm Manager use the wakelock mechanism wherever appropriate to maintain consistency between alarms and the rest of Android's wakelock-related behavior. Hence, when an alarm is fired, its consuming app gets the chance to do whatever operation is required before the system is allowed to suspend itself again, if need be.

At the time of this writing, Android's alarm driver is in the kernel's staging tree with upstreaming work pending.

Logger

Logging is another essential component of any Linux system, embedded ones included. Being able to analyze a system's logs for errors or warnings either postmortem or in real time can be vital to isolate fatal errors, especially transient ones. By default, most Linux distributions include two logging systems: the kernel's own log, typically accessed through the *dmesg* command, and the system logs, typically stored in files in the */var/log* directory. The kernel's log usually contains the messages printed out by the various `printk()` calls made within the kernel, either by core kernel code or by device drivers. For their part, the system logs contain messages coming from various daemons and utilities running in the system. In fact, you can use the *logger* command to send your own messages to the system log.

With regard to Android, the kernel's logging functionality is used as is. However, none of the usual system logging software packages typically found in most Linux distributions are found in Android. Instead, Android defines its own logging mechanisms based on the Android logger driver added to the kernel. The classic syslog relies on sending messages through sockets, and therefore generates a task switch. It also uses files to store its information, therefore generating writes to a storage device. In contrast, Android's logging functionality manages a handful of separate kernel-hosted buffers for logging data coming from user-space. Hence, no task-switches or file-writes are required for each event being logged. Instead, the driver maintains circular buffers in RAM where it logs every incoming event and returns immediately back to the caller.

There are numerous benefits to avoiding file-writes in the settings in which Android is used. For example, unlike in a desktop or server environment, it isn't necessarily desirable to have a log that grows indefinitely in an embedded system. It's also desirable to have a system that enables logging even though the filesystem types used may be read-only. Furthermore, most Android devices rely on solid-state storage devices, which have a limited number of erase cycles. Avoiding superfluous writes is crucial in those cases.

Because of its lightweight, efficient, and embedded-system-friendly design, Android's logger can actually be used by user-space components at runtime to regularly log events. In fact, the Log class available to app developers more or less directly invokes the logger driver to write to the main event buffer. Obviously, all good things can be abused, and it's preferable to keep the logging light, but still the level of use made possible by exposing Log through the app API, along with the level of use of logging within the AOSP itself, likely would have been very difficult to sustain had Android's logging been based on syslog.

Figure 2-2 describes Android's logging framework in more detail. As you can see, the logger driver is the core building block on which all other logging-related functionality relies. Each buffer it manages is exposed as a separate entry within /dev/log/. However, no user-space component directly interacts with that driver. Instead, they all rely on liblog, which provides a number of different logging functions. Depending on the functions being used and the parameters being passed, events will get logged to different buffers. The liblog functions used by the Log and Slog classes, for instance, will test whether the event being dispatched comes from a radio-related module. If so, the event is sent to the "radio" buffer. If not, the Log class will send the event to the "main" buffer, whereas the Slog class will send it to the "system" buffer. The "main" buffer is the one whose events are shown by the *logcat* command when it's issued without any parameters.

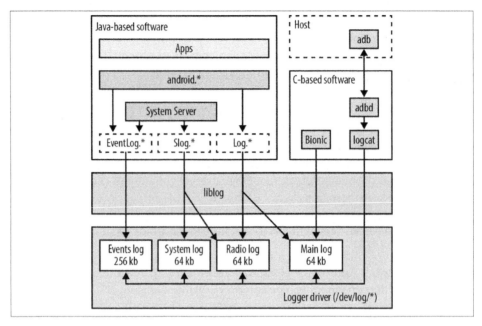

Figure 2-2. Android's logging framework

Both the Log and EventLog classes are exposed through the app development API, while Slog is for internal AOSP use only. Despite being available to app developers, though, EventLog is clearly identified in the documentation as mainly for system integrators, not app developers. In fact, the vast majority of code samples and examples provided as part of the developer documentation use the Log class. Typically, EventLog is used by system components to log binary events to the Android's "events" buffer. Some system components, especially System Server–hosted services, will use a combination of Log, Slog, and EventLog to log different events. An event that might be relevant to app developers, for instance, might be logged using Log, while an event relevant to platform developers or system integrators might be logged using either Slog or EventLog.

Note that the logcat utility, which is commonly used by app developers to dump the Android logs, also relies on liblog. In addition to providing access functions to the logger driver, liblog also provides functionality for formatting events for pretty printing and filtering. Another feature of liblog is that it requires every event being logged to have a priority, a tag, and data. The priority is either verbose, debug, info, warn, or error. The tag is a unique string that identifies the component or module writing to the log, and the data is the actual information that needs to be logged. This description should in fact sound fairly familiar to anyone exposed to the app development API, as this is exactly what's spelled out by the developer documentation for the Log class.

The final piece of the puzzle here is the *adb* command. As we'll discuss later, the AOSP includes an Android Debug Bridge (ADB) daemon that runs on the Android device and that is accessed from the host using the *adb* command-line tool. When you type *adb logcat* on the host, the daemon actually launches the *logcat* command locally on the target to dump its "main" buffer and then transfers that back to the host to be shown on the terminal.

At the time of this writing, the logger driver has been merged into the kernel's *drivers/staging/android/* directory. Have a look at the Mainline Android logger project (*http://elinux.org/Mainline_Android_logger_project*) for more information regarding the state of this driver's mainlining.

Other Notable Androidisms

A few other Androidisms, in addition to those already covered, are worth mentioning, even if I don't cover them in much detail.

Paranoid networking
  Usually in Linux, all processes are allowed to create sockets and interact with the network. Per Android's security model, however, access to network capabilities has to be controlled. Hence, an option is added to the kernel to gate access to socket creation and network interface administration based on whether the current process belongs to a certain group of processes or possesses certain capabilities. This applies to IPv4, IPv6, and Bluetooth.

  At the time of this writing, this functionality hasn't been merged into the mainline, and the path for its inclusion is unclear. You could run an AOSP on a kernel that doesn't have this functionality, but Android's permission system, especially with regard to socket creation, would be broken.

RAM console
  As I mentioned earlier, the kernel manages its own log, which you can access using the *dmesg* command. The content of this log is very useful, as it often contains critical messages from drivers and kernel subsystems. On a crash or a kernel panic, its content can be instrumental for postmortem analysis. Since this information is typically lost on reboot, Android adds a driver that registers a RAM-based console that survives reboots and makes its content accessible through */proc/last_kmsg*.

  At the time of this writing, the RAM console's functionality seems to have been merged into mainline within the pstore filesystem in the kernel's *fs/pstore/* directory.

Physical memory (pmem)
  Like ashmem, the pmem driver allows for sharing memory between processes. However, unlike ashmem, it allows the sharing of large chunks of physically contiguous memory regions, not virtual memory. In addition, these memory regions may be shared between processes and drivers. For the G1 handset, for instance,

pmem heaps are used for 2D hardware acceleration. Note, though, that pmem was used in very few devices. In fact, according to Brian Swetland, one of the Android kernel development team members, it was written to specifically address the MSM7201A's limitations, the MSM7201A being the SoC in the G1.

At the time of this writing, this driver is considered obsolete and has been dropped. It isn't found in the mainline kernel, and there are no plans to revive it. It appears that the ION memory allocator (*http://lwn.net/Articles/480055/*) is poised to replace whatever uses pmem had.

Hardware Support

Android's hardware support approach is significantly different from the classic approach typically found in the Linux kernel and Linux-based distributions. Specifically, the way hardware support is implemented, the abstractions built on that hardware support, and the mind-set surrounding the licensing and distribution of the resulting code are all different.

The Linux Approach

The usual way to provide support for new hardware in Linux is to create device drivers that are either built as part of the kernel or loaded dynamically at runtime through modules. The corresponding hardware is thereafter generally accessible in user-space through entries in */dev*. Linux's driver model defines three basic types of devices: character devices (devices that appear as a stream of bytes), block devices (essentially hard disks), and networking devices. Over the years, quite a few additional device and subsystem types have been added, such as for USB or Memory Technology Device (MTD) devices. Nevertheless, the APIs and methods for interfacing with the */dev* entry corresponding to a given type of device have remained fairly standardized and stable.

This has allowed various software stacks to be built on top of */dev* nodes either to interact with the hardware directly or to expose generic APIs that are used by user applications to provide access to the hardware. The vast majority of Linux distributions in fact ship with a similar set of core libraries and subsystems, such as the ALSA audio libraries and the X Window System, to interface with hardware devices exposed through */dev*.

With regard to licensing and distribution, the general "Linux" approach has always been that drivers should be merged and maintained as part of the mainline kernel and distributed with it under the terms of the GPL. So, while some device drivers are developed and maintained independently and some are even distributed under other licenses, the consensus has been that this isn't the preferred approach. In fact, with regard to licensing, non-GPL drivers have always been a contentious issue. Hence, the conventional wisdom is that users' and distributors' best bet for getting the latest drivers is usually to get the latest mainline kernel from *http://kernel.org*. This has been true since the kernel's early days and remains true despite some additions having been made to the kernel to allow the creation of user-space drivers.

Android's General Approach

Although Android builds on the kernel's hardware abstractions and capabilities, its approach is very different. On a purely technical level, the most glaring difference is that its subsystems and libraries don't rely on standard */dev* entries to function properly. Instead, the Android stack typically relies on shared libraries provided by manufacturers to interact with hardware. In effect, Android relies on what can be considered a Hardware Abstraction Layer (HAL), although, as we will see, the interface, behavior, and function of abstracted hardware components differ greatly from type to type.

In addition, most software stacks typically found in Linux distributions to interact with hardware are not found in Android. There is no X Window System, for instance, and while ALSA drivers are sometimes used—a decision left up to the hardware manufacturer who provides the shared library implementing audio support for the HAL—access to their functionality is different from that on standard Linux distributions. The ALSA libraries typically used in Linux desktop environments to interface with ALSA drivers, for example, aren't used in the official AOSP tree. Instead, recent Android releases include a BSD-licensed *tinyalsa* library as a replacement.

Figure 2-3 presents the typical way in which hardware is abstracted and supported in Android, along with the corresponding distribution and licensing. As you can see, Android still ultimately relies on the kernel to access the hardware. However, this is done through shared libraries that are either implemented by the device manufacturer or provided as part of the AOSP. Generally speaking, you can consider the HAL layer as being the hardware library loader shown in the diagram, along with the header files defining the various hardware types, with those same header files being used as the API definitions for the hardware library *.so* files.

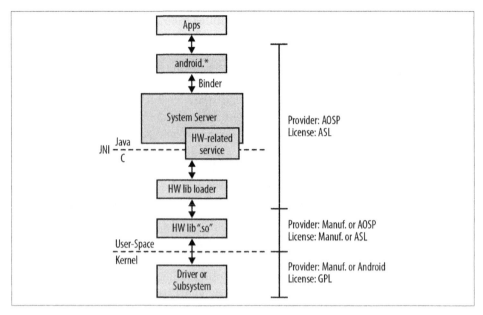

Figure 2-3. Android's "Hardware Abstraction Layer"

One of the main features of this approach is that the license under which the shared library is distributed is up to the hardware manufacturer. Hence, a device manufacturer can create a simplistic device driver that implements the most basic primitives to access a given piece of hardware and make that driver available under the GPL. Not much would be revealed about the hardware, since the driver wouldn't do anything fancy. That driver would then expose the hardware to user-space through mmap() or ioctl(), and the bulk of the intelligence would be implemented within a proprietary shared library in user-space that uses those functions to drive the hardware.

Android does not in fact specify how the shared library and the driver or kernel subsystem should interact. Only the API provided by the shared library to the upper layers is specified by the HAL. Hence, it's up to you to determine the specific driver interface that best fits your hardware, so long as the shared library you provide implements the appropriate API. Nevertheless, we will cover the typical methods used by Android to interface to hardware in the next section.

Where Android is relatively inconsistent is the way the hardware-supporting shared libraries are loaded by the upper layers. Remember for now that for most hardware types, there has to be a *.so* file that is either provided by the AOSP or that you must provide for Android to function properly.

No matter which mechanism is used to load a hardware-supporting shared library, a system service corresponding to the type of hardware is typically responsible for loading and interfacing with the shared library. That system service will be responsible for interacting and coordinating with the other system services to make the hardware behave coherently with the rest of the system and the APIs exposed to app developers. If you're adding support for a given type of hardware, it's therefore crucial that you try to understand in as much detail as possible the internals of the system service corresponding to your hardware. Usually, the system service will be split in two parts: one part in Java that implements most of the Android-specific intelligence, and another part in C/C++ whose main job is to interact with the HAL, the hardware-supporting shared library and other low-level functions.

Loading and Interfacing Methods

As I mentioned earlier, there are various ways in which system services and Android in general interact with the shared libraries implementing hardware support and hardware devices in general. It's difficult to fully understand why there is such a variety of methods, but I suspect that some of them evolved organically. Luckily, there seems to be a movement toward a more uniform way of doing things. Given that Android moves at a fairly rapid pace, this is one area that will require keeping an eye on for the foreseeable future, as it's likely to evolve.

Note that the methods described here are not necessarily mutually exclusive. Often a combination of these is used within the Android stack to load and interface with a shared library or some software layer before or after it. I'll cover specific hardware in the next section.

dlopen()-*loading through HAL*
Applies to: GPS, Lights, Sensors, and Display. Also applies to Audio and Camera starting from 4.0/Ice-Cream Sandwich.

Some hardware-supporting shared libraries are loaded by the *libhardware* library. This library is part of Android's HAL and exposes hw_get_module(), which is used by some system services and subsystems to explicitly load a given specific hardware-supporting shared library (a.k.a. a "module" in HAL terminology). hw_get_module() in turn relies on the classic dlopen() to load libraries into the caller's address space.

HAL "modules" shouldn't be confused with loadable kernel modules, which are a completely different and unrelated software construct, even though they share some similar properties.

Linker-loaded .so files
   Applies to: Audio, Camera, Wifi, Vibrator, and Power Management

   In some cases, system services are simply linked against a given *.so* file at build time. Hence, when the corresponding binary is run, the dynamic linker automatically loads the shared library into the process's address space.

Hardcoded `dlopen()`*s*
   Applies to: StageFright and Radio Interface Layer (RIL)

   In a few cases, the code invokes `dlopen()` directly instead of going through `libhardware` to fetch a hardware-enabling shared library. The rationale for using this method instead of the HAL is unclear.

Sockets
   Applies to: Bluetooth, Network Management, Disk Mounting, and Radio Interface Layer (RIL)

   Sockets are sometimes used by system services or framework components to talk to a remote daemon or service that actually interacts with the hardware.

Sysfs entries
   Applies to: Vibrator and Power Management

   Some entries in sysfs (*/sys*) can be used to control the behavior of hardware and/or kernel subsystems. In some cases, Android uses this method instead of */dev* entries to control the hardware. Use of sysfs entries instead of */dev* nodes makes sense, for instance, when defaults need to be set during system initialization when no part of the framework is yet running.

/dev nodes
   Applies to: Almost every type of hardware

   Arguably, any hardware abstraction must at some point communicate with an entry in */dev*, because that's how drivers are exposed to user-space. Some of this communication is likely hidden from Android itself because it interacts with a shared library instead, but in some corner cases AOSP components directly access device nodes. Such is the case of input libraries used by the Input Manager.

D-Bus
   Applies to: Bluetooth

   D-Bus is a classic messaging system found in most Linux distributions for facilitating communication between various desktop components. It's included in Android because it's the prescribed way for a non-GPL component to talk to the GPL-licensed BlueZ stack—Linux's default Bluetooth stack and the one used in Android—without being subject to the GPL's redistribution requirements; D-Bus itself being dual-licensed under the Academic Free License (AFL) and the GPL. Have a look at freedesktop.org's D-Bus page (*http://dbus.freedesktop.org*) for more information.

Given that BlueZ has been removed from the AOSP starting with 4.2/Jelly Bean, it's unclear what uses D-Bus will have, if any, in future Android releases.

Device Support Details

Table 2-1 summarizes the way in which each type of hardware is supported in Android. As you'll notice, there is a wide variety of combinations of mechanisms and interfaces. If you plan on implementing support for a specific type of hardware, the best way forward is to start from an existing sample implementation. The AOSP typically includes hardware support code for a few handsets, generally those that were used by Google to develop new Android releases and therefore served as lead devices. Sometimes the sources for hardware support are quite extensive, as was the case for the Samsung Nexus S (a.k.a. "Crespo," its code name) in Gingerbread, and the Galaxy Nexus (a.k.a. "Maguro") and the Nexus 7 (a.k.a. "Grouper") in Jelly Bean.

The only type of hardware for which you are unlikely to find publicly available implementations on which to base your own is the RIL. For various reasons, it's best not to let everyone be able to play with the airwaves. Hence, manufacturers don't make such implementations available. Instead, Google provides a reference RIL implementation in the AOSP should you want to implement a RIL.

Table 2-1. Android's hardware support methods and interfaces

Hardware	System Service	Interface to User-Space Hardware Support	Interface to Hardware
Audio	Audio Flinger	Linker-loaded[a] *libaudio.so*	Up to hardware manufacturer, though ALSA is typical
Bluetooth	Bluetooth Service	Socket/D-Bus to BlueZ[b]	BlueZ stack
Camera	Camera Service	Linker-loaded[c] *libcamera.so*	Up to hardware manufacturer, sometimes Video4Linux
Display	Surface Flinger	HAL-loaded *gralloc* module[d]	Up to hardware manufacturer, */dev/fb0* or */dev/graphics/fb0*
GPS	Location Manager	HAL-loaded *gps* module	Up to hardware manufacturer
Input	Input Manager	Native *libui.so* library[e]	Entries in */dev/input/*
Lights	Lights Service	HAL-loaded *lights* module	Up to hardware manufacturer
Media	N/A, StageFright framework within Media Service	dlopen on *libstagefrighthw.so*	Up to hardware manufacturer
Network interfaces[f]	Network Management Service	Socket to *netd*	ioctl() on interfaces
Power management	Power Manager Service	Linker-loaded *libhardware_legacy.so*	Entries in */sys/power/* or, in older days, */sys/android_power/*

Hardware	System Service	Interface to User-Space Hardware Support	Interface to Hardware
Radio (Phone)	Phone Service	Socket to *rild*, which itself does a dlopen() on manufacturer-provided *.so*	Up to hardware manufacturer
Storage	Mount Service	Socket to *vold*	System calls and /*sys* entries
Sensors	Sensor Service	HAL-loaded *sensors* module	Up to hardware manufacturer
Vibrator	Vibrator Service	Linker-loaded *libhardware_legacy.so*	Up to hardware manufacturer
WiFi	Wifi Service	Linker-loaded *libhardware_legacy.so*	Classic *wpa_supplicant*[g] in most cases

[a] This is HAL-loaded starting with 4.0/Ice-Cream Sandwich.

[b] BlueZ has been removed starting with 4.2/Jelly Bean. A Broadcom-supplied Bluetooth stack called *bluedroid* has replaced it. The new Bluetooth stack relies on HAL-loading like most other hardware types.

[c] This is HAL-loaded starting with 4.0/Ice-Cream Sandwich.

[d] The module used by the Surface Flinger is $hwcomposer$ starting with 4.0/Ice-Cream Sandwich

[e] This has been replaced by the $libinput.so$ library starting with 4.0/Ice-Cream Sandwich.

[f] This is for Tether, NAT, PPP, PAN, USB RNDIS (Windows). It isn't for WiFi.

[g] *wpa_supplicant* is the same software package used on any Linux desktop to manage WiFi networks and connections.

Native User-Space

Now that we've covered the low-level layers on which Android is built, let's start going up the stack. First off, we'll cover the native user-space environment in which Android operates. By "native user-space," I mean all the user-space components that run outside the Dalvik virtual machine. This includes quite a few binaries that are compiled to run natively on the target's CPU architecture. These are generally started either automatically or as needed by the init process according to its configuration files, or are available to be invoked from the command line once a developer shells into the device. Such binaries usually have direct access to the root filesystem and the native libraries included in the system. Their capabilities are restricted only by the filesystem rights granted to them and their effective UID and GID. They aren't subject to any of the restrictions imposed on a typical Android app by the Android Framework because they are running outside it.

Note that Android's user-space was designed pretty much from a blank slate and differs greatly from what you'd find in a standard Linux distribution. Hence, I will try as much as possible to explain where Android's user-space is different from or similar to what you'd usually find in a Linux-based system.

Filesystem Layout

Like any other Linux-based distribution, Android uses a root filesystem to store applications, libraries, and data. Unlike the vast majority of Linux-based distributions, however, the layout of Android's root filesystem does not adhere to the Filesystem Hierarchy Standard (FHS).[4] The kernel itself doesn't enforce the FHS, but most software packages built for Linux assume that the root filesystem they are running on conforms to the FHS. Hence, if you intend to port a standard Linux application to Android, you'll likely need to do some legwork to ensure that the filepaths it relies on are still valid, or use some form of "*chroot* jail" to isolate it and its supporting packages from the rest of the root filesystem (see *chroot*'s man page for details).

Given that most of the packages running in Android's user-space were written from scratch specifically for Android, this lack of conformity is of little to no consequence to Android itself. In fact, it has some benefits, as we'll see shortly. Still, it's important to learn how to navigate Android's root filesystem. If nothing else, you'll likely have to spend quite some time inside it as you bring Android up on your hardware or customize it for that hardware.

The two main directories in which Android operates are */system* and */data*. These directories do not emanate from the FHS. In fact, I can't think of any mainstream Linux distribution that uses either of these directories. Rather, they reflect the Android development team's own design. This is one of the first signs hinting that it might be possible to host Android side by side with a common Linux distribution on the same root filesystem. Have a look at Appendix A for more information on how to create such a hybrid.

/system is the main Android directory for storing immutable components generated by the build of the AOSP. This includes native binaries, native libraries, framework packages, and stock apps. It's usually mounted read-only from a separate image from the root filesystem, which is itself mounted from a RAM disk image. */data*, on the other hand, is Android's main directory for storing data and apps that change over time. This includes the data generated and stored by apps installed by the user alongside data generated by Android system components at runtime. It, too, is usually mounted from its own separate image, though in read-write mode.

Android also includes many directories commonly found in any Linux system, such as */dev*, */proc*, */sys*, */sbin*, */root*, */mnt*, and */etc*. These directories often serve similar if not identical purposes to the ones they serve on any Linux system, although they are very often trimmed down, as is the case of */sbin* and */etc*, and in some cases are empty, such as */root*.

---

4. The FHS (*http://www.pathname.com/fhs/*) is a community standard that describes the contents and use of the various directories within a Linux root filesystem.

Interestingly, Android doesn't include any /*bin* or /*lib* directories. These directories are typically crucial in a Linux system, containing, respectively, essential binaries and essential libraries. This is yet another artifact that opens the door for making Android coexist with standard Linux components.

There is of course more to be said about Android's root filesystem. The directories just mentioned, for instance, contain their own hierarchies. Also, Android's root filesystem contains other directories I haven't covered here. We will revisit the Android root filesystem and its makeup in more detail in Chapter 6.

Libraries

Android relies on about 100 dynamically loaded libraries, all stored in the /*system/lib* directory. A certain number of these come from external projects that were merged into Android's codebase to make their functionality available within the Android stack, but a large portion of the libraries in /*system/lib* are actually generated from within the AOSP itself. Table 2-2 lists the libraries included in the AOSP that come from external projects, whereas Table 2-3 summarizes the Android-specific libraries generated from within the AOSP.

Table 2-2. Libraries generated from external projects imported into the AOSP

Library(ies)	External Project	Original Location	License
audio.so, liba2dp, input.so, libbluetooth and *libbluetoothd*	BlueZ[a]	http://www.bluez.org	GPL
libcrypto.so and *libssl.so*	OpenSSL	http://www.openssl.org	Custom, BSD-like
libdbus.so	D-Bus	http://dbus.freedesktop.org	AFL and GPL
libexif.so[b]	Exif JPEG header manipulation tool	http://www.sentex.net/~mwandel/jhead/	Public Domain
libexpat.so	Expat XML Parser	http://expat.sourceforge.net	MIT
libFFTEm.so	neven face recognition library	N/A	ASL
libicui18n.so and *libicuuc.so*	International Components for Unicode	http://icu-project.org	MIT
libiprouteutil.so and *libnetlink.so*	iproute2 TCP/IP networking and traffic control	http://www.linuxfoundation.org/collaborate/workgroups/networking/iproute2	GPL
libjpeg.so	libjpeg	http://www.ijg.org	Custom, BSD-like
libnfc_ndef.so	NXP Semiconductor's NFC library	N/A	ASL
libskia.so and, in 2.3/Gingerbread, *libskiagl.so*	skia 2D graphics library	http://code.google.com/p/skia/	ASL
libsonivox	Sonic Network's Audio Synthesis library	N/A	ASL

Library(ies)	External Project	Original Location	License
libsqlite.so	SQLite database	http://www.sqlite.org	Public domain
libSR_AudioIn.so and, in 2.3/Gingerbread, libsrec_jni.so	Nuance Communications' Speech Recognition engine	N/A	ASL
libstlport.so	Implementation of the C++ Standard Template Library	http://stlport.sourceforge.net	Custom, BSD-like
libttspico.so	SVOX's Text-to-Speech speech synthesizer engine	N/A	ASL
libvorbisidec.so	Tremolo ARM-optimized Ogg Vorbis decompression library	http://wss.co.uk/pinknoise/tremolo/	Custom, BSD-like
libwebcore.so	WebKit Open Source Project	http://www.webkit.org	LGPL and BSD
libwpa_client.so	Library used by legacy HAL to talk to wpa_supplicant daemon	http://hostap.epitest.fi/wpa_supplicant/	GPL and BSD
libz.so	zlib compression library	http://zlib.net	Custom, BSD-like

[a] BlueZ has been replaced by an ASL-licensed, Broadcom-supplied Bluetooth stack called *bluedroid* that is also found in *external/*. The libraries generated by bluedroid are different from those listed here.

[b] Note that Android's *libexif.so*'s API is very different from that library's API as available in traditional Linux distributions.

Table 2-3. Android-specific libraries generated from within the AOSP

Category	Library(ies)	Description
Bionic	libc.so	C library
	libm.so	Math library
	libdl.so	Dynamic linking library
	libstdc++.so	C++ support library[a]
	libthread_db.so	Thread debugging library
Core[b]	libbinder.so	The Binder library
	libutils.so, libcutils.so, libnetutils.so, and libsysutils.so	Various utility libraries
	libsystem_server.so, libandroid_servers.so, libaudioflinger.so, libsurfaceflinger.so, libsensorservice.so, and libcameraservice.so	System-services-related libraries
	libcamera_client.so and, in 2.3/Gingerbread, libsurfaceflinger_client.so[c]	Client libraries for certain system services
	libpixelflinger.so	The PixelFlinger library
	libui.so	Low-level user-interface-related functionalities, such as user input events handling and dispatching and graphics buffer allocation and manipulation
	libgui.so	Library for functions related to sensors and, starting with 4.0/Ice-Cream Sandwich, client communication with the Surface Flinger

Category	Library(ies)	Description
	liblog.so	The logging library
	libandroid_runtime.so	The Android Runtime library
	libandroid.so	C interface to lifecycle management, input events, window management, assets, and Storage Manager
Dalvik	*libdvm.so*	The Dalvik VM library
	libnativehelper.so	JNI-related helper functions
Hardware	*libhardware.so*	The HAL library that provides hw_get_module() and uses dlopen() to load hardware support modules (i.e., shared libraries that provide hardware support to the HAL) on demand
	libhardware_legacy.so	Legacy HAL library providing hardware support for WiFi, power-management, and vibrator
	Various hardware-supporting shared libraries	Libraries that provide support for various hardware components; some are loaded through the HAL, while others are loaded automatically by the linker
Media	*libmediaplayerservice.so*	The Media Player service library
	libmedia.so	The low-level media functions used by the Media Player service
	libstagefright.so*	The many libraries that make up the StageFright media framework
	libeffects.so and the libraries in the *soundfx/* directory	The sound effects libraries
	libdrm1.so and *libdrm1_jni.so*	The DRM (Digital Rights Management) framework libraries
OpenGL	*libEGL.so, libETC1.so, libGLESv1_CM.so, libGLESv2.so,* and *egl/ligGLES_android.so*	Android's OpenGL implementation

[a] Some say that this library is similar in its role to the *libsupc++.a* found in standard Linux systems, while Android's *libstl port.so* is closer to traditional Linux systems' *libstdc++.so*.

[b] I'm using this category as catchall for many core Android functionalities.

[c] Starting with 4.0/Ice-Cream Sandwich, the functionality corresponding to *libsurfaceflinger_client.so* has been merged into *libgui.so*.

Since 2.3/Gingerbread, many libraries have been added to that AOSP. Tables 2-4 and 2-5 list some of the most notable additions you'll find in 4.1/Jelly Bean.

Table 2-4. *Important libraries from external projects found in 4.1/Jelly Bean*

Library(ies)	External Project	Original Location	License
libtinyalsa.so	tinyalsa	http://github.com/tinyalsa	ASL
libmtp.so	libmtp	http://libmtp.sourceforge.net/	LGPL
libchromium_net.so	WebKit	http://webkit.org/	LGPL and BSD
libmdnssd.so	mDNSResponder	http://www.opensource.apple.com/tarballs/mDNSResponder/	ASL

Table 2-5. Important Android-specific libraries found in 4.1/Jelly Bean

Category	Library(ies)	Description
Core	*libjnigraphics.so*	C interface to the 2D graphics system
	libcorkscrew.so	Debugging library
	libRS.so	Interface to RenderScript
Media	*libOpenMAXAL.so*	Native multimedia library, based on Khronos OpenMAX AL
	libOpenSLES.so	Khronos OpenSL EL compatible audio system
	libaudioutils.so	Echo cancellation and other audio tools

Init

One thing Android doesn't change is the kernel's boot process. Hence, whatever you know about the kernel's startup continues to apply just the same to Android's use of Linux. What changes in Android is what happens once the kernel finishes booting. Indeed, after it's finished initializing itself and the drivers it contains, the kernel starts just one user-space process, the *init* process. This process is then responsible for spawning all other processes and services in the system and for conducting critical operations such as reboots. Traditionally, Linux distributions have relied on SystemV init for the init process, although in recent years many distributions have created their own variants. Ubuntu, for instance, uses Upstart (*http://launchpad.net/upstart*). In embedded Linux systems, the classic package that provides init is BusyBox (*http://busybox.net*).

Android introduces its own custom init, which brings with it a few novelties.

Configuration language

Unlike traditional inits, which are predicated on the use of scripts that run per the current run-levels' configuration or on request, Android's init defines its own configuration semantics and relies on changes to global properties to trigger the execution of specific instructions.

The main configuration file for init is usually stored as */init.rc*, but there's also usually a device-specific configuration file stored as */init.<device_name>.rc*, where *<device_name>* is the name of the device. In some cases, such as the emulator, for example, there's also a device-specific script stored as */system/etc/init.<device_name>.sh*. You can get a high degree of control over the system's startup and its behavior by modifying those files. For instance, you can disable the Zygote—a key system component that we'll cover in greater detail later in this chapter and in Chapter 7—from starting up automatically and then starting it manually yourself after having used *adb* to shell into the device.

We'll discuss the init's configuration language in depth in Chapter 6.

Global properties

A very interesting aspect of Android's init is how it manages a global set of properties that can be accessed and set from many parts of the system, with the appropriate rights. Some of these properties are set at build time, while others are set in init's configuration files, and still others are set at runtime. Some properties are also persisted to storage for permanent use. Since init manages the properties, it can detect any changes and therefore trigger the execution of a set of commands based on its configuration.

The OOM adjustments mentioned earlier, for instance, are set on startup by the *init.rc* file. So are network properties. Some of the properties set at build time are stored in the */system/build.prop* file and include the build date and build system details. At runtime, the system will have over 100 different properties, ranging from IP and GSM configuration parameters to the battery's level. Use the *getprop* command to get the current list of properties and their values.

We'll discuss the init's global properties, the files used to provide its default values, and the relevant commands in greater detail in Chapter 6.

udev events

As I explained earlier, access to devices in Linux is done through nodes within the */dev* directory. In the old days, Linux distributions would ship with thousands of entries in that directory to accommodate all possible device configurations. Eventually, though, a few schemes were proposed to make the creation of such nodes dynamic. For some time now, the system in use has been *udev*, which relies on runtime events generated by the kernel every time hardware is added or removed from the system.

In most Linux distributions, the handling of udev hotplug events is done by the *udevd* daemon. In Android, these events are handled by the *ueventd* daemon built as part of Android's init and accessed through a symbolic link from */sbin/ueventd* to */init*. To know which entries to create in */dev*, ueventd relies on the */ueventd.rc* and */ueventd.<device_name>.rc* files.

We'll discuss the *ueventd* and its configuration files in detail in Chapter 6.

Toolbox

Much like the root filesystem's directory hierarchy, there are essential binaries on most Linux systems, listed by the FHS for the */bin* and */sbin* directories. In most Linux distributions, the binaries in those directories are built from separate packages coming from different projects available on the Internet. In an embedded system, it doesn't make sense to have to deal with so many packages, nor necessarily to have that many separate binaries.

The approach taken by the classic BusyBox package is to build a single binary that essentially has what amounts to a huge `switch-case`, which checks for the first parameter on the command line and executes the corresponding functionality. All commands are then made to be symbolic links to the *busybox* command. So when you type *ls*, for example, you're actually invoking BusyBox. But since BusyBox's behavior is predicated on the first parameter on the command line and that parameter is *ls*, it will behave as if you had run that command from a standard Linux shell.

Android doesn't use BusyBox but includes its own tool, Toolbox, that basically functions in the very same way, using symbolic links to the *toolbox* command. Unfortunately, Toolbox is nowhere as feature-rich as BusyBox. In fact, if you've ever used BusyBox, you're likely going to be very disappointed when using Toolbox. The rationale for creating a tool from scratch in this case seems to be the licensing angle, BusyBox being GPL licensed. In addition, some Android developers have stated that their goal was to create a minimal tool for shell-based debugging and not to provide a full replacement for shell tools, as BusyBox is. At any rate, Toolbox is BSD licensed, and manufacturers can therefore modify it and distribute it without having to track the modifications made by their developers or making any sources available to their customers.

You might still want to include BusyBox alongside Toolbox to benefit from its capabilities. If you don't want to ship it as part of your final product because of its licensing, you could include it temporarily during development and strip it from the final production release. I'll cover this in more detail in Appendix A.

Daemons

As part of the system startup, Android's init starts a few key daemons that continue to run throughout the lifetime of the system. Some daemons, such as *adbd*, are started on demand, depending on build options and changes to global properties. Table 2-6 provides a list of some of the more prominent daemons that Android runs. Many of these are discussed in much greater detail in Chapters 6 and 7.

Table 2-6. Android daemons

Daemon	Description
ueventd	Android's replacement for *udev*.
servicemanager	The Binder Context Manager. Acts as an index of all Binder services running in the system.
vold	The volume manager. Handles the mounting and formatting of mounted volumes and images.
netd	The network manager. Handles tethering, NAT, PPP, PAN, and USB RNDIS.
debuggerd	The debugger daemon. Invoked by Bionic's linker when a process crashes to do a postmortem analysis. Allows *gdb* to connect from the host.
rild	The RIL daemon. Mediates all communication between the Phone Service and the Baseband Processor.
Zygote	The Zygote process. It's responsible for warming up the system's cache and starting the System Server. We'll discuss it in more detail later in this chapter.

Daemon	Description
mediaserver	The Media server. Hosts most media-related services. We'll discuss it in more detail later in this chapter.
dbus-daemon	The D-Bus message daemon. Acts as an intermediary between D-Bus users. Have a look at its man page for more information.
bluetoothd	The Bluetooth daemon. Manages Bluetooth devices. Provides services through D-Bus. No longer in the AOSP as of 4.2/Jelly Bean, since the BlueZ stack has been removed.
installd	The *.apk* installation daemon. Takes care of installing and uninstalling *.apk* files and managing the related filesystem entries.
keystore	The KeyStore daemon. Manages an encrypted key-value pair store for cryptographic keys, SSL certs for instance.
system_server	Android's System Server. This daemon hosts the vast majority of system services that run in Android.
adbd	The ADB daemon. Manages all aspects of the connection between the target and the host's *adb* command.

Command-Line Utilities

More than 150 command-line utilities are scattered throughout Android's root filesystem. */system/bin* contains the majority of them, but some "extras" are in */system/xbin*, and a handful are in */sbin*. Around 50 of those in */system/bin* are actually symbolic links to */system/bin/toolbox*. The majority of the rest come from the Android base framework, from external projects merged into the AOSP, or from various other parts of the AOSP. We'll get the chance to cover the various binaries found in the AOSP in more detail in Chapters 6 and 7.

Dalvik and Android's Java

In a nutshell, Dalvik is Android's Java virtual machine. It allows Android to run the byte-code generated from Java-based apps and Android's own system components and provides both with the required hooks and environment to interface with the rest of the system, including native libraries and the rest of the native user-space. There's more to be said about Dalvik and Android's brand of Java, though. But before I can delve into that explanation, I must first cover some Java basics.

Without boring you with yet another history lesson on the Java language and its origins, suffice it to say that Java was created by James Gosling at Sun in the early '90s, that it rapidly became very popular, and that it was, in sum, more than well established before Android came around. From a developer perspective, two aspects are important to keep in mind with regard to Java: its differences from a traditional language such as C and C++, and the components that make up what we commonly refer to as "Java."

By design, Java is an interpreted language. Unlike C and C++, where the code you write gets compiled by a compiler into binary assembly instructions to be executed by a CPU matching the architecture targeted by the compiler, the code you write in Java gets compiled by a Java compiler into architecture-independent byte-code that is executed at runtime by a byte-code interpreter, also commonly referred to as a "virtual machine." This modus operandi, along with Java's semantics, enables the language to include quite a few features not traditionally found in previous languages, such as reflection and anonymous classes. Also, unlike C and C++, Java doesn't require you to keep track of objects you allocate. In fact, it requires you to lose track of all unused objects, since it has an integrated garbage collector that will ensure that all such objects are destroyed when no active code holds a reference to them any longer.

At a practical level, Java is actually made up of a few distinct things: the Java compiler, the Java byte-code interpreter—more commonly known as the Java Virtual Machine (JVM)—and the Java libraries commonly used by Java developers. Together, these are usually obtained by developers through the Java Development Kit (JDK) provided free of charge by Oracle. Android actually relies on the JDK for the Java compiler at build time, but it doesn't use the JVM or the libraries found in the JDK. Instead of the JVM it relies on Dalvik, and instead of the JDK libraries it relies on the Apache Harmony project, a clean-room implementation of the Java libraries hosted under the umbrella of the Apache project.

None of the JDK components are found in the images generated by the build of the AOSP. Hence, none of the JDK's components would be distributed by you when using Android for your embedded system.

Java Lingo

Java has its own specialized terminology. The following explanations should help you make sense of some of the terms being used in the text, if you aren't already familiar with them:

virtual machine
   This term was less ambiguous when Java came out, because "virtual machine" software products such as VMware and VirtualBox weren't as common or as popular as they are today. Such virtual machines do far more than interpret byte-code, as Java virtual machines do.

reflection
   The ability to ask an object whether it implements a certain method.

anonymous classes
   Snippets of code that are passed as a parameter to a method being invoked. An anonymous class might be used, for instance, as a callback registration method, thereby enabling the developer to see the code handling an event at the same location in the source code where she invokes the callback registration method.

> *.jar files*
>
> *.jar* files are actually Java ARchives (JAR) containing many *.class* files, each of which contains only a single class.

According to its developer, Dan Bornstein, Dalvik distinguishes itself from the JVM by being specifically designed for embedded systems. Namely, it targets systems that have slow CPUs and relatively little RAM, run OSes that don't use swap space, and are battery powered.

While the JVM munches on *.class* files, Dalvik prefers the *.dex* delicatessen. *.dex* files are actually generated by postprocessing the *.class* files generated by the Java compiler through Android's *dx* utility. Among other things, an uncompressed *.dex* file is 50% smaller than its originating *.jar* file.

For more information about the features and internals of Dalvik, I strongly encourage you to take a look at Dan Bornstein's Google I/O 2008 presentation entitled "Dalvik Virtual Machine Internals." It's about one hour long and available on YouTube (*http://www.youtube.com/watch?v=ptjedOZEXPM*). You can also just go to YouTube and search for "Dan Bornstein Dalvik."

> Another interesting factoid is that Dalvik is register-based, whereas the JVM is stack-based, though that is likely to have little to no meaning to you unless you're an avid student of VM theory, architecture, and internals.
>
> If you'd like to get the inside track on the benefits and trade-offs between stack-based VMs and register-based VMs, have a look at the paper entitled "Virtual Machine Showdown: Stack Versus Registers" by Shi et al. in proceedings of VEE'05, June 11–12, 2005, Chicago, p. 153–163.

A feature of Dalvik very much worth highlighting, though, is that since 2.2/Froyo it has included a Just-in-Time (JIT) compiler for ARM, with x86 and MIPS having been added since. Historically, JIT has been a defining feature for many VMs, helping them close the gap with noninterpreted languages. Indeed, having a JIT means that Dalvik converts apps' byte-codes to binary assembly instructions that run natively on the target's CPU instead of being interpreted one instruction at a time by the VM. The result of this conversion is then stored for future use. Hence, apps take longer to load the first time, but once they've been JIT'ed, they load and run much faster. The only caveat here is that JIT is available for a limited number of architectures only, namely ARM, x86, and MIPS.

As an embedded developer, you're unlikely to need to do anything specific to get Dalvik to work on your system. Dalvik was written to be architecture-independent. It has been reported that some of the early ports of Dalvik suffered from some endian issues. However, these issues seem to have subsided since.

Java Native Interface (JNI)

Despite its power and benefits, Java can't always operate in a vacuum, and code written in Java sometimes needs to interface with code coming from other languages. This is especially true in an embedded environment such as Android, where low-level functionality is never too far away. To that end, the Java Native Interface (JNI) mechanism is provided. It's essentially a call bridge to other languages such as C and C++. It's an equivalent to *P/Invoke* in the .NET/C# world.

App developers sometimes use JNI to call the native code they compile with the NDK from their regular Java code built using the SDK. Internally, though, the AOSP relies massively on JNI to enable Java-coded services and components to interface with Android's low-level functionality, which is mostly written in C and C++. Java-written system services, for instance, very often use JNI to communicate with matching native code that interfaces with a given service's corresponding hardware.

A large part of the heavy lifting to allow Java to communicate with other languages through JNI is actually done by Dalvik. If you go back to Table 2-3 in the previous section, for instance, you'll notice the *libnativehelper.so* library, which is provided as part of Dalvik for facilitating JNI calls.

Appendix B shows an example use of JNI to interface Java and C code. For the moment, keep in mind that JNI is central to platform work in Android and that it can be a relatively complex mechanism to use, especially to ensure that you use the appropriate call semantics and function parameters.

Unfortunately, JNI seems to be a dark art reserved for the initiated. In other words, it's rather difficult to find good documentation on it. There is one authoritative book on the topic, *The Java Native Interface Programmer's Guide and Specification* by Sheng Liang (Addison-Wesley, 1999).

System Services

System services are Android's man behind the curtain. Even if they aren't explicitly mentioned in Google's app development documentation, anything remotely interesting in Android goes through one of about 50 to 70 system services. These services cooperate to collectively provide what essentially amounts to an object-oriented OS built on top of Linux, which is exactly what Binder—the mechanism on which all system services are built—was intended for. The native user-space we just covered is actually designed very much as a support environment for Android's system services. It's therefore crucial to understand what system services exist and how they interact with one another and with the rest of the system. We've already covered some of this as part of discussing Android's hardware support.

Figure 2-4 illustrates in greater detail the system services first introduced in Figure 2-1. As you can see, there are in fact a couple of major processes involved. Most prominent is the System Server, whose components all run under the same process, *system_server*, and which is mostly made up of Java-coded services with two services written in C/C++. The System Server also includes some native code access through JNI to allow some of the Java-based services to interface to Android's lower layers. Another set of system services is housed within the Media Service, which runs as *mediaserver*. These services are all coded in C/C++ and are packaged alongside media-related components such as the StageFright multimedia framework and audio effects. Finally, the Phone application houses the Phone service separately from the rest. Since 4.0/Ice-Cream Sandwich, note that the Surface Flinger has been forked off into a separate standalone process.

The terminology here isn't my choosing, and it's unfortunately confusing. The "System Server" process houses **several** system services within the same process. So does the "Media Service." Both "System Server" and "Media Service" are spelled out as **singular** regardless of the number of system services they comprise. When this book refers to "system services," plural, it refers to all system services available in the system regardless of the process they run under. So, in short, neither "System Server" nor "Media Service" are part of the "system services." Instead, they are processes used to run the latter.

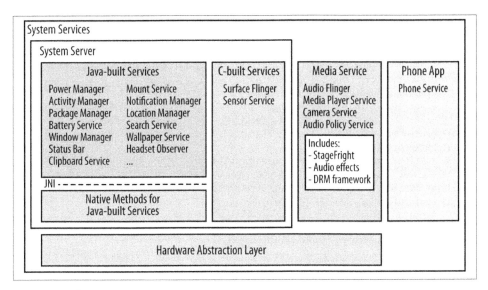

Figure 2-4. System services

Note that despite there being only a handful of processes to house the entirety of the Android's system services, they all appear to operate independently to anyone connecting to their services through Binder. Here's the output of the *service* utility on an Android 2.3/Gingerbread emulator:

```
# service list
Found 50 services:
0 phone: [com.android.internal.telephony.ITelephony]
1 iphonesubinfo: [com.android.internal.telephony.IPhoneSubInfo]
2 simphonebook: [com.android.internal.telephony.IIccPhoneBook]
3 isms: [com.android.internal.telephony.ISms]
4 diskstats: []
5 appwidget: [com.android.internal.appwidget.IAppWidgetService]
6 backup: [android.app.backup.IBackupManager]
7 uimode: [android.app.IUiModeManager]
8 usb: [android.hardware.usb.IUsbManager]
9 audio: [android.media.IAudioService]
10 wallpaper: [android.app.IWallpaperManager]
11 dropbox: [com.android.internal.os.IDropBoxManagerService]
12 search: [android.app.ISearchManager]
13 location: [android.location.ILocationManager]
14 devicestoragemonitor: []
15 notification: [android.app.INotificationManager]
16 mount: [IMountService]
17 accessibility: [android.view.accessibility.IAccessibilityManager]
18 throttle: [android.net.IThrottleManager]
19 connectivity: [android.net.IConnectivityManager]
20 wifi: [android.net.wifi.IWifiManager]
21 network_management: [android.os.INetworkManagementService]
```

```
22 netstat: [android.os.INetStatService]
23 input_method: [com.android.internal.view.IInputMethodManager]
24 clipboard: [android.text.IClipboard]
25 statusbar: [com.android.internal.statusbar.IStatusBarService]
26 device_policy: [android.app.admin.IDevicePolicyManager]
27 window: [android.view.IWindowManager]
28 alarm: [android.app.IAlarmManager]
29 vibrator: [android.os.IVibratorService]
30 hardware: [android.os.IHardwareService]
31 battery: []
32 content: [android.content.IContentService]
33 account: [android.accounts.IAccountManager]
34 permission: [android.os.IPermissionController]
35 cpuinfo: []
36 meminfo: []
37 activity: [android.app.IActivityManager]
38 package: [android.content.pm.IPackageManager]
39 telephony.registry: [com.android.internal.telephony.ITelephonyRegistry]
40 usagestats: [com.android.internal.app.IUsageStats]
41 batteryinfo: [com.android.internal.app.IBatteryStats]
42 power: [android.os.IPowerManager]
43 entropy: []
44 sensorservice: [android.gui.SensorServer]
45 SurfaceFlinger: [android.ui.ISurfaceComposer]
46 media.audio_policy: [android.media.IAudioPolicyService]
47 media.camera: [android.hardware.ICameraService]
48 media.player: [android.media.IMediaPlayerService]
49 media.audio_flinger: [android.media.IAudioFlinger]
```

Here's the same output on a 4.2/Jelly Bean emulator:

```
root@android:/ # service list
Found 68 services:
0 phone: [com.android.internal.telephony.ITelephony]
1 iphonesubinfo: [com.android.internal.telephony.IPhoneSubInfo]
2 simphonebook: [com.android.internal.telephony.IIccPhoneBook]
3 isms: [com.android.internal.telephony.ISms]
4 dreams: [android.service.dreams.IDreamManager]
5 commontime_management: []
6 samplingprofiler: []
7 diskstats: []
8 appwidget: [com.android.internal.appwidget.IAppWidgetService]
9 backup: [android.app.backup.IBackupManager]
10 uimode: [android.app.IUiModeManager]
11 serial: [android.hardware.ISerialManager]
12 usb: [android.hardware.usb.IUsbManager]
13 audio: [android.media.IAudioService]
14 wallpaper: [android.app.IWallpaperManager]
15 dropbox: [com.android.internal.os.IDropBoxManagerService]
16 search: [android.app.ISearchManager]
17 country_detector: [android.location.ICountryDetector]
18 location: [android.location.ILocationManager]
19 devicestoragemonitor: []
```

```
20 notification: [android.app.INotificationManager]
21 updatelock: [android.os.IUpdateLock]
22 throttle: [android.net.IThrottleManager]
23 servicediscovery: [android.net.nsd.INsdManager]
24 connectivity: [android.net.IConnectivityManager]
25 wifi: [android.net.wifi.IWifiManager]
26 wifip2p: [android.net.wifi.p2p.IWifiP2pManager]
27 netpolicy: [android.net.INetworkPolicyManager]
28 netstats: [android.net.INetworkStatsService]
29 textservices: [com.android.internal.textservice.ITextServicesManager]
30 network_management: [android.os.INetworkManagementService]
31 clipboard: [android.content.IClipboard]
32 statusbar: [com.android.internal.statusbar.IStatusBarService]
33 device_policy: [android.app.admin.IDevicePolicyManager]
34 lock_settings: [com.android.internal.widget.ILockSettings]
35 mount: [IMountService]
36 accessibility: [android.view.accessibility.IAccessibilityManager]
37 input_method: [com.android.internal.view.IInputMethodManager]
38 input: [android.hardware.input.IInputManager]
39 window: [android.view.IWindowManager]
40 alarm: [android.app.IAlarmManager]
41 vibrator: [android.os.IVibratorService]
42 battery: []
43 hardware: [android.os.IHardwareService]
44 content: [android.content.IContentService]
45 account: [android.accounts.IAccountManager]
46 user: [android.os.IUserManager]
47 permission: [android.os.IPermissionController]
48 cpuinfo: []
49 dbinfo: []
50 gfxinfo: []
51 meminfo: []
52 activity: [android.app.IActivityManager]
53 package: [android.content.pm.IPackageManager]
54 scheduling_policy: [android.os.ISchedulingPolicyService]
55 telephony.registry: [com.android.internal.telephony.ITelephonyRegistry]
56 display: [android.hardware.display.IDisplayManager]
57 usagestats: [com.android.internal.app.IUsageStats]
58 batteryinfo: [com.android.internal.app.IBatteryStats]
59 power: [android.os.IPowerManager]
60 entropy: []
61 sensorservice: [android.gui.SensorServer]
62 media.audio_policy: [android.media.IAudioPolicyService]
63 media.camera: [android.hardware.ICameraService]
64 media.player: [android.media.IMediaPlayerService]
65 media.audio_flinger: [android.media.IAudioFlinger]
66 drm.drmManager: [drm.IDrmManagerService]
67 SurfaceFlinger: [android.ui.ISurfaceComposer]
```

There is unfortunately not much documentation on how each of these services operates. You'll have to look at each service's source code to get a precise idea of how it works and how it interacts with other services.

> ### Reverse-Engineering Source Code
>
> Fully understanding the internals of Android's system services is like trying to swallow a whale. In 2.3/Gingerbread there were about 85,000 lines of Java code in the System Server alone, spread across 100 different files. And that didn't count any system service code written in C/C++. To add insult to injury, so to speak, the comments are few and far between and the design documents nonexistent. Arm yourself with a good dose of patience if you want to dig further here.
>
> One trick is to create a new Java project in Eclipse and import the System Server's code into that project. This won't compile in any way, but it'll allow you to benefit from Eclipse's Java browsing capabilities in trying to understand the code. For instance, you can open a single Java file, right-click the source browsing scrollbar area, and select Folding → Collapse All. This will essentially collapse all methods into a single line next to a plus sign (+) and will allow you to see the trees (the method names lined up one after another) instead of the leaves (the actual content of each method.) You'll very much still be in a forest, though.
>
> You can also try using one of the commercial source code analysis tools on the market from vendors such as Imagix, Rationale, Lattix, or Scitools. Although there are some open source analysis tools out there, most seem geared toward locating bugs, not reverse-engineering the code being analyzed. Still, some have reported that they've found Ctags and the open source AndroidXRef (*http://androidxref.com/*) projects helpful in their efforts.

Service Manager and Binder Interaction

As I explained earlier, the Binder mechanism used as system services' underlying fabric enables object-oriented RPC/IPC. For a process in the system to invoke a system service through Binder, though, it must first have a handle to it. For instance, Binder will enable an app developer to request a wakelock from the Power Manager by invoking the `acquire()` method of its `WakeLock` nested class. Before that call can be made, though, the developer must first get a handle to the Power Manager service. As we'll see in the next section, the app development API actually hides the details of how it gets this handle in an abstraction to the developer, but under the hood all system service handle lookups are done through the Service Manager, as illustrated in Figure 2-5.

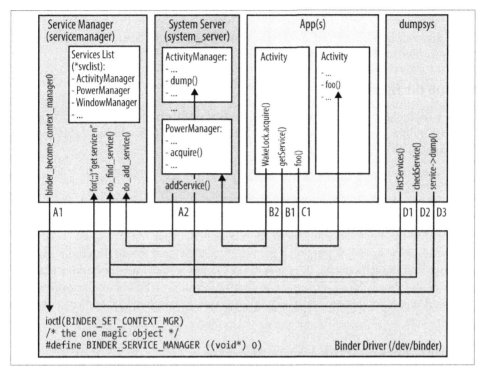

Figure 2-5. Service Manager and Binder interaction

Think of the Service Manager as a Yellow Pages book of all services available in the system. If a system service isn't registered with the Service Manager, then it's effectively invisible to the rest of the system. To provide this indexing capability, the Service Manager is started by *init* before any other service. It then opens */dev/binder* and uses a special ioctl() call to set itself as the Binder's *Context Manager* (A1 in Figure 2-5.) Thereafter, any process in the system that attempts to communicate with Binder ID 0 (a.k.a. the "magic" Binder or "magic object" in various parts of the code) is actually communicating through Binder to the Service Manager.

When the System Server starts, for instance, it registers every single service it instantiates with the Service Manager (A2). Later, when an app tries to talk to a system service, such as the Power Manager service, it first asks the Service Manager for a handle to the service (B1) and then invokes that service's methods (B2). In contrast, a call to a service component running within an app goes directly through Binder (C1) and is not looked up through the Service Manager.

The Service Manager is also used in a special way by a number of command-line utilities such as the *dumpsys* utility, which allows you to dump the status of a single or all system services. To get the list of all services, *dumpsys* loops around to get every system service (D1), requesting the nth plus one at every iteration until there aren't any more. To get each service, *dumpsys* just asks the Service Manager to locate that specific one (D2). With a service handle in hand, *dumpsys* invokes that service's dump() function to dump its status (D3) and displays that on the terminal.

Calling on Services

All of what I just explained is, as I said earlier, almost invisible to regular app developers. Here's a snippet, for instance, that allows us to grab a wakelock within an app using the regular application development API:

```
PowerManager pm = (PowerManager) getSystemService(POWER_SERVICE);
PowerManager.WakeLock wakeLock =
    pm.newWakeLock(PowerManager.FULL_WAKE_LOCK, "myPreciousWakeLock");
wakeLock.acquire(100);
```

Notice that we don't see any hint of the Service Manager here. Instead, we're using getSystemService() and passing it the POWER_SERVICE parameter. Internally, though, the code that implements getSystemService() does actually use the Service Manager to locate the Power Manager service so that we create a wakelock and acquire it. Appendix B shows you how to add a system service and make it available through getSystemService().

A Service Example: the Activity Manager

Although covering each and every system service is outside the scope of this book, let's have a quick look at the Activity Manager, one of the key system services. In 2.3/Gingerbread, the Activity Manager's sources actually span over 30 files and 20,000 lines of code. If there's a core to Android's internals, this service is very much near it. It takes care of the starting of new components, such as Activities and Services, along with the fetching of Content Providers and intent broadcasting. If you ever got the dreaded ANR (Application Not Responding) dialog box, know that the Activity Manager was behind it. It's also involved in the maintenance of OOM adjustments used by the in-kernel low-memory handler, permissions, task management, etc.

For instance, when the user clicks an icon to start an app from his home screen, the first thing that happens is the Launcher's onClick() callback is called (the Launcher being the default app packaged with the AOSP that takes care of the main interface with the user, the home screen). To deal with the event, the Launcher will then call, through Binder, the startActivity() method of the Activity Manager service. The service will then call the startViaZygote() method, which will open a socket to the Zygote and ask it to start the Activity. All this may make more sense after you read the final section of this chapter.

If you're familiar with Linux's internals, a good way to think of the Activity Manager is that it's to Android what the content of the *kernel/* directory in the kernel's sources is to Linux. It's that important.

Stock AOSP Packages

The AOSP ships with a certain number of default packages that are found in most Android devices. As I mentioned in the previous chapter, though, some apps such as Maps, YouTube, and Gmail aren't part of the AOSP. Let's take a look at some of the most notable packages included by default; as we'll see below, the AOSP includes many more packages. Table 2-7 lists the most important stock apps included in the 2.3/Gingerbread AOSP; Table 2-8 lists that AOSP's main content providers; and Table 2-9 lists the corresponding IMEs (input method editors).

While stock apps are coded very much like standard apps, most won't build outside the AOSP using the standard SDK. Hence, if you'd like to create your own version of one of these apps (i.e., fork it), you'll either have to do it inside the AOSP or invest some time in getting the app to build outside the AOSP with the standard SDK. For one thing, these apps sometimes use APIs that are accessible within the AOSP but aren't exported through the standard SDK.

Table 2-7. Stock AOSP apps

App in AOSP	Name Displayed in Launcher	Description
AccountsAndSyncSettings	N/A	Accounts management app
Bluetooth	N/A	Bluetooth manager
Browser	Browser	Default Android browser, includes bookmark widget
Calculator	Calculator	Calculator app
Calendar	Calendar	Calendar app
Camera	Camera	Camera app
CertInstaller	N/A	UI for installing certificates
Contacts	Contacts	Contacts manager app
DeskClock	Clock	Clock and alarm app, including the clock widget
DownloadProviderUi	Downloads	UI for DownloadProvider
Development	Dev Tools	Miscellaneous dev tools
Email	Email	Default Android email app
Gallery	Gallery	Default gallery app for viewing pictures
Gallery3D	Gallery	Fancy gallery with "sexier" UI
HTMLViewer	N/A	App for viewing HTML files
Launcher2	N/A	Default home screen
Mms	Messaging	SMS/MMS app
Music	Music	Music player

App in AOSP	Name Displayed in Launcher	Description
Nfc	N/A	NFC configuration UI and NFC system service
PackageInstaller	N/A	App install/uninstall UI
Phone	Phone	Default phone dialer/UI and phone system service
Protips	N/A	Home screen tips
Provision	N/A	App for setting a flag indicating whether a device was provisioned
QuickSearchBox	Search	Search app and widget
Settings	Settings	Settings app, also accessible through home screen menu
SoundRecorder	N/A	Sound recording app; activated when recording intent is sent, not by user
SpeechRecorder	Speech Recorder	Speech recording app
SystemUI	N/A	Status bar

Table 2-8. Stock AOSP providers

Provider	Description
ApplicationsProvider	Provider to search installed apps
CalendarProvider	Main Android calendar storage and provider
ContactsProvider	Main Android contacts storage and provider
DownloadProvider[a]	Download management, storage, and access
DrmProvider	Management and access of DRM-protected storage
MediaProvider	Media storage and provider
TelephonyProvider	Carrier and SMS/MMS storage and provider
UserDictionaryProvider	Storage and provider for user-defined words dictionary

[a] Interestingly, this package's source code includes a design document, a rarity in the AOSP.

Table 2-9. Stock AOSP input methods

Input Method	Description
LatinIME	Latin keyboard
OpenWnn	Japanese keyboard
PinyinIME	Chinese keyboard

The AOSP contains a lot more packages than those listed in the above tables. Indeed, if you search the sources, you'll find that a 4.2/Jelly Bean release can generate about 500 apps. A large number of those are either tests or samples and aren't worth focusing on in the current discussion. Roughly a quarter of these apps are worth putting into a final product, and they are mostly found in the following directories of the AOSP:

- *packages/apps/*
- *packages/inputmethods/*
- *packages/providers/*
- *packages/screensavers/* (new to 4.2/Jelly Bean)
- *packages/wallpapers/*
- *frameworks/base/packages/*
- *development/apps/*

You'll probably want to look at the content of those directories in conjunction with the above tables to determine which packages are worth further investigation in the context of your project. Like many other things in the AOSP, of course, the packages it contains change over time, as do their locations. Here's a summary of some of the location changes that have occurred between 2.3.4/Gingerbread and 4.2/Jelly Bean:

- AccountAndSyncSettings and Gallery3D have been removed from *packages/apps/*, and the following packages have been added: CellBroadcastReceiver, SmartCardService, BasicSmsReceiver, Exchange, Gallery2, KeyChain, MusicFX, SpareParts, VideoEditor, and LegacyCamera.
- TtsService and VpnServices have been removed from *frameworks/base/packages/*, and the following packages have been added: BackupRestoreConfirmation, SharedStorageBackup, VpnDialogs, WAPPushManager, FakeOemFeatures, FusedLocation, and InputDevices.

System Startup

The best way to bring together everything we've discussed is to look at Android's startup. As you can see in Figure 2-6, the first cog to turn is the CPU. It typically has a hardcoded address from which it fetches its first instructions. That address usually points to a chip that has the bootloader programmed on it. The bootloader then initializes the RAM, puts basic hardware in a quiescent state, loads the kernel and RAM disk, and jumps into the kernel. More recent SoC devices, which include a CPU and a slew of peripherals in a single chip, can actually boot straight from a properly formatted SD card or SD-card-like chip. The PandaBoard and recent editions of the BeagleBoard, for instance, don't have any onboard flash chips because they boot straight from an SD card.

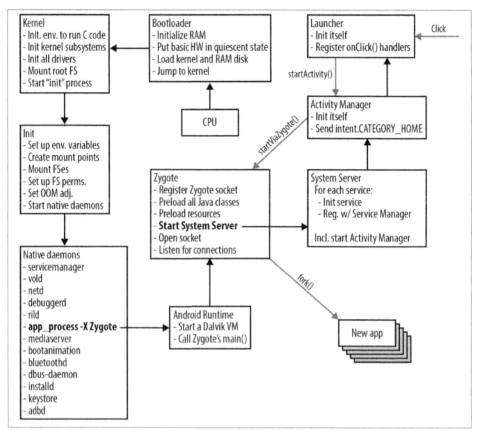

Figure 2-6. Android's boot sequence

The initial kernel startup is very hardware dependent, but its purpose is to set things up so that the CPU can start executing C code as early as possible. Once that's done, the kernel jumps to the architecture-independent `start_kernel()` function, initializes its various subsystems, and proceeds to call the "init" functions of all built-in drivers. The majority of messages printed out by the kernel at startup come from these steps. The kernel then mounts its root filesystem and starts the init process.

That's when Android's init kicks in and executes the instructions stored in its */init.rc* file to set up environment variables such as the system path, create mount points, mount filesystems, set OOM adjustments, and start native daemons. We've already covered the various native daemons active in Android, but it's worth focusing a little on the Zygote. The Zygote is a special daemon whose job is to launch apps. Its functionality is centralized here in order to unify the components shared by all apps and to shorten their start-up time. The init doesn't actually start the Zygote directly; instead it uses the *app_process* command to get Zygote started by the Android Runtime. The runtime then starts the first Dalvik VM of the system and tells it to invoke the Zygote's main().

Zygote is active only when a new app needs to be launched. To achieve a speedier app launch, the Zygote starts by preloading all Java classes and resources that an app may potentially need at runtime. This effectively loads those into the system's RAM. The Zygote then listens for connections on its socket (*/dev/socket/zygote*) for requests to start new apps. When it gets a request to start an app, it forks itself and launches the new app. The beauty of having all apps fork from the Zygote is that it's a "virgin" VM that has all the system classes and resources an app may need preloaded and ready to be used. In other words, new apps don't have to wait until those are loaded to start executing.

All of this works because the Linux kernel implements a copy-on-write (COW) policy for forks. As you may know, forking in Unix involves creating a new process that is an exact copy of the parent process. With COW, Linux doesn't actually copy anything. Instead, it maps the pages of the new process over to those of the parent process and makes copies only when the new process writes to a page. But in fact the classes and resources loaded are never written to, because they're the default ones and are pretty much immutable within the lifetime of the system. So all processes directly forking from the Zygote are essentially using its own mapped copies. Therefore, regardless of the number of apps running on the system, only one copy of the system classes and the resources is ever loaded in RAM.

Although the Zygote is designed to listen to connections for requests to fork new apps, there is one "app" that the Zygote actually starts explicitly: the System Server. This is the first app started by the Zygote, and it continues to live on as an entirely separate process from its parent. The System Server then starts initializing each system service it houses and registering it with the previously started Service Manager. One of the services it starts, the Activity Manager, will end its initialization by sending an intent of type Intent.CATEGORY_HOME. This starts the Launcher app, which then displays the home screen familiar to all Android users.

When the user clicks an icon on the home screen, the process I described in "A Service Example: the Activity Manager" on page 70 takes place. The Launcher asks the Activity Manager to start the process, which in turn "forwards" that request on to the Zygote, which itself forks and starts the new app, which is then displayed to the user.

Once the system has finished starting up, the process list will look something like this:

```
# ps
USER     PID   PPID  VSIZE  RSS   WCHAN     PC                NAME
root     1     0     268    180   c009b74c  0000875c  S       /init
root     2     0     0      0     c004e72c  00000000  S       kthreadd
root     3     2     0      0     c003fdc8  00000000  S       ksoftirqd/0
root     4     2     0      0     c004b2c4  00000000  S       events/0
root     5     2     0      0     c004b2c4  00000000  S       khelper
root     6     2     0      0     c004b2c4  00000000  S       suspend
```

```
root         7    2      0     0  c004b2c4 00000000 S kblockd/0
root         8    2      0     0  c004b2c4 00000000 S cqueue
root         9    2      0     0  c018179c 00000000 S kseriod
root        10    2      0     0  c004b2c4 00000000 S kmmcd
root        11    2      0     0  c006fc74 00000000 S pdflush
root        12    2      0     0  c006fc74 00000000 S pdflush
root        13    2      0     0  c0079750 00000000 D kswapd0
root        14    2      0     0  c004b2c4 00000000 S aio/0
root        22    2      0     0  c017ef48 00000000 S mtdblockd
root        23    2      0     0  c004b2c4 00000000 S kstriped
root        24    2      0     0  c004b2c4 00000000 S hid_compat
root        25    2      0     0  c004b2c4 00000000 S rpciod/0
root        26    1    232   136  c009b74c 0000875c S /sbin/ueventd
system      27    1    804   216  c01a94a4 afd0b6fc S /system/bin/servicemanager
root        28    1   3864   308  ffffffff afd0bdac S /system/bin/vold
root        29    1   3836   304  ffffffff afd0bdac S /system/bin/netd
root        30    1    664   192  c01b52b4 afd0c0cc S /system/bin/debuggerd
radio       31    1   5396   440  ffffffff afd0bdac S /system/bin/rild
root        32    1  60832 16348  c009b74c afd0b844 S zygote
media       33    1  17976  1104  ffffffff afd0b6fc S /system/bin/mediaserver
bluetooth   34    1   1256   280  c009b74c afd0c59c S /system/bin/dbus-daemon
root        35    1    812   232  c02181f4 afd0b45c S /system/bin/installd
keystore    36    1   1744   212  c01b52b4 afd0c0cc S /system/bin/keystore
root        38    1    824   272  c00b8fec afd0c51c S /system/bin/qemud
shell       40    1    732   204  c0158eb0 afd0b45c S /system/bin/sh
root        41    1   3368   172  ffffffff 00008294 S /sbin/adbd
system      65   32 123128 25232  ffffffff afd0b6fc S system_server
app_15     115   32  77232 17576  ffffffff afd0c51c S com.android.inputmethod.
                                                       latin
radio      120   32  86060 17952  ffffffff afd0c51c S com.android.phone
system     122   32  73160 17656  ffffffff afd0c51c S com.android.systemui
app_27     125   32  80664 22900  ffffffff afd0c51c S com.android.launcher
app_5      173   32  74404 18024  ffffffff afd0c51c S android.process.acore
app_2      212   32  73112 17032  ffffffff afd0c51c S android.process.media
app_19     284   32  70336 16672  ffffffff afd0c51c S com.android.bluetooth
app_22     292   32  72752 17844  ffffffff afd0c51c S com.android.email
app_23     320   32  70276 15792  ffffffff afd0c51c S com.android.music
app_28     328   32  70744 16444  ffffffff afd0c51c S com.android.quicksearchbox
app_14     345   32  69708 15404  ffffffff afd0c51c S com.android.protips
app_21     354   32  70912 17152  ffffffff afd0c51c S com.cooliris.media
root       366   41   2128   292  c003da38 00110c84 S /bin/sh
root       367  366    888   324  00000000 afd0b45c R /system/bin/ps
```

This output actually comes from a 2.3/Gingerbread Android emulator, so it contains some emulator-specific artifacts such as the *qemud* daemon. Notice that the apps running all bear their fully qualified package names despite being forked from the Zygote. This is a neat trick that can be pulled in Linux by using the prctl() system call with PR_SET_NAME to tell the kernel to change the calling process's name. Have a look at prctl()'s man page if you're interested in it. Note also that the first process started by init is actually *ueventd*. All processes prior to that are actually started from within the kernel by subsystems or drivers.

Most importantly, notice that the Zygote's process identifier (PID) is 32 and the the parent PID (PPID) of all apps is 32. This illustrates the earlier explanations that the Zygote is the parent of all apps in the system.

CHAPTER 3
AOSP Jump-Start

Now that you have a solid understanding of the basics, let's start getting our hands dirty with the Android Open Source Project (AOSP). We'll start by covering how to get the AOSP distribution from *http://android.googlesource.com/*. Before actually building and running the AOSP, we'll spend some time exploring the AOSP's contents and explain how the sources reflect what we just saw in the previous chapter. Finally, we'll close the chapter by covering the use of *adb* and the emulator, two very important tools when doing any sort of platform work.

Above all, this chapter is meant to be fun. The AOSP is an exciting piece of software with a tremendous amount of innovation. OK, I'll admit it's not all rosy, and some parts do have rough edges. Still, other parts are pure genius. The most amazing thing of all, obviously, is that we can all download it, modify it, and ship our own custom products based on it. So roll up your sleeves and let's get started.

Development Host Setup

As we discussed in "Development Setup and Tools" on page 22, you'll need an Ubuntu-based desktop in order to work on the AOSP. Even though other systems can be made to work, that's the one Google documents as being supported. I suggest you flip back and reread that section to review the basic host setup required for AOSP work. Also, I suggest you have a look at the Initializing a Build Environment (*http://source.android.com/source/initializing.html*) section of Google's *http://source.android.com* website for the latest information on how to set up your host for building Android's sources. That page also covers configuring *udev* to ensure permissions are properly set to let you access an Android device connected to your host.

Getting the AOSP

As I mentioned earlier, the official AOSP is available at *http://android.google source.com*, which sports the Gitweb interface (*git*'s Web interface) shown in Figure 3-1. When you visit the site, you will see a fairly large number of git repositories you can pull. Needless to say, pulling each and every one of these manually would be rather tedious; there are over 100. And, in fact, pulling them all would be quite useless because only a subset of these projects is needed. The right way to pull the AOSP is to use the *repo* tool, which is available at the very same location. First, though, you'll need to get *repo* itself:

```
$ sudo apt-get install curl
$ curl https://dl-ssl.google.com/dl/googlesource/git-repo/repo > ~/bin/repo
$ chmod a+x ~/bin/repo
```

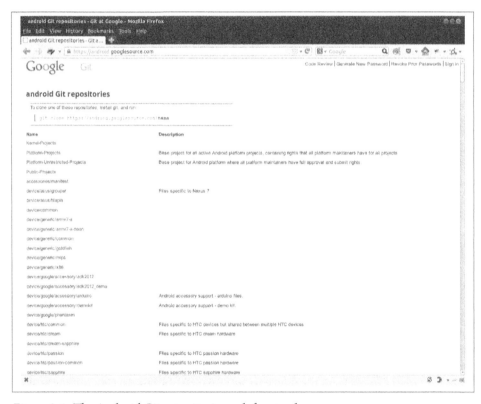

Figure 3-1. The Android Git repositories web frontend

Under Ubuntu, *~/bin* is automatically added to your path when you log in, *if it already exists*. So, if you don't have a *bin/* directory in your home directory, create it, and then log out and log back in to make it part of your path. Otherwise, the shell won't be able to find *repo*, even if you fetch it as I just showed.

If this doesn't work, either in Ubuntu or any other distribution you may be using, add a PATH=$PATH:~/bin to your *~/.profile* manually, and then log out and log back in.

You don't have to put *repo* in *~/bin*, but it has to be in your path. So regardless of where you put it, just make sure it's available to you in all locations in the filesystem from the command line.

Despite its structure as a single shell script, *repo* is actually quite an intricate tool. It can simultaneously pull from multiple *git* repositories to create an Android distribution. The repositories it pulls from are given to it through a *manifest* file, which is an XML file describing the projects that need to be pulled from and their location. *repo* is in fact layered on top of *git*, and each project it pulls from is an independent *git* repository. You can find out more about what pushed Google to create *repo* from the blog post Gerrit and Repo, the Android Source Management Tools (*http://google-opensource.blogspot.com/2008/11/gerrit-and-repo-android-source.html*), published in November 2008, soon after Android's first open source release.

Confusing as it may be, note that *repo*'s "manifest" file has absolutely **nothing** to do with "manifest" files (*AndroidManifest.xml*) used by app developers to describe their apps to the system. Their formats and uses are completely different. Fortunately, they rarely have to be used within the same context, so while you should keep this fact in mind, we won't need to worry too much about it in the coming explanations.

Before you can use *repo*, you'll need to make sure that *git* is installed on your system, as it may not have been there by default:

```
$ sudo apt-get install git
```

Now that we've got *repo* and *git*, let's get ourselves a copy of the AOSP:

```
$ mkdir -p ~/android/aosp-2.3.x
$ cd ~/android/aosp-2.3.x
$ repo init -u https://android.googlesource.com/platform/manifest.git
  -b gingerbread
$ repo sync
```

The last command should run for quite some time as it goes and fetches the sources of all the projects described in the manifest file. After all, the AOSP is several gigabytes in size uncompiled, as mentioned in "Development Setup and Tools" on page 22. Keep in mind that network bandwidth and latencies will play a big role in how long this takes. Note also that we are fetching a specific branch of the tree, Gingerbread. That's the `-b gingerbread` part of the third command. If you omit that part, you will be getting the *master* branch. It's been the experience of many people that the master branch doesn't always build or run properly, because it contains the tip of the open development branch. Tagged branches, on the other hand, mostly work out of the box. If you're planning to make contributions back to the AOSP, however, note that Google accepts contributions to the master branch only.

You can get more information about *repo*'s capabilities by using its online help:

```
$ repo help
usage: repo COMMAND [ARGS]

The most commonly used repo commands are:

  abandon      Permanently abandon a development branch
  branch       View current topic branches
  branches     View current topic branches
  checkout     Checkout a branch for development
  cherry-pick  Cherry-pick a change.
  diff         Show changes between commit and working tree
  download     Download and checkout a change
  grep         Print lines matching a pattern
  init         Initialize repo in the current directory
  list         List projects and their associated directories
  overview     Display overview of unmerged project branches
  prune        Prune (delete) already merged topics
  rebase       Rebase local branches on upstream branch
  smartsync    Update working tree to the latest known good revision
  stage        Stage file(s) for commit
  start        Start a new branch for development
  status       Show the working tree status
  sync         Update working tree to the latest revision
  upload       Upload changes for code review

See 'repo help <command>' for more information on a specific command.
See 'repo help --all' for a complete list of recognized commands.
```

As the above output indicates, you can also ask for more information about any of *repo*'s subcommands:

```
$ repo help init
Summary
-------
Initialize repo in the current directory

Usage: repo init [options]
```

```
Options:
  -h, --help            show this help message and exit

  Logging options:
    -q, --quiet         be quiet

  Manifest options:
    -u URL, --manifest-url=URL
                        manifest repository location
    -b REVISION, --manifest-branch=REVISION
                        manifest branch or revision
    -m NAME.xml, --manifest-name=NAME.xml
                        initial manifest file
    --mirror            create a replica of the remote repositories rather
                        than a client working directory
    --reference=DIR     location of mirror directory
    --depth=DEPTH       create a shallow clone with given depth; see git clone
    -g GROUP, --groups=GROUP
                        restrict manifest projects to ones with a specified
                        group
    -p PLATFORM, --platform=PLATFORM
                        restrict manifest projects to ones with a specified
                        platform group [auto|all|none|linux|darwin|...]

  repo Version options:
    --repo-url=URL      repo repository location
    --repo-branch=REVISION
                        repo branch or revision
    --no-repo-verify    do not verify repo source code

  Other options:
    --config-name       Always prompt for name/e-mail

Description
-----------
The 'repo init' command is run once to install and initialize repo. The
latest repo source code and manifest collection is downloaded from the
server and is installed in the .repo/ directory in the current working
directory.

The optional -b argument can be used to select the manifest branch to
checkout and use. If no branch is specified, master is assumed.

The optional -m argument can be used to specify an alternate manifest to
be used. If no manifest is specified, the manifest default.xml will be
used.

The --reference option can be used to point to a directory that has the
content of a --mirror sync. This will make the working directory use as
much data as possible from the local reference directory when fetching
from the server. This will make the sync go a lot faster by reducing
```

```
data traffic on the network.

Switching Manifest Branches
---------------------------
To switch to another manifest branch, `repo init -b otherbranch` may be
used in an existing client. However, as this only updates the manifest,
a subsequent `repo sync` (or `repo sync -d`) is necessary to update the
working directory files.
```

When you look at *repo sync*'s online help, for instance, one of the flags you will likely want to investigate further is *-j*, since it permits syncing several git trees in parallel. This is especially useful if you've got a generous corporate net connection and would like to speed up your downloading of the AOSP—by default, *repo* does four parallel downloads:

```
$ repo sync -j8
```

Getting other branches and tags is also relatively simple. Here's getting 4.2/Jelly Bean:

```
$ mkdir -p ~/android/aosp-4.2
$ cd ~/android/aosp-4.2
$ repo init -u https://android.googlesource.com/platform/manifest
-b android-4.2_r1

$ repo sync
```

In contrast to the earlier command, I'm using a specific version number instead of a version name. Codenames, Tags, and Build Numbers (*http://source.android.com/source/build-numbers.html*) provides a full list of the official tags and version numbers. You can find the available tags and branches for yourself by doing something like this:
[1]

```
$ mkdir ~/android/aosp-branches-tags
$ cd ~/android/aosp-branches-tags
$ git clone https://android.googlesource.com/platform/manifest.git
$ cd manifest
$ git tag
android-1.6_r1.1_
android-1.6_r1.2_
android-1.6_r1.3_
android-1.6_r1.4_
android-1.6_r1.5_
android-1.6_r1_
android-1.6_r2_
android-2.0.1_r1_
android-2.0_r1_
android-2.1_r1_
android-2.1_r2.1p2_
android-2.1_r2.1p_
```

---

1. Thanks to Linaro's Bernhard Rosenkränzer for pointing out this really useful trick.

```
...
android-4.1.1_r6
android-4.1.1_r6.1
android-4.1.2_r1
android-4.2.1_r1__
android-4.2_r1___
android-cts-2.2_r8
android-cts-2.3_r10
android-cts-2.3_r11
...
$ git branch -a
* master
  remotes/origin/HEAD -> origin/master
  remotes/origin/android-1.6_r1
  remotes/origin/android-1.6_r1.1
  remotes/origin/android-1.6_r1.2
  remotes/origin/android-1.6_r1.3
  remotes/origin/android-1.6_r1.4
  remotes/origin/android-1.6_r1.5
  remotes/origin/android-1.6_r2
  remotes/origin/android-2.0.1_r1
  remotes/origin/android-2.0_r1
  remotes/origin/android-2.1_r1
  remotes/origin/android-2.1_r2
  remotes/origin/android-2.1_r2.1p
  remotes/origin/android-2.1_r2.1p2
...
  remotes/origin/android-4.1.1_r6.1
  remotes/origin/android-4.1.2_r1
  remotes/origin/android-4.2.1_r1
  remotes/origin/android-4.2_r1
  remotes/origin/android-cts-2.2_r8
  remotes/origin/android-cts-2.3_r10
  remotes/origin/android-cts-2.3_r11
...
  remotes/origin/android-sdk-support_r11
  remotes/origin/froyo
  remotes/origin/gingerbread
  remotes/origin/gingerbread-release
  remotes/origin/ics-mr0
  remotes/origin/ics-mr1
  remotes/origin/ics-plus-aosp
  remotes/origin/jb-dev
  remotes/origin/jb-mr1-dev
  remotes/origin/jumper-stable
  remotes/origin/master
  remotes/origin/master-dalvik
  remotes/origin/tools_r20
  remotes/origin/tools_r21
  remotes/origin/tools_r21.1
  remotes/origin/tradefed
```

All of the above is, of course, limited to the official AOSP. Have a look at Appendix E for a list of other AOSP trees that may be relevant to your work, such as those maintained by Linaro and CynogenMod. Interestingly, most of these alternative trees also rely on *repo*, which is all the more reason to learn how to master this tool.

Inside the AOSP

Now that we've got a copy of the AOSP, let's start looking at what's inside and, most importantly, connect that to what we just saw in the previous chapter. Feel free to skip over this section and come back to it after the next section if you're too eager to get your own custom Android running. For those of you still reading, have a look at Table 3-1 for a summary of the AOSP's top-level directory for 2.3.7/Gingerbread and 4.2/Jelly Bean. Where "N/A" is listed in one of the Size columns for a directory, that directory doesn't exist in that version. Also, the sizes given don't include the *.git* directories that might have been included underneath any of the given entries.

Table 3-1. AOSP content summary

Directory	Content	Size (in MB) in 2.3.7	Size (in MB) in 4.2
abi	Minimal C++ Run-Time Type Information support	N/A	0.1
bionic	Android's custom C library	14	18
bootable	OTA, recovery mechanism and reference bootloader	4	4
build	Build system	4	5
cts	Comptability Test Suite	77	136
dalvik	Dalvik VM	35	40
development	Development tools	64	87
device	Device-specific files and components	17	43
docs	Content of http://source.android.com	N/A	6
external	External projects imported into the AOSP	849	1,595
frameworks	Core components such as system services	360	1,150
gdk	Unknown[a]	N/A	5
hardware	HAL and hardware support libraries	27	52
libcore	Apache Harmony	54	40
libnativehelper[b]	Helper functions for use with JNI	N/A	0.1
ndk	Native Development Kit	13	31
packages	Stock Android apps, providers, and IMEs	115	278
pdk	Platform Development Kit	N/A	0.3
prebuilt	Prebuilt binaries, including toolchains	1,389	N/A
prebuilts	Replacement for *prebuilt*	N/A	2,387

Directory	Content	Size (in MB) in 2.3.7	Size (in MB) in 4.2
sdk	Software Development Kit	14	54
system	"Embedded Linux" platform that houses Android	32	9
tools	Various IDE tools	N/A	34

[a] Despite several attempts, the author has been unable to identify what purpose this directory serves, apart from it having something to do with the NDK and LLVM. Even the git logs don't hint at what the acronym stands for. It's possibly experimental code for future use.

[b] This was a subdirectory of *dalvik/* in 2.3.7.

As you can see, *prebuilt* (*prebuilts* in 4.2/Jelly Bean) and *external* are the two largest directories in the tree, accounting for close to 75% of its size in 2.3.7/Gingerbread and above 65% of its size in 4.2/Jelly Bean. Interestingly, both of these directories are mostly made up of content from other open source projects and include things like various GNU toolchain versions, kernel images, common libraries, and frameworks such as OpenSSL and WebKit, etc. *libcore* is also from another open source project, Apache Harmony. In essence, this is further evidence of how much Android relies on the rest of the open source ecosystem to exist. Still, Android contains a fair bit of "original" (or nearly) code: about 800 MB of it in 2.3.7/Gingerbread and about 2 GB in 4.2/Jelly Bean.

To best understand Android's sources, it's useful to refer back to Figure 2-1, which illustrated Android's architecture. Figure 3-2 is a variant of that figure, which illustrates the location of each Android component in the AOSP sources. Obviously, a lot of key components come from *frameworks/base/*, which is where the bulk of Android's "brains" are located. It's in fact worth taking a closer look at that directory and at *system/core/*, in Tables 3-2 and 3-3 respectively, as they contain a large chunk of the moving parts you'll likely be interested in interfacing with or modifying as an embedded developer.

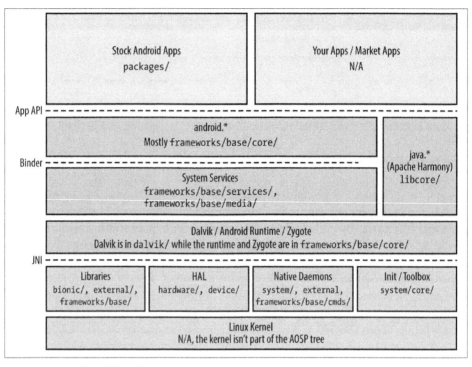

Figure 3-2. Android's architecture

Table 3-2. Content summary for frameworks/base/ in 2.3/Gingerbread

Directory	Content
cmds	Framework-related commands and daemons
core	The android.* packages
data	Fonts and sounds
graphics	2D graphics and Renderscript
include	C-language include files
keystore	Security key store
libs	C libraries
location	Location provider
media	Media Service, StageFright, codecs, etc.
native	Native code for some framework components
obex	Bluetooth Obex
opengl	OpenGL library and Java code
packages	A few core packages such as the Status Bar
services	System services

Directory	Content
telephony	Telephony API, which talks to the *rild* radio layer interface
tools	A few core tools such as *aapt* and *aidl*
voip	RTP and SIP APIs
vpn	VPN Manager
wifi	Wifi Manager and API

In addition to *base/*, *frameworks/* contained few other directories at the time of 2.3/Gingerbread. In between that version and 4.2/Jelly Bean, *frameworks/base/* has gone through a number of cleanups, and several parts of it have been moved up a directory level and into *frameworks/* (Table 3-4). *frameworks/base/media/* for instance, is now *frameworks/av/media/* instead. Also, *frameworks/native/* now contains several native libraries and system services that were previously in *frameworks/base/*.

Table 3-3. Content summary for system/core/ in 2.3/Gingerbread

Directory	Content
adb[a]	The ADB daemon and client
cpio	*mkbootfs* tool used to generate RAM disk images[b]
debuggerd	*debuggerd* command mentioned in Chapter 2 and covered in Chapter 6
fastboot	*fastboot* utility used to communicate with Android bootloaders using the "fastboot" protocol
include	C-language headers for all things "system"
init	Android's *init*
libacc	"Almost" C Compiler library for compiling C-like code; used by RenderScript in 2.3/Gingerbread[c]
libcutils	Various C utility functions not part of the standard C library; used throughout the tree
libdiskconfig	For reading and configuring disks; used by *vold*
liblinenoise	BSD-licensed readline() replacement from http://github.com/antirez/linenoise; used by Android's shell
liblog	Logging library that interfaces with the Android kernel logger as seen in Figure 2-2; used throughout the tree
libmincrypt	Basic RSA and SHA functions; used by the recovery mechanism and *mkbootimg* utility
libnetutils	Network configuration library; used by *netd*
libpixelflinger	Low-level graphic rendering functions
libsysutils	Utility functions for talking with various components of the system, including the framework; used by *netd* and *vold*
libzipfile	Wrapper around zlib for dealing with ZIP files
logcat	The *logcat* utility

Directory	Content
logwrapper	Utility that forks and runs the command passed to it while redirecting stdout and stderr to Android's logger
mkbootimg	Utility for creating a boot image using a RAM disk and a kernel
netcfg	Network configuration utility
rootdir	Default Android root directory structure and content
run-as	Utility for running a program as a given user ID
sdcard	Emulates FAT using FUSE
sh	Android shell
toolbox	Android's Toolbox (BusyBox replacement)

[a] Some entries have been omitted because they aren't currently used by any part of the AOSP. They are likely legacy components.
[b] This is used to create both the default RAM disk image used to boot Android and the recovery image.
[c] This description might not make any sense to you unless you know what RenderScript is. Have a look at Google's documentation for RenderScript; the relevance of *libacc* in that context should be clearer.

Table 3-4. Major additions made to system/core/ between 2.3/Gingerbread and 4.2/Jelly Bean.

Directory	Content
charger	Full-screen battery state display
fs_mgr	Filesystem manager
gpttool	Tool for dealing with GPT (UEFI) partition table
libcorkscrew	Debugging/backtrace library
libion	Library for interfacing with the ION driver
libnl_2	Library for handling NetLink sockets
libsuspend	Library for interfacing with the kernel's power management functionality, including autosleep
libsync	Library for interface with */dev/sw_sync*
libusbhost	Library for USB host mode handling

Apart from *core/*, *system/* also includes a few more directories, such as *netd/* and *vold/*, which contain the *netd* and *vold* daemons, respectively.

In addition to the top-level directories, the root directory also includes a single *Makefile*. That file is, however, mostly empty, its main use being to include the entry point to Android's build system:

```
### DO NOT EDIT THIS FILE ###
include build/core/main.mk
### DO NOT EDIT THIS FILE ###
```

As you've likely figured out already, there's far more to the AOSP than what I just presented to you. There are, after all, more than 14,000 directories and 100,000 files in 2.3.x/

Gingerbread, and more than 40,000 directories and 265,000 files in 4.2/Jelly Bean. By most standards, it's a fairly large project. In comparison, early 3.0.x releases of the Linux kernel have about 2,000 directories and 35,000 files. We will certainly get the chance to explore more parts of the AOSP's functionality and sources as we move forward. I highly recommend, though, that you start exploring and experimenting with the sources in earnest, as it will likely take several months before you can comfortably navigate your way through.

Build Basics

So now we have an AOSP and a general idea of what's inside, so let's get it up and running. There's one last thing we need to do before we can build it, though. We need to make sure we've got all the packages necessary on our Ubuntu install. Here are the instructions for 64-bit Ubuntu 11.04, assuming we're building 2.3/Gingerbread. Even if you are using an older or newer version of some Debian-based Linux distribution, the instructions will be fairly similar. (See also "Building on Virtual Machines or Non-Linux Systems" on page 98 for other systems on which you can build the AOSP.) As I mentioned earlier, refer to Google's Initializing a Build Environment (*http://source.android.com/source/initializing.html*) for the latest version of packages required to build recent AOSPs on more recent Ubuntu versions.

Build System Setup

First, let's get some of the basic packages installed on our development system. You might have some of these already installed as part of other development work you've been doing, and that's fine. Ubuntu's package management system will ignore your request to install those packages.

> Note that the following commands are broken down on several lines to fit this book's width. The use of the \ character at the end of a line on the shell forces it to start over on another line (the one starting with the > character) to give you the chance to continue entering your command. As such, you're expected to type the \ characters at the end of the lines in the following commands, but the > at the beginning of the subsequent lines isn't something you type; it's inserted by the shell. Other commands in this book use the same trick for presentation purposes.

```
$ sudo apt-get install bison flex gperf git-core gnupg zip tofrodos \
> build-essential g++-multilib libc6-dev libc6-dev-i386 ia32-libs mingw32 \
> zlib1g-dev lib32z1-dev x11proto-core-dev libx11-dev \
> lib32readline5-dev libgl1-mesa-dev lib32ncurses5-dev
```

You might also need to fix a few symbolic links:

```
$ sudo ln -s /usr/lib32/libstdc++.so.6 /usr/lib32/libstdc++.so
$ sudo ln -s /usr/lib32/libz.so.1 /usr/lib32/libz.so
```

Finally, you need to install Sun's JDK; it's "officially" discouraged to use the OpenJDK with the AOSP (see this posting (*https://groups.google.com/forum/?fromgroups=#!topic/android-building/IGCVGp9huLg*) by Google's Jean-Baptiste Queru), though some people are able to use it successfully (see sidebar below) and *gcj* won't do. In Ubuntu, you used to be able to get the JDK by using the following sequence of commands:

```
$ sudo add-apt-repository "deb http://archive.canonical.com/ natty partner"
$ sudo apt-get update
$ sudo apt-get install sun-java6-jdk
```

Unfortunately there seems to have been some disagreement between Canonical (the company behind Ubuntu) and Oracle, and these instructions no longer work at the time of this writing. Instead, you should refer to Ubuntu's instructions (*https://help.ubuntu.com/community/Java#Oracle_.28Sun.29_Java_6*) for getting the JDK version 6 working on your host. Note that version 7 doesn't work at the time of this writing for the AOSP. Essentially, the Ubuntu instructions explain that you need to get the JDK binary from Oracle's site (*http://www.oracle.com/technetwork/java/javase/downloads/index.html*) and install it. Here's a slightly modified version of the currently published instructions, which you're likely going to have to adapt to the latest version of the JDK:

```
$ chmod u+x jdk-6u38-linux-x64.bin
$ ./jdk-6u38-linux-x64.bin
$ sudo mkdir -p /usr/lib/jvm
$ sudo mv jdk1.6.0_38 /usr/lib/jvm/
$ sudo update-alternatives --install "/usr/bin/java" "java" \
> "/usr/lib/jvm/jdk1.6.0_38/bin/java" 1
$ sudo update-alternatives --install "/usr/bin/javac" "javac" \
> "/usr/lib/jvm/jdk1.6.0_38/bin/javac" 1
$ sudo update-alternatives --install "/usr/bin/javah" "javah" \
> "/usr/lib/jvm/jdk1.6.0_38/bin/javah" 1
$ sudo update-alternatives --install "/usr/bin/javadoc" "javadoc" \
> "/usr/lib/jvm/jdk1.6.0_38/bin/javadoc" 1
$ sudo update-alternatives --install "/usr/bin/jar" "jar" \
> "/usr/lib/jvm/jdk1.6.0_38/bin/jar" 1
```

You'll then have to run the following commands and select the version you just installed:

```
$ sudo update-alternatives --config java
There are 2 choices for the alternative java (providing /usr/bin/java).

  Selection    Path                                              Priority   Status
------------------------------------------------------------
* 0            /usr/lib/jvm/java-6-openjdk-amd64/jre/bin/java     1061      auto mode
  1            /usr/lib/jvm/java-6-openjdk-amd64/jre/bin/java     1061      manual mode
  2            /usr/lib/jvm/jdk1.6.0_38/bin/java                  1         manual mode

Press enter to keep the current choice[*], or type selection number: 2
$ sudo update-alternatives --display java
```

```
    java - manual mode
      link currently points to /usr/lib/jvm/jdk1.6.0_38/bin/java
    ...
$ sudo update-alternatives --config javac
    ...
$ sudo update-alternatives --config javah
    ...
$ sudo update-alternatives --config javadoc
    ...
$ sudo update-alternatives --config jar
    ...
```

As you can see, Oracle's JDK and the OpenJDK can coexist on the same Ubuntu installation. You just need to make sure the defaults point to the right JDK as needed. The above instructions have you installing Oracle's JDK systemwide and changing the defaults of some commands to use the binaries in that package instead of whatever was installed by default in Ubuntu. Nothing precludes you from installing Oracle's JDK somewhere into your home directory and changing the PATH variable to point to the *bin/* directory extracted by the running of Oracle's installation binary.

> ## Using the OpenJDK instead of Oracle's JDK
>
> Following the rules can sometimes be boring. Despite the official recommendations to stick to Oracle's JDK, many have actually successfully used the OpenJDK to build the AOSP. Here's a patch from Linaro's Bernhard Rosenkränzer that allows you to build the AOSP with the OpenJDK:
>
> ```
> diff --git a/core/main.mk b/core/main.mk
> index 87488f4..32e3aec 100644
> --- a/core/main.mk
> +++ b/core/main.mk
> @@ -125,7 +125,14 @@ endif
>  # Check for the correct version of java
>   java_version := $(shell java -version 2>&1 | head -n 1 | grep '^java .*[
>   "]1\.6[\. "$$]')
>   ifneq ($(shell java -version 2>&1 | grep -i openjdk),)
>  -java_version :=
>  +$(warning ************************************************************)
>  +$(warning AOSP errors out when using OpenJDK, saying you need to use)
>  +$(warning Java SE 1.6 instead.)
>  +$(warning A build with OpenJDK seems to work fine though - if you)
>  +$(warning run into any Java errors, you may want to try using the)
>  +$(warning version required by AOSP though.)
>  +$(warning ************************************************************)
>  +#java_version :=
>   endif
>   ifeq ($(strip $(java_version)),)
>   $(info ************************************************************)
> ```

A few Linaro engineers report they have no problems either compiling the AOSP this way or running the resulting images. Others seem to report javadoc issues, as Google's Jean-Baptiste Queru (*https://groups.google.com/forum/?fromgroups=#!topic/android-building/IGCVGp9huLg*) hints. We can hope that future efforts will provide further evidence as to the viability of using the OpenJDK.

Your system is now ready to build Android. Obviously you don't need to do this package installation process every time you build Android. You'll need to do it only once for every Android development system you set up.

Building Android

We are now ready to build Android. Let's go to the directory where we downloaded Android and configure the build system:

```
$ cd ~/android/aosp-2.3.x
$ . build/envsetup.sh
including device/acme/coyotepad/vendorsetup.sh
including device/htc/passion/vendorsetup.sh
including device/samsung/crespo4g/vendorsetup.sh
including device/samsung/crespo/vendorsetup.sh
$ lunch

You're building on Linux

Lunch menu... pick a combo:
     1. generic-eng
     2. simulator
     3. full_passion-userdebug
     4. full_crespo4g-userdebug
     5. full_crespo-userdebug

Which would you like? [generic-eng] ENTER

============================================
PLATFORM_VERSION_CODENAME=REL
PLATFORM_VERSION=2.3.4
TARGET_PRODUCT=generic
TARGET_BUILD_VARIANT=eng
TARGET_SIMULATOR=false
TARGET_BUILD_TYPE=release
TARGET_BUILD_APPS=
TARGET_ARCH=arm
HOST_ARCH=x86
HOST_OS=linux
HOST_BUILD_TYPE=release
BUILD_ID=GINGERBREAD
============================================
```

For 4.2/Jelly Bean, the same operations on Ubuntu 12.04 would yield this instead:

```
$ cd ~/android/aosp-4.2
$ . build/envsetup.sh
including device/asus/grouper/vendorsetup.sh
including device/asus/tilapia/vendorsetup.sh
including device/generic/armv7-a-neon/vendorsetup.sh
including device/generic/armv7-a/vendorsetup.sh
including device/generic/mips/vendorsetup.sh
including device/generic/x86/vendorsetup.sh
including device/lge/mako/vendorsetup.sh
including device/samsung/maguro/vendorsetup.sh
including device/samsung/manta/vendorsetup.sh
including device/samsung/toroplus/vendorsetup.sh
including device/samsung/toro/vendorsetup.sh
including device/ti/panda/vendorsetup.sh
including sdk/bash_completion/adb.bash
$ lunch

You're building on Linux

Lunch menu... pick a combo:
     1. full-eng
     2. full_x86-eng
     3. vbox_x86-eng
     4. full_mips-eng
     5. full_grouper-userdebug
     6. full_tilapia-userdebug
     7. mini_armv7a_neon-userdebug
     8. mini_armv7a-userdebug
     9. mini_mips-userdebug
    10. mini_x86-userdebug
    11. full_mako-userdebug
    12. full_maguro-userdebug
    13. full_manta-userdebug
    14. full_toroplus-userdebug
    15. full_toro-userdebug
    16. full_panda-userdebug

Which would you like? [full-eng] ENTER
============================================
PLATFORM_VERSION_CODENAME=REL
PLATFORM_VERSION=4.2
TARGET_PRODUCT=full
TARGET_BUILD_VARIANT=eng
TARGET_BUILD_TYPE=release
TARGET_BUILD_APPS=
TARGET_ARCH=arm
TARGET_ARCH_VARIANT=armv7-a
HOST_ARCH=x86
HOST_OS=linux
HOST_OS_EXTRA=Linux-3.2.0-35-generic-x86_64-with-Ubuntu-12.04-precise
HOST_BUILD_TYPE=release
```

```
BUILD_ID=JOP40C
OUT_DIR=out
=============================================
```

In both cases, note that we typed a period, a space, and then build/envsetup.sh. This forces the shell to run the *envsetup.sh* script within the current shell. If we were to just run the script, the shell would spawn a new shell and run the script in that new shell. That would be useless since *envsetup.sh* defines new shell commands, such as *lunch*, and sets up environment variables required for the rest of the build.

We will explore *envsetup.sh* and *lunch* in more detail later. For the moment, though, note that the generic-eng combo in 2.3/Gingerbread and full-eng combo in 4.2/Jelly Bean means that we configured the build system to create images for running in the Android emulator. This is the same QEMU emulator software used by developers to test their apps when developing using the SDK on a workstation. Here it will be running our own custom images instead of the default ones shipped with the SDK. It's also the same emulator that was used by the Android development team to develop Android while there were no devices for it yet. So while it's not real hardware and is therefore by no means a perfect target, it's still more than sufficient to cover most of the terrain we need to cover. Once you know your specific target, you should be able to adapt the instructions found in the rest of this book, possibly with some help from the book *Building Embedded Linux Systems*, to get your custom Android images loaded on your device and your hardware to boot them.

Now that the environment has been set up, we can actually build Android:

```
$ make -j16
=============================================
PLATFORM_VERSION_CODENAME=REL
PLATFORM_VERSION=2.3.4
TARGET_PRODUCT=generic
TARGET_BUILD_VARIANT=eng
TARGET_SIMULATOR=false
TARGET_BUILD_TYPE=release
TARGET_BUILD_APPS=
TARGET_ARCH=arm
HOST_ARCH=x86
HOST_OS=linux
HOST_BUILD_TYPE=release
BUILD_ID=GINGERBREAD
=============================================
Checking build tools versions...
find: `frameworks/base/frameworks/base/docs/html': No such file or directory
find: `out/target/common/docs/gen': No such file or directory
find: `frameworks/base/frameworks/base/docs/html': No such file or directory
find: `out/target/common/docs/gen': No such file or directory
find: `frameworks/base/frameworks/base/docs/html': No such file or directory
find: `out/target/common/docs/gen': No such file or directory
find: `frameworks/base/frameworks/base/docs/html': No such file or directory
```

```
find: `out/target/common/docs/gen': No such file or directory
find: `frameworks/base/frameworks/base/docs/html': No such file or directory
find: `out/target/common/docs/gen': No such file or directory
host Java: apicheck (out/host/common/obj/JAVA_LIBRARIES/apicheck_intermediates/c
lasses)
Header: out/host/linux-x86/obj/include/libexpat/expat.h
Header: out/host/linux-x86/obj/include/libexpat/expat_external.h
Header: out/target/product/generic/obj/include/libexpat/expat.h
Header: out/target/product/generic/obj/include/libexpat/expat_external.h
Header: out/host/linux-x86/obj/include/libpng/png.h
Header: out/host/linux-x86/obj/include/libpng/pngconf.h
Header: out/host/linux-x86/obj/include/libpng/pngusr.h
Header: out/target/product/generic/obj/include/libpng/png.h
Header: out/target/product/generic/obj/include/libpng/pngconf.h
Header: out/target/product/generic/obj/include/libpng/pngusr.h
Header: out/target/product/generic/obj/include/libwpa_client/wpa_ctrl.h
Header: out/target/product/generic/obj/include/libsonivox/eas_types.h
Header: out/target/product/generic/obj/include/libsonivox/eas.h
Header: out/target/product/generic/obj/include/libsonivox/eas_reverb.h
Header: out/target/product/generic/obj/include/libsonivox/jet.h
Header: out/target/product/generic/obj/include/libsonivox/ARM_synth_constants_gn
u.inc
host Java: clearsilver (out/host/common/obj/JAVA_LIBRARIES/clearsilver_intermedi
ates/classes)
target Java: core (out/target/common/obj/JAVA_LIBRARIES/core_intermediates/class
es)
host Java: dx (out/host/common/obj/JAVA_LIBRARIES/dx_intermediates/classes)
Notice file: frameworks/base/libs/utils/NOTICE -- out/host/linux-x86/obj/NOTICE_
FILES/src//lib/libutils.a.txt
Notice file: system/core/libcutils/NOTICE -- out/host/linux-x86/obj/NOTICE_FILES
/src//lib/libcutils.a.txt
...
```

Note that several lines, especially at the end of the output, are wrapped around to the following line because they wouldn't fit in the width permitted by this book's pages. You will see this occurring in several of the output screens printed throughout this book. I've tried to keep the line-wrap at 80 characters, though sometimes I could get away with a little more without it being too obvious.

In sum, make sure you keep an eye out for wrapped lines in output in the rest of the book.

Now is a good time to go for a snack or to watch tonight's hockey game—it's a Canadian thing, I can't help it. On a more serious note, your build time will obviously depend on your system's capabilities. On a laptop with a quad-core CORE i7 Intel processor with hyperthreading enabled and 8GB of RAM, this actual command will take about 20 minutes to build 2.3/Gingerbread and 80 minutes to build 4.2/Jelly Bean. On an older laptop with a dual-core Centrino 2 Intel processor and 2GB of RAM, a *make -j4* would take about an hour to build 2.3/Gingerbread—I wouldn't try building 4.2/Jelly Bean on such a machine. Note that the *-j* parameter of *make* allows you to specify how many jobs to run in parallel. Some say that it's best to use your number of processors times 2, which is what I'm doing here. Others say it's best to add 2 to the number of processors you have. Following that advice, I would have used 10 and 4 instead of 16 and 4.

Generally speaking, the AOSP is a very heavy piece of software to build. I highly recommend you use the most powerful system you can get your hands on, no holds barred. Having lots of RAM is also very highly recommended. In fact, if the entire AOSP tree can fit in the filesystem cache maintained by the kernel in RAM, then you'll minimize your build times. You can also use solid-state drives instead of regular hard drives. They've been shown to significantly reduce the AOSP's build times.

> ## Building on Virtual Machines or Non-Linux Systems
>
> I often get asked about building the AOSP in virtual machines; most often because the development team, or their IT department, is standardized on Windows. While I've seen this work and have put together images to do that myself, your results will vary. It'll usually take more than twice as much time to build in a VM than building natively on the same system. So if you're going to do a lot of work on the AOSP, I strongly suggest you build it natively. And, yes, this involves having a Linux machine at hand.
>
> An increasing number of developers also prefer Mac OS X over Linux and Windows, including many at Google itself. Hence, the official instructions at *http://source.android.com* also describe how to build on a Mac. These instructions, though, tend to break after Mac OS updates. Fortunately for Mac-based developers, they are many and they are rather zealous. Hence, you'll eventually find updated instructions on the web or on the various Google Groups about how to build the AOSP on your new version of OS X. Here's one posting explaining how to build Gingerbread on OS X Lion: Building Gingerbread on OS X Lion (*http://groups.google.com/group/android-building/msg/4b9e6168ecae68a5*). Bear in mind, though, that as I mentioned in Chapter 1, Google's own Android build farms are Ubuntu based. If you choose to build on OS X, you'll likely always be playing catch-up. At worst, you can use a VM as in the Windows case.
>
> If you do choose to go the VM route, make sure you configure the VM to use as many CPUs as there are available in your system. Most BIOSes I've seen seem to disable by default the instruction sets that allow multiple-CPU virtualization. VirtualBox, for instance, will complain about some obscure error if you try to allocate more than one CPU to an image while those instruction sets are disabled. You must go to the BIOS and enable those options for your VM software to be able to grant the guest OS multiple CPUs.

There are a few other things to consider regarding the build. First, note that in between printing out the build configuration and the printing of the first output of the actual build (where it prints out: `host Java: apicheck (out/host/common/o...`), there will be a rather long delay where nothing will get printed out, save for the "No such file or directory" warnings. I'll explain this delay in more detail later, but suffice it to say that the build system is taking that time to figure out the rules of how to build every part of the AOSP.

Note also that you'll see plenty of warning statements. These are rather "normal," not so much in terms of maintaining software quality, but in that they are pervasive in Android's build. They usually won't have an impact on the final product being compiled. So, contrary to the best of my software engineering instincts, I have to recommend you completely ignore warnings and stick to fixing errors only. Unless, of course, those warnings stem from software you added yourself. By all means, make sure you get rid of **those** warnings.

Running Android

With the build completed, all you need to do is start the emulator to run your own custom-built images:

```
$ emulator &
```

This will start the emulator window that will boot into a full Android environment as illustrated in Figure 3-3 (showing 2.3/Gingerbread).

Figure 3-3. Android emulator running custom images

You can then interact with the AOSP you just built as if it were running on a real device. Since your monitor is likely not a touch screen, however, you will need to use your mouse as if it were your finger. A single touch is a click, and swiping is done by holding down the mouse button, moving around, and letting go of the mouse button to signify that your finger has been removed from the touch screen. You also have a full keyboard at your disposal, with all the buttons you would find on a phone equipped with a QWERTY keyboard, although you can use your regular keyboard to input text in text boxes.

Despite its features and realism, the emulator does have its issues. For one thing, it takes some time to boot. It will take longest to boot the first time, because Dalvik is creating a JIT cache for the apps running on the phone. Note that the creation of the Dalvik cache isn't unique to the emulator. No matter what type of device you run Android on, modern Dalvik needs a JIT cache, whether it be created at boot time or, as we'll see in Chapter 7, at build time.

Even after the first boot, though, you might find the emulator heavy, especially if you're in a modify-compile-test loop. Also, it doesn't perfectly emulate everything. For instance, it traditionally has a hard time firing off rotation change events when it's made to rotate using F11 or F12. This, though, is mostly an issue for app developers.

If for any reason you close the shell where you had configured, built, and started Android —or if you need to start a new one and have access to all the tools and binaries created from the build, you must invoke the *envsetup.sh* script and the *lunch* commands again in order to set up environment variables. Here are commands from a new shell, for instance:

```
$ cd ~/android/aosp-2.3.x
$ emulator &
No command 'emulator' found, did you mean:
 Command 'qemulator' from package 'qemulator' (universe)
emulator: command not found
$ . build/envsetup.sh
$ lunch

You're building on Linux

Lunch menu... pick a combo:
     1. generic-eng
     2. simulator
     3. full_passion-userdebug
     4. full_crespo4g-userdebug
     5. full_crespo-userdebug

Which would you like? [generic-eng] ENTER


============================================
PLATFORM_VERSION_CODENAME=REL
PLATFORM_VERSION=2.3.4
```

```
TARGET_PRODUCT=generic
TARGET_BUILD_VARIANT=eng
...
==============================================
$ emulator &
$
```

Note that the second time we issued *emulator*, the shell didn't complain that the command was missing anymore. The same goes for a lot of other Android tools, such as the *adb* command we're about to look at. Note also that we didn't need to issue any *make* commands, because we had already built Android. In this case, we just needed to make sure the environment variables were properly set in order for the results of the previous build to be available to us again.

Using the Android Debug Bridge (ADB)

One of the most interesting aspects of the development environment put together by the Android development team is that you can shell into the running emulator, or any real device connected through USB for that matter, using the *adb* tool:

```
$ adb shell   ❶
* daemon not running. starting it now on port 5037 *
* daemon started successfully *
# cat /proc/cpuinfo   ❷
Processor       : ARM926EJ-S rev 5 (v5l)
BogoMIPS        : 405.50
Features        : swp half thumb fastmult vfp edsp java
CPU implementer : 0x41
CPU architecture: 5TEJ
CPU variant     : 0x0
CPU part        : 0x926
CPU revision    : 5

Hardware        : Goldfish
Revision        : 0000
Serial          : 0000000000000000
```

❶ This is issued in the same shell where you started the emulator.

❷ This is the target's shell, and *cat* is actually running on the "target" (i.e., the emulator).

As you can see, the kernel running in the emulator reports that it's seeing an ARM processor, which is in fact the predominant platform used with Android. Also, the kernel says it's running on a platform called *Goldfish*. This is the code name for the emulator, and you will see it in quite a few places.

Now that you've got a shell into the emulator and you are root, which is the default in the emulator, you can run any command much as if you had shelled into a remote

machine or a traditional, network-connected embedded Linux system. The Android
Debug Bridge (ADB) is what makes this possible. To exit an ADB shell session, all you
need to do is type Ctrl-D:

```
# CTRL-D ❶
$ ❷
```

❶ This is in the target shell.
❷ This is back on the host.

When you start *adb* for the first time on the host, it starts a server in the background
whose job is to manage the connections to all Android devices connected to the host.
That was the part of the earlier output that said a daemon was being started on port
5037. You can actually ask that daemon what devices it sees:

```
$ adb devices
List of devices attached
emulator-5554   device
0000021459584822   device
emulator-5556   offline
```

This is the output with one emulator instance running, one device connected through
USB, and another emulator instance starting up. If there are multiple devices connected,
you can tell it which device you want to talk to using the -s flag to identify the serial
number of the device:

```
$ adb -s 0000021459584822 shell
$ id
uid=2000(shell) gid=2000(shell) groups=1003(graphics),1004(input), ...
$ su
su: permission denied
```

Note that in this case, I'm getting a $ for my shell prompt instead of a #. This means that
contrary to the earlier interaction, I'm not running as root, as can also be seen from the
output of the *id* command. This is actually a real commercial Android phone, and my
inability above to gain root privileges using the *su* command is typical. Hence, my ability
to make any modifications to this device will be fairly limited. Unless, of course, I find
some way to "root" the phone (i.e., gain root access).

Historically, device manufacturers have been very reluctant for various reasons to give
root access to their devices and have put in a number of provisions to make that as
difficult as possible, if not impossible. That's why "rooting" devices is held up as a holy
grail by many power users and hackers. As of early 2013, some manufacturers, including
Motorola, HTC, and Sony Mobile, have spelled out policy changes that seem to be aimed
at making it easier for users to root their devices, with caveats of course. But this isn't
mainstream yet. And, unfortunately, it's subject to the whims of network operators, who
can still decide to lock down devices left unlocked by the handset manufacturer.

You may be tempted to try to root a commercial phone or device for experimenting with Android platform development. I would suggest you think this through carefully. While there are plenty of instructions out there explaining how to replace your standard images with what is often referred to as "custom ROMs" such as CyanogenMod and others, you need to be aware that any false step could well result in "bricking" the device (i.e., rendering it unbootable or erasing critical boot-time code). You then have an expensive paperweight (hence the term "bricking") instead of a phone.

If you want to experiment with running custom AOSP builds on real hardware, I suggest you get yourself something like a BeagleBoard xM or a PandaBoard. These boards are made for tinkering. If nothing else, they don't have a built-in flash chip that you may risk damaging. Instead, the SoCs on those devices boot straight from SD cards. Hence, fixing a broken image is simply a matter of unplugging the SD card from the board, connecting it to your workstation, reprogramming it, and plugging it back into the board.

Some commercial phones and devices allow you to "unlock" the firmware, often with the *fastboot oem unlock* command, and therefore you can burn your own images with less risk of bricking your device. Still, the bootloader in those cases becomes the single point of failure; if you damage it for some reason, you could still end up with a bricked device. The best configuration is one where you can reprogram all storage devices no matter what commands you mistype.

adb can of course do a lot more than just give you a shell, and I encourage you to start it without any parameters to look at its usage output:

```
$ adb
Android Debug Bridge version 1.0.26

 -d                            - directs command to the only connected USB device
                                 returns an error if more than one USB device is
                                 present.
 -e                            - directs command to the only running emulator.
                                 returns an error if more than one emulator is
                                 running.
 -s <serial number>            - directs command to the USB device or emulator
                                 with the given serial number. Overrides
                                 ANDROID_SERIAL
...
device commands:
  adb push <local> <remote>    - copy file/dir to device
  adb pull <remote> [<local>]  - copy file/dir from device
  adb sync [ <directory> ]     - copy host->device only if changed
                                 (-l means list but don't copy)
                                 (see 'adb help all')
  adb shell                    - run remote shell interactively
```

```
adb shell <command>            - run remote shell command
adb emu <command>              - run emulator console command
...
```

You can, for instance, use *adb* to dump the data contained in the main logger buffer:

```
$ adb logcat
I/DEBUG   (   30): debuggerd: Sep 10 2011 13:44:19
I/Netd    (   29): Netd 1.0 starting
I/Vold    (   28): Vold 2.1 (the revenge) firing up
D/qemud   (   38): entering main loop
D/Vold    (   28): USB mass storage support is not enabled in the kernel
D/Vold    (   28): usb_configuration switch is not enabled in the kernel
D/Vold    (   28): Volume sdcard state changing -1 (Initializing) -> 0 (No-Media
)
D/qemud   (   38): fdhandler_accept_event: accepting on fd 9
D/qemud   (   38): created client 0xe078 listening on fd 10
D/qemud   (   38): client_fd_receive: attempting registration for service 'boot-
properties'
D/qemud   (   38): client_fd_receive:    -> received channel id 1
D/qemud   (   38): client_registration: registration succeeded for client 1
I/qemu-props(   54): connected to 'boot-properties' qemud service.
I/qemu-props(   54): receiving..
I/qemu-props(   54): received: qemu.sf.lcd_density=160
I/qemu-props(   54): receiving..
I/qemu-props(   54): received: dalvik.vm.heapsize=16m
I/qemu-props(   54): receiving..
D/qemud   (   38): fdhandler_event: disconnect on fd 10
I/qemu-props(   54): exiting (2 properties set).
D/AndroidRuntime(   32):
D/AndroidRuntime(   32): >>>>>> AndroidRuntime START com.android.internal.os.Zyg
oteInit <<<<<<
D/AndroidRuntime(   32): CheckJNI is ON
I/        (   33): ServiceManager: 0xad50
...
```

This is very useful for observing the runtime behavior of key system components, including services run by the System Server.

You can also copy files to and from the device:

```
$ adb push data.txt /data/local
1 KB/s (87 bytes in 0.043s)
$ adb pull /proc/config.gz
95 KB/s (7087 bytes in 0.072s)
```

Again, given its centrality to Android development, I invite you to read up on *adb*'s use. We will continue using it throughout the book and cover it in much greater detail in Chapter 6. Keep in mind, though, that *adb* can have its quirks. First and foremost, many have found its host-side daemon to be somewhat flaky. For some reason or another, it sometimes doesn't correctly identify the state of connected devices and continues to state that they are offline while you try connecting to them. Or *adb* might just hang on the command line waiting for the device while the device is clearly active and able to receive ADB commands. The solution to those issues is almost invariably to kill the host-side daemon:[2]

```
$ adb kill-server
```

Not to worry—the next time you issue any *adb* command, the daemon will automatically be restarted. It's unclear what causes this behavior, and maybe this problem will get resolved at some point in the future. In the meantime, keep in mind that if you see some odd behavior when using ADB, killing the host-side daemon is usually something you want to try before investigating other potential issues.

As I said above, we'll discuss ADB in much greater detail in Chapter 6. Still, another source of information on *adb* is the Android Debug Bridge (*http://developer.android.com/tools/help/adb.html*) part of Google's Android Developers Guide. As Tim Bird[3] recommends, you want to print a copy and put it under your pillow.

Mastering the Emulator

As I said earlier, you can go a long way in platform development by simply using the emulator. It effectively emulates an ARM target, and more recently an x86 target, too, with a minimal set of hardware. We'll spend some time here going through some more advanced aspects of dealing with the emulator. As with many Android pieces, the emulator is quite a complex piece of software in and of itself. Still, we can get a very good idea of its capabilities by surveying a few key features.

Earlier we started the emulator by simply typing:

```
$ emulator &
```

But the *emulator* command can also take quite a few parameters. You can see the online help by adding the -help flag on the command line:

```
$ emulator -help
Android Emulator usage: emulator [options] [-qemu args]
  options:
    -sysdir <dir>              search for system disk images in <dir>
    -system <file>             read initial system image from <file>
    -datadir <dir>             write user data into <dir>
    -kernel <file>             use specific emulated kernel
    -ramdisk <file>            ramdisk image (default <system>/ramdisk.img
    -image <file>              obsolete, use -system <file> instead
    -init-data <file>          initial data image (default <system>/
```

---

2. It's actually somewhat interesting that the Android development team felt the need to build such functionality right into *adb*. Clearly they were encountering issues with that daemon themselves.

3. Tim is the maintainer of *http://elinux.org*, the guy behind the Embedded Linux Conference, and the chair of the Linux Foundation's CE Workgroup, and he's been doing a lot of cool Android stuff at Sony.

```
        -initdata <file>            userdata.img
                                    same as '-init-data <file>'
        -data <file>                data image (default <datadir>/userdata-
                                    qemu.img
        -partition-size <size>      system/data partition size in MBs
...
```

One especially useful flag is -kernel. It allows you to tell the emulator to use another kernel than the default prebuilt one found in *prebuilt/android-arm/kernel/*:

```
$ emulator -kernel path_to_your_kernel_image/zImage
```

If you want to use a kernel that has module support, for instance, you'll need to build your own, because the prebuilt one doesn't have module support enabled by default. Also, by default, the emulator won't show you the kernel's boot messages. You can, however, pass the -show-kernel flag to see them:

```
$ emulator -show-kernel
Uncompressing Linux.......................................................
............................ done, booting the kernel.
Initializing cgroup subsys cpu
Linux version 2.6.29-00261-g0097074-dirty (digit@digit.mtv.corp.google.com) (gcc
  version 4.4.0 (GCC) ) #20 Wed Mar 31 09:54:02 PDT 2010
CPU: ARM926EJ-S [41069265] revision 5 (ARMv5TEJ), cr=00093177
CPU: VIVT data cache, VIVT instruction cache
Machine: Goldfish
Memory policy: ECC disabled, Data cache writeback
Built 1 zonelists in Zone order, mobility grouping on.  Total pages: 24384
Kernel command line: qemu=1 console=ttyS0 android.checkjni=1 android.qemud=ttyS1
  android.ndns=3
Unknown boot option `android.checkjni=1': ignoring
Unknown boot option `android.qemud=ttyS1': ignoring
Unknown boot option `android.ndns=3': ignoring
PID hash table entries: 512 (order: 9, 2048 bytes)
Console: colour dummy device 80x30
Dentry cache hash table entries: 16384 (order: 4, 65536 bytes)
Memory: 96MB = 96MB total
Memory: 91548KB available (2616K code, 681K data, 104K init)
Calibrating delay loop... 403.04 BogoMIPS (lpj=2015232)
Mount-cache hash table entries: 512
Initializing cgroup subsys debug
Initializing cgroup subsys cpuacct
Initializing cgroup subsys freezer
CPU: Testing write buffer coherency: ok
...
```

You can also have the emulator print out information about its own execution using the -verbose flag, thereby allowing you to see, for example, which images files it's using:

```
$ emulator -verbose
emulator: found Android build root: /home/karim/android/aosp-2.3.x
emulator: found Android build out:  /home/karim/android/aosp-2.3.x/out/target/pr
oduct/generic
```

```
emulator:    locking user data image at /home/karim/android/aosp-2.3.x/out/targ
et/product/generic/userdata-qemu.img
emulator: selecting default skin name 'HVGA'
emulator: found skin-specific hardware.ini: /home/karim/android/aosp-2.3.x/sdk/e
mulator/skins/HVGA/hardware.ini
emulator: autoconfig: -skin HVGA
emulator: autoconfig: -skindir /home/karim/android/aosp-2.3.x/sdk/emulator/skins
emulator: keyset loaded from: /home/karim/.android/default.keyset
emulator: trying to load skin file '/home/karim/android/aosp-2.3.x/sdk/emulator/
skins/HVGA/layout'
emulator: skin network speed: 'full'
emulator: skin network delay: 'none'
emulator: no SD Card image at '/home/karim/android/aosp-2.3.x/out/target/product
/generic/sdcard.img'
emulator: registered 'boot-properties' qemud service
emulator: registered 'boot-properties' qemud service
emulator: Adding boot property: 'qemu.sf.lcd_density' = '160'
emulator: Adding boot property: 'dalvik.vm.heapsize' = '16m'
emulator: argv[00] = "emulator"
emulator: argv[01] = "-kernel"
emulator: argv[02] = "/home/karim/android/aosp-2.3.x/prebuilt/android-arm/kernel
/kernel-qemu"
emulator: argv[03] = "-initrd"
emulator: argv[04] = "/home/karim/android/aosp-2.3.x/out/target/product/generic/
ramdisk.img"
emulator: argv[05] = "-nand"
emulator: argv[06] = "system,size=0x4200000,initfile=/home/karim/android/aosp-2.
3.x/out/target/product/generic/system.img"
emulator: argv[07] = "-nand"
emulator: argv[08] = "userdata,size=0x4200000,file=/home/karim/android/aosp-2.3.
x/out/target/product/generic/userdata-qemu.img"
emulator: argv[09] = "-nand"
...
```

Up to this point, I've used the terms QEMU and emulator interchangeably. The reality, though, is that the *emulator* command isn't actually QEMU: It's a custom wrapper around it created by the Android development team. You can, however, interact with the emulator's QEMU by using the -qemu flag. Anything you pass after that flag is passed on to QEMU and not the *emulator* wrapper:

```
$ emulator -qemu -h
QEMU PC emulator version 0.10.50Android, Copyright (c) 2003-2008 Fabrice Bellard
usage: qemu [options] [disk_image]

'disk_image' is a raw hard image image for IDE hard disk 0

Standard options:
-h or -help     display this help and exit
-version        display version information and exit
-M machine      select emulated machine (-M ? for list)
-cpu cpu        select CPU (-cpu ? for list)
-smp n          set the number of CPUs to 'n' [default=1]
```

```
-numa node[,mem=size][,cpus=cpu[-cpu]][,nodeid=node]
-fda/-fdb file  use 'file' as floppy disk 0/1 image
-hda/-hdb file  use 'file' as IDE hard disk 0/1 image
...
$ emulator -qemu -...
```

We saw earlier how we can use *adb* to interact with the AOSP running within the emulator, and we just saw how we can use various options to change the way the emulator is started. Interestingly, we can also control the emulator's behavior at runtime by *telnet*ing into it. Every emulator instance that starts is assigned a port number on the host. Look again at Figure 3-3 and check the top-left corner of the emulator's window. The number up there (5554 in this case) is the port number at which that emulator instance is listening. The next emulator that starts simultaneously will get 5556, the next 5558, and so on. To gain access to the emulator's special console, you can use the regular *telnet* command:

```
$ telnet localhost 5554
Trying 127.0.0.1...
Connected to localhost.
Escape character is '^]'.
Android Console: type 'help' for a list of commands
OK
help
Android console command help:

        help|h|?        print a list of commands
        event           simulate hardware events
        geo             Geo-location commands
        gsm             GSM related commands
        kill            kill the emulator instance
        network         manage network settings
        power           power related commands
        quit|exit       quit control session
        redir           manage port redirections
        sms             SMS related commands
        avd             manager virtual device state
        window          manage emulator window

try 'help <command>' for command-specific help
OK
```

Using that console, you can do some nifty tricks like redirecting a port from the host to the target:

```
redir add tcp:8080:80
OK
redir list
tcp:8080  => 80
OK
```

From here on, anything accessing 8080 on your host will actually be speaking to whatever is listening to port 80 on that emulated Android. Nothing listens to that port by default on Android, but you can, for example, have BusyBox's *httpd* running on Android and connect to it in this way.

The emulator also exposes a few "magic" IPs to the emulated Android. IP address 10.0.2.2, for instance, is an alias to your workstation's 127.0.0.1. If you have Apache running on your workstation, you can open the emulator's browser and type `http://10.0.2.2` and you'll be able to browse whatever content is served up by Apache.

For more information on how to operate the emulator and its various options, have a look at the Using the Android Emulator (*http://developer.android.com/tools/devices/emulator.html*) section of Google's Android Developers Guide (*http://developer.android.com*). It's written for an app developer audience, but it will still be very useful to you even if you're doing platform work.

CHAPTER 4
The Build System

The goal of the previous chapter was to get you up and running as quickly as possible with custom AOSP development. There's nothing precluding you from closing this book at this point and starting to dig in and modify your AOSP tree to fit your needs. All you need to do to test your modifications is to rebuild the AOSP, start the emulator again, and, if need be, shell back into it using ADB. If you want to maximize your efforts, however, you'll likely want some insight into Android's build system.

Despite its modularity, Android's build system is fairly complex and doesn't resemble any of the mainstream build systems out there; none that are used for most open source projects, at least. Specifically, it uses *make* in a fairly unconventional way and doesn't provide any sort of *menuconfig*-based configuration (or equivalent for that matter). Android very much has its own build paradigm that takes some time to get used to. So grab yourself a good coffee or two—things are about to get serious.

Like the rest of the AOSP, the build system is a moving target. So while the following information should remain valid for a long time, you should be on the lookout for changes in the AOSP version you're using.

Comparison with Other Build Systems

Before I start explaining how Android's build system works, allow me to begin by emphasizing how it differs from what you might already know. First and foremost, unlike most *make*-based build systems, the Android build system doesn't rely on recursive makefiles. Unlike the Linux kernel, for instance, there isn't a top-level makefile that will recursively invoke subdirectories' makefiles. Instead, there is a script that explores all directories and subdirectories until it finds an *Android.mk* file, whereupon it stops and doesn't explore the subdirectories underneath that file's location—unless the *Android.mk* found instructs the build system otherwise. Note that Android doesn't rely on makefiles called *Makefile*. Instead, it's the *Android.mk* files that specify how the local "module" is built.

 Android build "modules" have nothing to do with kernel "modules." Within the context of Android's build system, a "module" is any component of the AOSP that needs to be built. This might be a binary, an app package, a library, etc., and it might have to be built for the target or the host, but it's still a "module" with regards to the build system.

How Many Build Modules?

Just to give you an idea of how many modules can be built by the AOSP, try running this command in your tree:

```
$ find . -name Android.mk | wc -l
```

This will look for all *Android.mk* files and count how many there are. In 2.3.7/Gingerbread there are 1,143 and in 4.2/Jelly Bean, 2,037.

Another Android specificity is the way the build system is configured. While most of us are used to systems based on kernel-style menuconfig or GNU autotools (i.e., autoconf, automake, etc.), Android relies on a set of variables that are either set dynamically as part of the shell's environment by way of *envsetup.sh* and *lunch* or are defined statically ahead of time in a *buildspec.mk* file. Also—always seeming to be a surprise to newcomers—the level of configurability made possible by Android's build system is fairly limited. So while you can specify the properties of the target for which you want the AOSP to be built and, to a certain extent, which apps should be included by default in the resulting AOSP, there is no way for you to enable or disable most features, as is possible à la menuconfig. You can't, for instance, decide that you don't want power management support or that you don't want the Location Service to start by default.

Also, the build system doesn't generate object files or any sort of intermediate output within the same location as the source files. You won't find the *.o* files alongside their *.c* source files within the source tree, for instance. In fact, none of the existing AOSP directories are used in any of the output. Instead, the build system creates an *out/* directory where it stores everything it generates. Hence, a *make clean* is very much the same thing as an *rm -rf out/*. In other words, removing the *out/* directory wipes out anything that was built.

The last thing to say about the build system before we start exploring it in more detail is that it's heavily tied to *GNU make*. And, more to the point, version 3.81; even the newer 3.82 won't work with many AOSP versions without patching. The build system in fact heavily relies on many *GNU make*-specific features such as the `define`, `include`, and `ifndef` directives.

> ### Some Background on the Design of Android's Build System
>
> If you would like to get more insight into the design choices that were made when Android's build system was put together, have a look at the *build/core/build-system.html* file in the AOSP. It's dated May 2006 and seems to have been the document that went around within the Android dev team to get consensus on a rework of the build system. Some of the information and the hypothesis are out of date or have been obsoleted, but most of the nuggets of the current build system are there. In general, I've found that the further back the document was created by the Android dev team, the more insightful it is regarding raw motivations and technical background. Newer documents tend to be "cleaned up" and abstract, if they exist at all.
>
> If you want to understand the technical underpinnings of why Android's build system doesn't use recursive *make*, have a look at the paper entitled "Recursive Make Considered Harmful" (*http://aegis.sourceforge.net/auug97.pdf*) by Peter Miller in AUUGN Journal of AUUG Inc., 19(1), pp. 14–25. The paper explores the issues surrounding the use of recursive makefiles and explains a different approach involving the use of a single global makefile for building the entire project based on module-provided *.mk* files, which is exactly what Android does.

Architecture

As illustrated in Figure 4-1, the entry point to making sense of the build system is the *main.mk* file found in the *build/core/* directory, which is invoked through the top-level makefile, as we saw earlier. The *build/core/* directory actually contains the bulk of the build system, and we'll cover key files from there. Again, remember that Android's build system pulls everything into a single makefile; it isn't recursive. Hence, each *.mk* file you see eventually becomes part of a single huge makefile that contains the rules for building all the pieces in the system.

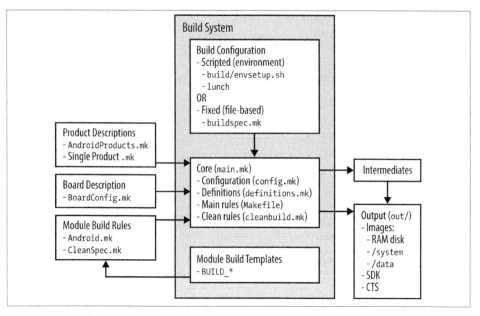

Figure 4-1. Android's build system

Why Does make Hang?

Every time you type *make*, you witness the aggregation of the *.mk* files into a single set through what might seem like an annoying build artifact: The build system prints out the build configuration and seems to hang for quite some time without printing anything to the screen. After these long moments of screen silence, it actually starts proceeding again and builds every part of the AOSP, at which point you see regular output to your screen as you'd expect from any regular build system. Anyone who's built the AOSP has wondered what in the world the build system is doing during that time. What it's doing is incorporating every *Android.mk* file it can find in the AOSP.

If you want to see this in action, edit *build/core/main.mk* and replace this line:

```
include $(subdir_makefiles)
```

with this:

```
$(foreach subdir_makefile, $(subdir_makefiles), \
   $(info Including $(subdir_makefile)) \
   $(eval include $(subdir_makefile)) \
)
subdir_makefile :=
```

The next time you type *make*, you'll actually see what's happening:

```
$ make -j16
===============================================
PLATFORM_VERSION_CODENAME=REL
PLATFORM_VERSION=2.3.4
TARGET_PRODUCT=generic
...
===============================================
Including ./bionic/Android.mk
Including ./development/samples/Snake/Android.mk
Including ./libcore/Android.mk
Including ./external/elfutils/Android.mk
Including ./packages/apps/Camera/Android.mk
Including ./device/htc/passion-common/Android.mk
...
```

Configuration

One of the first things the build system does is pull in the build configuration through the inclusion of *config.mk*. The build can be configured either by the use of the *envsetup.sh* and *lunch* commands or by providing a *buildspec.mk* file at the top-level directory. In either case, some of the following variables need to be set.

TARGET_PRODUCT
    Android flavor to be built. Each recipe can, for instance, include a different set of apps or locales or build different parts of the tree. Have a look at the various single product *.mk* files included by the *AndroidProducts.mk* files in *build/target/product/*, *device/samsung/crespo/*, and *device/htc/passion/* for examples in 2.3/Gingerbread. In case of 4.2/Jelly Bean, look at *device/asus/grouper/* and *device/samsung/amgnuro/* instead of Crespo and Passion. Values include the following:

generic
    The "vanilla" kind, the most basic build of the AOSP parts you can have.

full
    The "all dressed" kind, with most apps and the major locales enabled.

full_crespo
    Same as full but for Crespo (Samsung Nexus S).

full_grouper
    Same as full but for Grouper (Asus Nexus 7).

sim
    Android simulator (see "The Simulator: A Piece of Android's History" on page 117). Even though this is available in 2.3/Gingerbread, this target has since been removed and isn't in 4.2/Jelly Bean.

sdk
: The SDK; includes a vast number of locales.

TARGET_BUILD_VARIANT
: Selects which modules to install. Each module is supposed to have a LOCAL_MOD ULE_TAGS variable set in its *Android.mk* to at least one of the following:[1] user, debug, eng, tests, optional, or samples. By selecting the variant, you will tell the build system which module subsets should be included—the only exception to this is packages (i.e., modules that generate *.apk* files) for which these rules don't apply. Specifically:

eng
: Includes all modules tagged as user, debug, or eng.

userdebug
: Includes both modules tagged as user and debug.

user
: Includes only modules tagged as user.

TARGET_BUILD_TYPE
: Dictates whether or not special build flags are used or DEBUG variables are defined in the code. The possible values here are either release or debug. Most notably, the *frameworks/base/Android.mk* file chooses between either *frameworks/base/core/config/debug* or *frameworks/base/core/config/ndebug*, depending on whether or not this variable is set to debug. The former causes the ConfigBuildFlags.DEBUG Java constant to be set to true, whereas the latter causes it to be set to false. Some code in parts of the system services, for instance, is conditional on DEBUG. Typically, TARGET_BUILD_TYPE is set to release.

TARGET_TOOLS_PREFIX
: By default, the build system will use one of the cross-development toolchains shipped with it underneath the *prebuilt/* directory — *prebuilts/* as of 4.2/Jelly Bean. However, if you'd like it to use another toolchain, you can set this value to point to its location.

OUT_DIR
: By default, the build system will put all build output into the *out/* directory. You can use this variable to provide an alternate output directory.

---

1. If you do not provide a value, defaults will be used. For instance, all apps are set to optional by default. Also, some modules are part of GRANDFATHERED_USER_MODULES in *user_tags.mk*. No LOCAL_MODULE_TAGS need be specified for those; they're always included.

BUILD_ENV_SEQUENCE_NUMBER

If you use the template *build/buildspec.mk.default* to create your own *build spec.mk* file, this value will be properly set. However, if you create a *buildspec.mk* with an older AOSP release and try to use it in a future AOSP release that contains important changes to its build system and, hence, a different value, this variable will act as a safety net. It will cause the build system to inform you that your *build spec.mk* file doesn't match your build system.

The Simulator: A Piece of Android's History

If you go back to the menu printed by 2.3/Gingerbread's *lunch* in "Building Android" on page 94, you'll notice an entry called simulator. In fact you'll find references to the simulator at a number of locations in 2.3/Gingerbread, including quite a few *Android.mk* files and subdirectories in the tree. The most important thing you need to know about the simulator is that it has *nothing* to do with the emulator. They are two completely different things.

That said, the simulator appears to be a remnant of the Android team's early work to create Android. Since at the time they didn't even have Android running in QEMU, they used their desktop OSes and the LD_PRELOAD mechanism to simulate an Android device, hence the term "simulator." It appears that they stopped using it as soon as running Android on QEMU became possible. It continued being in the AOSP up until 4.0/Ice-Cream Sandwich, though, and was potentially useful for building parts of the AOSP for development and testing on developer workstations. 4.2/Jelly Bean, for instance, doesn't have a simulator target.

The presence of the simulator build target in 2.3/Gingerbread and before didn't mean that you could run the AOSP on your desktop. In fact you couldn't, if only because you needed a kernel that had Binder included and you would've needed to be using Bionic instead of your system's default C library. But, if you wanted to run parts of what's built from the AOSP on your desktop, this product target allowed you to do so.

In 2.3/Gingerbread, various parts of the code build very differently if the target is the simulator. When browsing the code, for example, you'll sometimes find conditional builds around the HAVE_ANDROID_OS C macro, which is only defined when compiling for the simulator. The code that talks to the Binder is one of these. If HAVE_AN DROID_OS is not defined, that code will return an error to its caller instead of trying to actually talk to the Binder driver.

For the full story behind the simulator, have a look at Android developer Andrew McFadden's response to a post entitled "Android Simulator Environment" (*http://groups.google.com/group/android-porting/msg/9f27c8d072c1b112*) on the android-porting mailing list in April 2009.

In addition to selecting which parts of the AOSP to build and which options to build them with, the build system also needs to know about the target it's building for. This is provided through a *BoardConfig.mk* file, which will specify things such as the command line to be provided to the kernel, the base address at which the kernel should be loaded, or the instruction set version most appropriate for the board's CPU (TARGET_ARCH_VARIANT). Have a look at *build/target/board/* for a set of per-target directories that each contain a *BoardConfig.mk* file. Also have a look at the various *device/*/TARGET_DEVICE/BoardConfig.mk* files included in the AOSP. The latter are much richer than the former because they contain a lot more hardware-specific information. The device name (i.e., TARGET_DEVICE) is derived from the PRODUCT_DEVICE specified in the product *.mk* file provided for the TARGET_PRODUCT set in the configuration. In 2.3/Gingerbread, for example, *device/samsung/crespo/AndroidProducts.mk* includes *device/samsung/crespo/full_crespo.mk*, which sets PRODUCT_DEVICE to crespo. Hence, the build system looks for a *BoardConfig.mk* in *device/*/crespo/*, and there happens to be one at that location. The same goes on in 4.2/Jelly Bean for the PRODUCT_DEVICE set in *device/asus/grouper/full_grouper.mk* to grouper, thereby pointing the build system to *device/*/grouper/BoardConfig.mk*.

The final piece of the puzzle with regard to configuration is the CPU-specific options used to build Android. For ARM, those are contained in *build/core/combo/arch/arm/armv*.mk*, with TARGET_ARCH_VARIANT determining the actual file to use. Each file lists CPU-specific cross-compiler and cross-linker flags used for building C/C++ files. They also contain a number of ARCH_ARM_HAVE_* variables that enable others parts of the AOSP to build code conditionally based on whether a given ARM feature is found in the target's CPU.

envsetup.sh

Now that you understand the kinds of configuration input the build system needs, we can discuss the role of *envsetup.sh* in more detail. As its name implies, *envsetup.sh* actually is for setting up a build environment for Android. It does only part of the job, though. Mainly, it defines a series of shell commands that are useful to any sort of AOSP work:

```
$ cd ~/android/aosp-2.3.x
$ . build/envsetup.sh
$ help
Invoke ". build/envsetup.sh" from your shell to add the following functions to
your environment:
- croot:   Changes directory to the top of the tree.
- m:       Makes from the top of the tree.
- mm:      Builds all of the modules in the current directory.
- mmm:     Builds all of the modules in the supplied directories.
- cgrep:   Greps on all local C/C++ files.
- jgrep:   Greps on all local Java files.
```

```
- resgrep:  Greps on all local res/*.xml files.
- godir:    Go to the directory containing a file.

Look at the source to view more functions. The complete list is:
add_lunch_combo cgrep check_product check_variant choosecombo chooseproduct choo
setype choosevariant cproj croot findmakefile gdbclient get_abs_build_var getbug
reports get_build_var getprebuilt gettop godir help isviewserverstarted jgrep lu
nch m mm mmm pgrep pid printconfig print_lunch_menu resgrep runhat runtest set_j
ava_home setpaths set_sequence_number set_stuff_for_environment settitle smokete
st startviewserver stopviewserver systemstack tapas tracedmdump
```

In 4.2/Jelly Bean, *hmm* has replaced *help*, and the command set made available to you has been expanded:

```
$ cd ~/android/aosp-4.2
$ . build/envsetup.sh
$ hmm
Invoke ". build/envsetup.sh" from your shell to add the following functions to y
our environment:
- lunch:    lunch <product_name>-<build_variant>
- tapas:    tapas [<App1> <App2> ...] [arm|x86|mips] [eng|userdebug|user]
- croot:    Changes directory to the top of the tree.
- m:        Makes from the top of the tree.
- mm:       Builds all of the modules in the current directory.
- mmm:      Builds all of the modules in the supplied directories.
- cgrep:    Greps on all local C/C++ files.
- jgrep:    Greps on all local Java files.
- resgrep:  Greps on all local res/*.xml files.
- godir:    Go to the directory containing a file.

Look at the source to view more functions. The complete list is:
addcompletions add_lunch_combo cgrep check_product check_variant choosecombo cho
oseproduct choosetype choosevariant cproj croot findmakefile gdbclient get_abs_b
uild_var getbugreports get_build_var getlastscreenshot getprebuilt getscreenshot
path getsdcardpath gettargetarch gettop godir hmm isviewserverstarted jgrep key_
back key_home key_menu lunch _lunch m mm mmm pid printconfig print_lunch_menu re
sgrep runhat runtest set_java_home setpaths set_sequence_number set_stuff_for_en
vironment settitle smoketest startviewserver stopviewserver systemstack tapas tr
acedmdump
```

You'll likely find the *croot* and *godir* commands quite useful for traversing the tree. Some parts of it are quite deep, given the use of Java and its requirement that packages be stored in directory trees bearing the same hierarchy as each subpart of the corresponding fully qualified package name. For instance, a file part of the com.foo.bar package must be stored under the *com/foo/bar/* directory. Hence, it's not rare to find yourself 7 to 10 directories underneath the AOSP's top-level directory, and it rapidly becomes tedious to type something like *cd ../../../* ... to return to an upper part of the tree.

m and *mm* are also quite useful since they allow you to, respectively, build from the top level regardless of where you are or just build the modules found in the current directory. For example, if you made a modification to the Launcher and are in *packages/apps/*

Launcher2, you can rebuild just that module by typing *mm* instead of *cd*'ing back to the top level and typing *make*. Note that *mm* doesn't rebuild the entire tree and, therefore, won't regenerate AOSP images even if a dependent module has changed. *m* will do that, though. Still, *mm* can be useful to test whether your local changes break the build or not until you're ready to regenerate the full AOSP.

Although the online help doesn't mention *lunch*, it is one of the commands defined by *envsetup.sh*. When you run *lunch* without any parameters, it shows you a list of potential choices. This is the list from 2.3/Gingerbread:

```
$ lunch

You're building on Linux

Lunch menu... pick a combo:
     1. generic-eng
     2. simulator
     3. full_passion-userdebug
     4. full_crespo4g-userdebug
     5. full_crespo-userdebug

Which would you like? [generic-eng]
```

This is the list from 4.2/Jelly Bean:

```
$ lunch

You're building on Linux

Lunch menu... pick a combo:
     1. full-eng
     2. full_x86-eng
     3. vbox_x86-eng
     4. full_mips-eng
     5. full_grouper-userdebug
     6. full_tilapia-userdebug
     7. mini_armv7a_neon-userdebug
     8. mini_armv7a-userdebug
     9. mini_mips-userdebug
    10. mini_x86-userdebug
    11. full_mako-userdebug
    12. full_maguro-userdebug
    13. full_manta-userdebug
    14. full_toroplus-userdebug
    15. full_toro-userdebug
    16. full_panda-userdebug

Which would you like? [full-eng]
```

These choices are not static. Most depend on what's in the AOSP at the time *envsetup.sh* runs. They're in fact individually added using the add_lunch_combo() function that the script defines. In 2.3/Gingerbread, for instance, *envsetup.sh* adds generic-eng and simulator by default:

```
# add the default one here
add_lunch_combo generic-eng

# if we're on linux, add the simulator.  There is a special case
# in lunch to deal with the simulator
if [ "$(uname)" = "Linux" ] ; then
    add_lunch_combo simulator
fi
```

In 4.2/Jelly Bean, simulator is no longer a valid target and *envsetup.sh* does this instead:

```
# add the default one here
add_lunch_combo full-eng
add_lunch_combo full_x86-eng
add_lunch_combo vbox_x86-eng
add_lunch_combo full_mips-eng
```

envsetup.sh also includes all the vendor-supplied scripts it can find. Here's how it's done in 2.3/Gingerbread:

```
# Execute the contents of any vendorsetup.sh files we can find.
for f in `/bin/ls vendor/*/vendorsetup.sh vendor/*/build/vendorsetup.sh device/*/*/vendorsetup.sh 2> /dev/null`
do
    echo "including $f"
    . $f
done
unset f
```

Here's how it's done in 4.2/Jelly Bean:

```
# Execute the contents of any vendorsetup.sh files we can find.
for f in `/bin/ls vendor/*/vendorsetup.sh vendor/*/*/vendorsetup.sh device/*/*/vendorsetup.sh 2> /dev/null`
do
    echo "including $f"
    . $f
done
unset f
```

In 2.3/Gingerbread the *device/samsung/crespo/vendorsetup.sh* file, for instance, does this:

```
add_lunch_combo full_crespo-userdebug
```

Similarly, in 4.2/Jelly Bean the *device/asus/grouper/vendorsetup.sh* file does this:

```
add_lunch_combo full_grouper-userdebug
```

So that's how you end up with the menu we saw earlier. Note that the menu asks you to choose a *combo*. Essentially, this is a combination of a TARGET_PRODUCT and TAR

GET_BUILD_VARIANT, with the exception of the simulator in 2.3/Gingerbread. The menu provides the default combinations, but the others remain valid and can be passed to *lunch* as parameters on the command line. In 2.3/Gingerbread, for instance, you can do something like this:

```
$ lunch generic-user

============================================
PLATFORM_VERSION_CODENAME=REL
PLATFORM_VERSION=2.3.4
TARGET_PRODUCT=generic
TARGET_BUILD_VARIANT=user
TARGET_SIMULATOR=false
TARGET_BUILD_TYPE=release
TARGET_BUILD_APPS=
TARGET_ARCH=arm
HOST_ARCH=x86
HOST_OS=linux
HOST_BUILD_TYPE=release
BUILD_ID=GINGERBREAD
============================================
```

```
$ lunch full_crespo-eng

============================================
PLATFORM_VERSION_CODENAME=REL
PLATFORM_VERSION=2.3.4
TARGET_PRODUCT=full_crespo
TARGET_BUILD_VARIANT=eng
TARGET_SIMULATOR=false
TARGET_BUILD_TYPE=release
TARGET_BUILD_APPS=
TARGET_ARCH=arm
HOST_ARCH=x86
HOST_OS=linux
HOST_BUILD_TYPE=release
BUILD_ID=GINGERBREAD
============================================
```

Once *lunch* has finished running for a generic-eng combo, it will set up environment variables described in Table 4-1 in your current shell to provide the build system with the required configuration information.

Table 4-1. Environment variables set by lunch (in no particular order) for the default build target (i.e., generic-eng) in 2.3/Gingerbread

Variable	Value
PATH	$ANDROID_JAVA_TOOLCHAIN:$PATH:$ANDROID_BUILD_PATHS
ANDROID_EABI_TOOLCHAIN	*aosp-root*/prebuilt/linux-x86/toolchain/arm-eabi-4.4.3/bin

Variable	Value
ANDROID_TOOLCHAIN	$ANDROID_EABI_TOOLCHAIN
ANDROID_QTOOLS	*aosp-root*/development/emulator/qtools
ANDROID_BUILD_PATHS	*aosp-root*/out/host/linux-x86:$ANDROID_TOOLCHAIN:$ANDROID_QTOOLS:$ANDROID_TOOLCHAIN:$ANDROID_EABI_TOOLCHAIN
ANDROID_BUILD_TOP	*aosp-root*
ANDROID_JAVA_TOOLCHAIN	$JAVA_HOME/bin
ANDROID_PRODUCT_OUT	*aosp-root*/out/target/product/generic
OUT	ANDROID_PRODUCT_OUT
BUILD_ENV_SEQUENCE_NUMBER	10
OPROFILE_EVENTS_DIR	*aosp-root*/prebuilt/linux-x86/oprofile
TARGET_BUILD_TYPE	release
TARGET_PRODUCT	generic
TARGET_BUILD_VARIANT	eng
TARGET_BUILD_APPS	empty
TARGET_SIMULATOR	false
PROMPT_COMMAND	\"\033]0;[${TARGET_PRODUCT}-${TARGET_BUILD_VARIANT}] ${USER}@${HOSTNAME}: ${PWD}\007\"
JAVA_HOME	/usr/lib/jvm/java-6-sun

Using ccache

If you've already done any AOSP building while reading these pages, you've noticed how long the process is. Obviously, unless you can construct yourself a bleeding-edge build farm, any sort of speedup on your current hardware would be greatly appreciated. As a sign that the Android development team might itself also feel the pain of the rather long builds, they've added support for ccache. ccache stands for *Compiler Cache* and is part of the Samba Project (*http://ccache.samba.org/*). It's a mechanism that caches the object files generated by the compiler based on the preprocessor's output. Hence, if under two separate builds the preprocessor's output is identical, use of ccache will result in the second build not actually using the compiler to build the file. Instead, the cached object file will be copied to the destination where the compiler's output would have been.

To enable the use of ccache, all you need to do is make sure that the USE_CCACHE environment variable is set to 1 before you start your build:

```
$ export USE_CCACHE=1
```

You won't gain any acceleration the first time you run, since the cache will be empty at that time. Every other time you build from scratch, though, the cache will help accelerate the build process. The only downside is that ccache is for C/C++ files only. Hence, it can't accelerate the build of any Java file, I must add sadly. In 2.3/Gingerbread, there are about 15,000 C/C++ files and 18,000 Java files in the AOSP. Those numbers are 27,000 and 29,000 in 4.2/Jelly Bean. So, while the cache isn't a panacea, it's better than nothing.

If you'd like to learn more about ccache, have a look at the article titled "Improve collaborative build times with ccache" (*http://www.ibm.com/developerworks/linux/library/l-ccache/index.html*) by Martin Brown on IBM's developerWorks site. The article also explores the use of *distcc*, which allows you to distribute builds over several machines, so you can pool your team's workstation caches together.

For all its benefits, some developers have reported weird errors in some cases when using *ccache*. For instance, I ran into such issues while maintaining my own AOSP fork. First, I got a version of the AOSP on my workstation and built it, creating a warm cache. I then proceeded to upload that tree to *http://github.com*. Finally, I did a *repo sync* on the tree I had just uploaded but from another directory on my workstation than the original one uploaded. Using *diff* to compare both trees showed both trees were identical. Yet, the original built fine with the warm cache while the second continued to fail building until the cache was erased.

Of course, if you get tired of always typing *build/envsetup.sh* and *lunch*, all you need to do is copy the *build/buildspec.mk.default* into the top-level directory, rename it to *build spec.mk*, and edit it to match the configuration that would have otherwise been set by running those commands. The file already contains all the variables you need to provide; it's just a matter of uncommenting the corresponding lines and setting the values appropriately. Once you've done that, all you have to do is go to the AOSP's directory and invoke *make* directly. You can skip *envsetup.sh* and *lunch*.

Function Definitions

Because the build system is fairly large—there are more than 40 .mk files in *build/core/* alone—there are benefits in being able to reuse as much code as possible. This is why the build system defines a large number of *functions* in the *definitions.mk* file. That file is actually the largest one in the build system at about 60KB, with about 140 functions on about 1,800 lines of makefile code in 2.3/Gingerbread. It's still the largest file in the build system in 4.2/Jelly Bean at about 73KB, 170 functions, and about 2,100 lines of makefile code. Functions offer a variety of operations, including file lookup (e.g., all-makefiles-under and all-c-files-under), transformation (e.g., transform-c-to-o and transform-java-to-classes.jar), copying (e.g., copy-file-to-target), and utility (e.g., my-dir).

Not only are these functions used throughout the rest of the build system's components, acting as its core library, but they're sometimes also directly used in modules' *Android.mk* files. Here's an example snippet from the Calculator app's *Android.mk*:

```
LOCAL_SRC_FILES := $(call all-java-files-under, src)
```

Although thoroughly describing *definitions.mk* is outside the scope of this book, it should be fairly easy for you to explore it on your own. If nothing else, most of the functions in it are preceded with a comment explaining what they do. Here's an example from 2.3/Gingerbread:

```
###########################################################
## Find all of the java files under the named directories.
## Meant to be used like:
##     SRC_FILES := $(call all-java-files-under,src tests)
###########################################################

define all-java-files-under
$(patsubst ./%,%, \
  $(shell cd $(LOCAL_PATH) ; \
        find $(1) -name "*.java" -and -not -name ".*") \
 )
endef
```

Main Make Recipes

At this point you might be wondering where any of the goodies are actually generated. How are the various images such as RAM disk generated or how is the SDK put together, for example? Well, I hope you won't hold a grudge, but I've been keeping the best for last. So without further ado, have a look at the *Makefile* in *build/core/* (not the top-level one). The file starts with an innocuous-looking comment:

```
# Put some miscellaneous rules here
```

But don't be fooled. This is where some of the best meat is. Here's the snippet that takes care of generating the RAM disk, for example, in 2.3/Gingerbread:

```
# ------------------------------------------------------------------
# the ramdisk
INTERNAL_RAMDISK_FILES := $(filter $(TARGET_ROOT_OUT)/%, \
$(ALL_PREBUILT) \
$(ALL_COPIED_HEADERS) \
$(ALL_GENERATED_SOURCES) \
$(ALL_DEFAULT_INSTALLED_MODULES))

BUILT_RAMDISK_TARGET := $(PRODUCT_OUT)/ramdisk.img

# We just build this directly to the install location.
INSTALLED_RAMDISK_TARGET := $(BUILT_RAMDISK_TARGET)
$(INSTALLED_RAMDISK_TARGET): $(MKBOOTFS) $(INTERNAL_RAMDISK_FILES) | $(MINIGZIP)
$(call pretty,"Target ram disk: $@")
$(hide) $(MKBOOTFS) $(TARGET_ROOT_OUT) | $(MINIGZIP) > $@
```

And here's the snippet that creates the certs packages for checking over-the-air (OTA) updates in the same AOSP version:

```
#  -------------------------------------------------------------------
# Build a keystore with the authorized keys in it, used to verify the
# authenticity of downloaded OTA packages.
#
# This rule adds to ALL_DEFAULT_INSTALLED_MODULES, so it needs to come
# before the rules that use that variable to build the image.
ALL_DEFAULT_INSTALLED_MODULES += $(TARGET_OUT_ETC)/security/otacerts.zip
$(TARGET_OUT_ETC)/security/otacerts.zip: KEY_CERT_PAIR :=
$(DEFAULT_KEY_CERT_PAIR)
$(TARGET_OUT_ETC)/security/otacerts.zip: $(addsuffix .x509.pem,
$(DEFAULT_KEY_CERT_PAIR))
    $(hide) rm -f $@
    $(hide) mkdir -p $(dir $@)
    $(hide) zip -qj $@ $<

.PHONY: otacerts
otacerts: $(TARGET_OUT_ETC)/security/otacerts.zip
```

Obviously there's a lot more than I can fit here, but have a look at *Makefile* for information on how any of the following are created:

- Properties (including the target's */default.prop* and */system/build.prop*).
- RAM disk.
- Boot image (combining the RAM disk and a kernel image).
- *NOTICE* files: These are files required by the AOSP's use of the Apache Software License (ASL). Have a look at the ASL for more information about *NOTICE* files.
- OTA keystore.
- Recovery image.
- System image (the target's */system* directory).
- Data partition image (the target's */data* directory).
- OTA update package.
- SDK.

Nevertheless, some things **aren't** in this file:

Kernel images
    Don't look for any rule to build these. There is no kernel part of the official AOSP releases—some of the third-party projects listed in Appendix E, however, actually do package kernel sources directly into the AOSPs they distribute. Instead, you need to find an Androidized kernel for your target, build it separately from the AOSP, and feed it to the AOSP. You can find a few examples of this in the devices in the *device/* directory. In 2.3/Gingerbread, for example, *device/samsung/crespo/* includes a kernel image (file called *kernel*) and a loadable module for the Crespo's WiFi

(*bcm4329.ko* file). Both of these are built outside the AOSP and copied in binary form into the tree for inclusion with the rest of the build.

NDK

While the code to build the NDK is in the AOSP, it's entirely separate from the AOSP's build system in *build/*. Instead, the NDK's build system is in *ndk/build/*. We'll discuss how to build the NDK shortly.

CTS

The rules for building the CTS are in *build/core/tasks/cts.mk*.

Cleaning

As I mentioned earlier, a *make clean* is very much the equivalent of wiping out the *out/* directory. The `clean` target itself is defined in *main.mk*. There are, however, other cleanup targets. Most notably, `installclean`, which is defined in *cleanbuild.mk*, is automatically invoked whenever you change `TARGET_PRODUCT`, `TARGET_BUILD_VARIANT` or `PRODUCT_LOCALES`. For instance, if I had first built 2.3/Gingerbread for the `generic-eng` combo and then used *lunch* to switch the combo to `full-eng`, the next time I started *make*, some of the build output would be automatically pruned using `installclean`:

```
$ make -j16
============================================
PLATFORM_VERSION_CODENAME=REL
PLATFORM_VERSION=2.3.4
TARGET_PRODUCT=full
TARGET_BUILD_VARIANT=eng
...
============================================
*** Build configuration changed: "generic-eng-{mdpi,nodpi}" -> "full-eng-{en_US,
en_GB,fr_FR,it_IT,de_DE,es_ES,mdpi,nodpi}"
*** Forcing "make installclean"...
*** rm -rf out/target/product/generic/data/* out/target/product/generic/data-qem
u/* out/target/product/generic/userdata-qemu.img out/host/linux-x86/obj/NOTICE_F
ILES out/host/linux-x86/sdk out/target/product/generic/*.img out/target/product/
generic/*.txt out/target/product/generic/*.xlb out/target/product/generic/*.zip
out/target/product/generic/data out/target/product/generic/obj/APPS out/target/p
roduct/generic/obj/NOTICE_FILES out/target/product/generic/obj/PACKAGING out/tar
get/product/generic/recovery out/target/product/generic/root out/target/product/
generic/system out/target/product/generic/dex_bootjars out/target/product/generi
c/obj/JAVA_LIBRARIES
*** Done with the cleaning, now starting the real build.
```

In contrast to `clean`, `installclean` doesn't wipe out the entirety of *out/*. Instead, it only nukes the parts that need rebuilding given the combo configuration change. There's also a `clobber` target which is essentially the same thing as a `clean`.

Module Build Templates

What I just described is the build system's architecture and the mechanics of its core components. Having read that, you should have a much better idea of how Android is built from a top-down perspective. Very little of that, however, permeates down to the level of AOSP modules' *Android.mk* files. The system has in fact been architected so that module build recipes are pretty much independent from the build system's internals. Instead, build templates are provided so that module authors can get their modules built appropriately. Each template is tailored for a specific type of module, and module authors can use a set of documented variables, all prefixed by LOCAL_, to modulate the templates' behavior and output. Of course, the templates and underlying support files (mainly *base_rules.mk*) closely interact with the rest of the build system to deal properly with each module's build output. But that's invisible to the module's author.

The templates are themselves found in the same location as the rest of the build system in *build/core/*. *Android.mk* gets access to them through the include directive. Here's an example:

```
include $(BUILD_PACKAGE)
```

As you can see, *Android.mk* files don't actually include the *.mk* templates by name. Instead, they include a variable that is set to the corresponding *.mk* file. Table 4-2 provides the full list of available module templates.

Table 4-2. Module build templates list

Variable	Template	What It Builds	Most Notable Use
BUILD_EXECUTABLE	*executable.mk*	Target binaries	Native commands and daemons
BUILD_HOST_EXECUTABLE	*host_executable.mk*	Host binaries	Development tools
BUILD_RAW_EXECUTABLE	*raw_executable.mk*	Target binaries that run on bare metal	Code in the *bootloader/* directory
BUILD_JAVA_LIBRARY	*java_library.mk*	Target Java libaries	Apache Harmony and Android Framework
BUILD_STATIC_JAVA_LIBRARY	*static_java_library.mk*	Target static Java libraries	N/A, few modules use this
BUILD_HOST_JAVA_LIBRARY	*host_java_library.mk*	Host Java libraries	Development tools
BUILD_SHARED_LIBRARY	*shared_library.mk*	Target shared libraries	A vast number of modules, including many in *external/* and *frameworks/base/*
BUILD_STATIC_LIBRARY	*static_library.mk*	Target static libraries	A vast number of modules, including many in *external/*

Variable	Template	What It Builds	Most Notable Use
BUILD_HOST_SHARED_LIBRARY	host_shared_library.mk	Host shared libraries	Development tools
BUILD_HOST_STATIC_LIBRARY	host_static_library.mk	Host static libraries	Development tools
BUILD_RAW_STATIC_LIBRARY	raw_static_library.mk	Target static libraries that run on bare metal	Code in bootloader/
BUILD_PREBUILT	prebuilt.mk	Copies prebuilt target files	Configuration files and binaries
BUILD_HOST_PREBUILT	host_prebuilt.mk	Copies prebuilt host files	Tools in prebuilt/ and configuration files
BUILD_MULTI_PREBUILT	multi_prebuilt.mk	Copies prebuilt modules of multiple but known types, like Java libraries or executables	Rarely used
BUILD_PACKAGE	package.mk	Built-in AOSP apps (i.e., anything that ends up being an .apk)	All apps in the AOSP
BUILD_KEY_CHAR_MAP	key_char_map.mk	Device character maps	All device character maps in AOSP

These build templates allow *Android.mk* files to be usually fairly lightweight:

```
LOCAL_PATH := $(call my-dir) ❶
include $(CLEAR_VARS) ❷

LOCAL_VARIABLE_1 := value_1 ❸

LOCAL_VARIABLE_2 := value_2

...

include $(BUILD_MODULE_TYPE) ❹
```

❶ Tells the build template where the current module is located.

❷ Clears all previously set LOCAL_* variables that might have been set for other modules.

❸ Sets various LOCAL_* variables to module-specific values.

❹ Invokes the build template that corresponds to the current module's type.

Note that CLEAR_VARS, which is provided by *clear_vars.mk*,[2] is very important. Recall that the build system includes all *Android.mk* into what amounts to a single huge makefile. Including CLEAR_VARS ensures that the LOCAL_* values set for modules preceding yours are zeroed out by the time your *Android.mk* is included. Also, a single *Android.mk* can describe multiple modules one after the other. Hence, CLEAR_VARS ensures that previous module recipes don't pollute subsequent ones.

Here's the Service Manager's *Android.mk* in 2.3/Gingerbread, for instance (*frameworks/base/cmds/servicemanager/*):[3]

```
LOCAL_PATH:= $(call my-dir)
include $(CLEAR_VARS)

LOCAL_SHARED_LIBRARIES := liblog
LOCAL_SRC_FILES := service_manager.c binder.c
LOCAL_MODULE := servicemanager
ifeq ($(BOARD_USE_LVMX),true)
    LOCAL_CFLAGS += -DLVMX
endif

include $(BUILD_EXECUTABLE)
```

And here's the one[4] from 2.3/Gingerbread's Desk Clock app (*packages/app/DeskClock/*):

```
LOCAL_PATH:= $(call my-dir)
include $(CLEAR_VARS)

LOCAL_MODULE_TAGS := optional
LOCAL_SRC_FILES := $(call all-java-files-under, src)
LOCAL_PACKAGE_NAME := DeskClock
LOCAL_OVERRIDES_PACKAGES := AlarmClock
LOCAL_SDK_VERSION := current

include $(BUILD_PACKAGE)

include $(call all-makefiles-under,$(LOCAL_PATH))
```

As you can see, essentially the same structure is used in both modules, even though they provide very different input and result in very different output. Notice also the last line from the Desk Clock's *Android.mk*, which basically includes all subdirectories'

---

2. This file contains a set list of variables starting with the string LOCAL_. If a variable isn't specifically listed in this file, it won't be taken into account by CLEAR_VARS.
3. This version is cleaned up a little (removed commented code, for instance) and slightly reformatted.
4. Also slightly modified to remove white space and comments.

Android.mk files. As I said earlier, the build system looks for the first makefile in a hierarchy and doesn't look in any subdirectories underneath the directory where one was found, hence the need to manually invoke those. Obviously, the code here just goes out and looks for all makefiles underneath. However, some parts of the AOSP either explicitly list subdirectories or conditionally select them based on configuration.

The documentation at *http://source.android.com* used to provide an exhaustive list of all the LOCAL_* variables with their meaning and use. Unfortunately, at the time of this writing, this list is no longer available. The *build/core/build-system.html* file, however, contains an earlier version of that list, and you should refer to that one until up-to-date lists become available again. Here are some of the most frequently encountered LOCAL_* variables:

LOCAL_PATH
: The path of the current module's sources, typically provided by invoking $(call my-dir).

LOCAL_MODULE
: The name to attribute to this module's build output. The actual filename or output and its location will depend on the build template you include. If this is set to foo, for example, and you build an executable, then the final executable will be a command called *foo* and it will be put in the target's */system/bin/*. If LOCAL_MODULE is set to libfoo and you include BUILD_SHARED_LIBRARY instead of BUILD_EXECUTABLE, the build system will generate *libfoo.so* and put it in */system/lib/*.

  Note that the name you provide here must be unique for the particular module class (i.e., build template type) you are building. There can't be two *libfoo.so* libraries, for instance. It's expected that the module name will have to be globally unique (i.e., across all module classes) at some point in the future.

LOCAL_SRC_FILES
: The source files used to build the module. You may provide those by using one of the build system's defined functions, as the Desk Clock uses all-java-files-under, or you may list the files explicitly, as the Service Manager does.

LOCAL_PACKAGE_NAME
: Unlike all other modules, apps use this variable instead of LOCAL_MODULE to provide their names, as you can witness by comparing the two *Android.mk* files shown earlier.

LOCAL_SHARED_LIBRARIES
: Use this to list all the libraries your module depends on. As mentioned earlier, the Service Manager's dependency on liblog is specified using this variable.

LOCAL_MODULE_TAGS
: As I mentioned earlier, this allows you to control under which TARGET_BUILD_VARIANT this module is built. Usually, this should just be set to optional.

LOCAL_MODULE_PATH
: Use this to override the default install location for the type of module you're building.

A good way to find out about more LOCAL_* variables is to look at existing *Android.mk* files in the AOSP. Also, *clear_vars.mk* contains the full list of variables that are cleared. So while it doesn't give you the meaning of each, it certainly lists them all.

Also, in addition to the cleaning targets that affect the AOSP globally, each module can define its own cleaning rules by providing a *CleanSpec.mk*, much like modules provide *Android.mk* files. Unlike the latter, though, the former aren't required. By default, the build system has cleaning rules for each type of module. But you can specify your own rules in a *CleanSpec.mk* in case your module's build does something the build system doesn't generate by default and, therefore, wouldn't typically know how to clean up.

Output

Now that we've looked at how the build system works and how module build templates are used by modules, let's look at the output it creates in *out/*. At a fairly high level, the build output operates in three stages and in two modes, one for the host and one for the target:

1. *Intermediates* are generated using the module sources. These intermediates' format and location depend on the module's sources. They may be *.o* files for C/C++ code, for example, or *.jar* files for Java-based code.
2. Intermediates are used by the build system to create actual binaries and packages: taking *.o* files, for example, and linking them into an actual binary.
3. The binaries and packages are assembled together into the final output requested of the build system. Binaries, for instance, are copied into directories containing the root and */system* filesystems, and images of those filesystems are generated for use on the actual device.

out/ is mainly separated into two directories, reflecting its operating modes: *host/* and *target/*. In each directory, you will find a couple of *obj/* directories that contain the various intermediates generated during the build. Most of these are stored in subdirectories named like the one that the BUILD_* macros presented earlier and serve a specific complementary purpose during the build system's operation:

- *EXECUTABLES/*
- *JAVA_LIBRARIES/*
- *SHARED_LIBRARIES/*
- *STATIC_LIBRARIES/*
- *APPS/*
- *DATA/*
- *ETC/*
- *KEYCHARS/*
- *PACKAGING/*
- *NOTICE_FILES/*
- *include/*
- *lib/*

The directory you'll likely be most interested in is *out/target/product/PRODUCT_DE VICE/*. That's where the output images will be located for the PRODUCT_DEVICE defined in the corresponding product configuration's *.mk*. Table 4-3 explains the content of that directory.

Table 4-3. Product output

Entry	Description
android-info.txt	Contains the code name for the board for which this product is configured
clean_steps.mk	Contains a list of steps that must be executed to clean the tree, as provided in *CleanS pec.mk* files by calling the add-clean-step function
data/	The target's */data* directory
installed-files.txt	A list of all the files installed in *data/* and *system/* directories
obj/	The target product's intermediaries
previous_build_con fig.mk	The last build target; will be used on the next *make* to check if the config has changed, thereby forcing an installclean
ramdisk.img	The RAM disk image generated based on the content of the *root/* directory
root/	The content of the target's root filesystem
symbols/	Unstripped versions of the binaries put in the root filesystem and */system* directory
system/	The target's */system* directory
system.img	The */system* image, based on the content of the *system/* directory
userdata.img	The */data* image, based on the content of the *data/* directory

Architecture | 133

Have a look back at Chapter 2 for a refresher on the root filesystem, /system, and /data. Essentially, though, when the kernel boots, it will mount the RAM disk image and execute the /init found inside. That binary, in turn, will run the /init.rc script that will mount both the /system and /data images at their respective locations. We'll come back to the root filesystem layout and the system's operation at boot time in Chapter 6.

Build Recipes

With the build system's architecture and functioning in mind, let's take a look at some of the most common, and some slightly uncommon, build recipes. We'll only lightly touch on using the results of each recipe, but you should have enough information to get started.

The Default droid Build

Earlier, we went through a number of plain *make* commands but never really explained the default target. When you run plain *make*, it's as if you had typed:[5]

```
$ make droid
```

droid is in fact the default target as defined in *main.mk*. You don't usually need to specify this target manually. I'm providing it here for completeness, so you know it exists.

Seeing the Build Commands

When you build the AOSP, you'll notice that it doesn't actually show you the commands it's running. Instead, it prints out only a summary of each step it's at. If you want to see everything it does, like the *gcc* command lines for example, add the showcommands target to the command line:

```
$ make showcommands
...
host Java: apicheck (out/host/common/obj/JAVA_LIBRARIES/apicheck_intermediates/c
lasses)
for f in ; do if [ ! -f $f ]; then echo Missing file $f; exit 1; fi; unzip -qo $
f -d  out/host/common/obj/JAVA_LIBRARIES/apicheck_intermediates/classes; (cd  ou
t/host/common/obj/JAVA_LIBRARIES/apicheck_intermediates/classes && rm -rf META-I
NF); done
javac -J-Xmx512M -target 1.5 -Xmaxerrs 9999999 -encoding ascii -g     -extdirs ""
 -d out/host/common/obj/JAVA_LIBRARIES/apicheck_intermediates/classes \@out/host
/common/obj/JAVA_LIBRARIES/apicheck_intermediates/java-source-list-uniq || ( rm
-rf out/host/common/obj/JAVA_LIBRARIES/apicheck_intermediates/classes ; exit 41
)
rm -f out/host/common/obj/JAVA_LIBRARIES/apicheck_intermediates/java-source-list
rm -f out/host/common/obj/JAVA_LIBRARIES/apicheck_intermediates/java-source-list
```

---

5. This assumes you had already run *envsetup.sh* and *lunch*.

```
        -uniq
jar -cfm out/host/common/obj/JAVA_LIBRARIES/apicheck_intermediates/javalib.jar b
uild/tools/apicheck/src/MANIFEST.mf   -C out/host/common/obj/JAVA_LIBRARIES/apich
eck_intermediates/classes .
Header: out/host/linux-x86/obj/include/libexpat/expat.h
cp -f external/expat/lib/expat.h out/host/linux-x86/obj/include/libexpat/expat.h
Header: out/host/linux-x86/obj/include/libexpat/expat_external.h
cp -f external/expat/lib/expat_external.h out/host/linux-x86/obj/include/libexpa
t/expat_external.h
Header: out/target/product/generic/obj/include/libexpat/expat.h
cp -f external/expat/lib/expat.h out/target/product/generic/obj/include/libexpat
/expat.h
...
```

Illustrating what I explained in the previous section, this is the same as:

```
$ make droid showcommands
```

As you'll rapidly notice when using this, it generates a lot of output and is therefore hard to follow. You may, however, want to save the standard output and standard error into files if you'd like to analyze the actual commands used to build the AOSP:

```
$ make showcommands > aosp-build-stdout 2> aosp-build-stderr
```

You can also do something like this to merge all output into a single file:

```
$ make showcommands 2>&1 | tell build.log
```

Some also report that they prefer using the *nohup* command instead:

```
$ nohup make showcommands
```

Building the SDK for Linux and Mac OS

The official Android SDK is available at *http://developer.android.com*. You can, however, build your own SDK using the AOSP if, for instance, you extended the core APIs to expose new functionality and would like to distribute the result to developers so they can benefit from your new APIs. To do so, you'll need to select a special combo:

```
$ . build/envsetup.sh
$ lunch sdk-eng
$ make sdk
```

Once this is done, the SDK will be in *out/host/linux-x86/sdk/* when built on Linux and in *out/host/darwin-x86/sdk/* when built on a Mac. There will be two copies, one a ZIP file, much like the one distributed at *http://developer.android.com*, and one uncompressed and ready to use.

Assuming you had already configured Eclipse for Android development using the instructions at *http://developer.android.com*, you'll need to carry out two additional steps to use your newly built SDK. First, you'll need to tell Eclipse the location of the new SDK. To do so, go to Window→Preferences→Android, enter the path to the new SDK in the SDK Location box, and click OK. Also, for reasons that aren't entirely clear to me at the time of this writing, you also need to go to Window→Android SDK Manager, deselect all the items that might be selected except the first two under Tools, and then click "Install 2 packages..." Once that is done, you'll be able to create new projects using the new SDK and access any new APIs you expose in it. If you don't do that second step, you'll be able to create new Android projects, but none of them will resolve Java libraries properly and will, therefore, never build.

Building the SDK for Windows

The instructions for building the SDK for Windows are slightly different from Linux and Mac OS:

```
$ . build/envsetup.sh
$ lunch sdk-eng
$ make win_sdk
```

The resulting output will be in *out/host/windows/sdk/*.

Building the CTS

If you want to build the CTS, you don't need to use *envsetup.sh* or *lunch*. You can go right ahead and type:

```
$ make cts
...
Generating test description for package android.sax
Generating test description for package android.performance
Generating test description for package android.graphics
Generating test description for package android.database
Generating test description for package android.text
Generating test description for package android.webkit
Generating test description for package android.gesture
Generating test plan CTS
Generating test plan Android
Generating test plan Java
Generating test plan VM
Generating test plan Signature
Generating test plan RefApp
Generating test plan Performance
Generating test plan AppSecurity
Package CTS: out/host/linux-x86/cts/android-cts.zip
Install: out/host/linux-x86/bin/adb
```

The *cts* command includes its own online help. Here's the corresponding sample output from 2.3/Gingerbread:

```
$ cd out/host/linux-x86/bin/
$ ./cts
Listening for transport dt_socket at address: 1337
```

```
Android CTS version 2.3_r3
$ cts_host > help
Usage: command options
Available commands and options:
  Host:
    help: show this message
    exit: exit cts command line
  Plan:
    ls --plan: list available plans
    ls --plan plan_name: list contents of the plan with specified name
    add --plan plan_name: add a new plan with specified name
    add --derivedplan plan_name -s/--session session_id -r/--result result_type:
 derive a plan from the given session
    rm --plan plan_name/all: remove a plan or all plans from repository
    start --plan test_plan_name: run a test plan
    start --plan test_plan_name -d/--device device_ID: run a test plan using the
 specified device
    start --plan test_plan_name -t/--test test_name: run a specific test
...
$ cts_host > ls --plan
List of plans (8 in total):
Signature
RefApp
VM
Performance
AppSecurity
Android
Java
CTS
```

Once you have a target up and running, such as the emulator, you can launch the test suite and it will use *adb* to run tests on the target:

```
$ ./cts start --plan CTS
Listening for transport dt_socket at address: 1337
Android CTS version 2.3_r3
Device(emulator-5554) connected
cts_host > start test plan CTS

CTS_INFO >>> Checking API...

CTS_INFO >>> This might take several minutes, please be patient...
...
```

Building the NDK

As I had mentioned earlier, the NDK has its own separate build system, with its own setup and help system, which you can invoke like this:

```
$ cd ndk/build/tools
$ export ANDROID_NDK_ROOT=aosp-root/ndk
$ ./make-release --help
```

```
Usage: make-release.sh [options]

Valid options (defaults are in brackets):
    --help                      Print this help.
    --verbose                   Enable verbose mode.
    --release=name              Specify release name [20110921]
    --prefix=name               Specify package prefix [android-ndk]
    --development=path          Path to development/ndk directory [/home/karim/
                                opersys-dev/android/aosp-2.3.4/development/ndk]
    --out-dir=path              Path to output directory [/tmp/ndk-release]
    --force                     Force build (do not ask initial question) [no]
    --incremental               Enable incremental packaging (debug only). [no]
    --darwin-ssh=hostname       Specify Darwin hostname to ssh to for the build.
    --systems=list              List of host systems to build for [linux-x86]
    --toolchain-src-dir=path    Use toolchain sources from path
```

When you are ready to build the NDK, you can invoke *make-release* as follows, and witness its rather emphatic warning:

```
$ ./make-release
IMPORTANT WARNING !!

This script is used to generate an NDK release package from scratch
for the following host platforms: linux-x86

This process is EXTREMELY LONG and may take SEVERAL HOURS on a dual-core
machine. If you plan to do that often, please read docs/DEVELOPMENT.TXT
that provides instructions on how to do that more easily.

Are you sure you want to do that [y/N]
y
Downloading toolchain sources...
...
```

Updating the API

The build systems has safeguards in case you modify the AOSP's core API. If you do, the build will fail by default with a warning such as this:

```
*******************************
You have tried to change the API from what has been previously approved.

To make these errors go away, you have two choices:
   1) You can add "@hide" javadoc comments to the methods, etc. listed in the
      errors above.

   2) You can update current.xml by executing the following command:
         make update-api

      To submit the revised current.xml to the main Android repository,
      you will need approval.
```

```
********************************
make: *** [out/target/common/obj/PACKAGING/checkapi-current-timestamp] Error 38
make: *** Waiting for unfinished jobs....
```

As the error message suggests, to get the build to continue, you'll need to do something like this:

```
$ make update-api
...
Install: out/host/linux-x86/framework/apicheck.jar
Install: out/host/linux-x86/framework/clearsilver.jar
Install: out/host/linux-x86/framework/droiddoc.jar
Install: out/host/linux-x86/lib/libneo_util.so
Install: out/host/linux-x86/lib/libneo_cs.so
Install: out/host/linux-x86/lib/libneo_cgi.so
Install: out/host/linux-x86/lib/libclearsilver-jni.so
Copying: out/target/common/obj/JAVA_LIBRARIES/core_intermediates/emma_out/lib/cl
asses-jarjar.jar
Install: out/host/linux-x86/framework/dx.jar
Install: out/host/linux-x86/bin/dx
Install: out/host/linux-x86/bin/aapt
Copying: out/target/common/obj/JAVA_LIBRARIES/bouncycastle_intermediates/emma_ou
t/lib/classes-jarjar.jar
Copying: out/target/common/obj/JAVA_LIBRARIES/ext_intermediates/emma_out/lib/cla
sses-jarjar.jar
Install: out/host/linux-x86/bin/aidl
Copying: out/target/common/obj/JAVA_LIBRARIES/core-junit_intermediates/emma_out/
lib/classes-jarjar.jar
Copying: out/target/common/obj/JAVA_LIBRARIES/framework_intermediates/emma_out/l
ib/classes-jarjar.jar
Copying current.xml
```

The next time you start *make*, you won't get any more errors regarding API changes. Obviously at this point you're no longer compatible with the official APIs and are therefore unlikely to be able to get certified as an "Android" device by Google.

Building a Single Module

Up to now, we've looked at building the entire tree. You can also build individual modules. Here's how you can ask the build system to build the Launcher2 module (i.e., the Home screen):

```
$ make Launcher2
```

You can also clean modules individually:

```
$ make clean-Launcher2
```

If you'd like to force the build system to regenerate the system image to include your updated module, you can add the snod target to the command line:

```
$ make Launcher2 snod
============================================
PLATFORM_VERSION_CODENAME=REL
PLATFORM_VERSION=2.3.4
TARGET_PRODUCT=generic
...
target Package: Launcher2 (out/target/product/generic/obj/APPS/Launcher2_interme
diates/package.apk)
  'out/target/common/obj/APPS/Launcher2_intermediates//classes.dex' as 'classes.d
ex'...
Install: out/target/product/generic/system/app/Launcher2.apk
Install: out/host/linux-x86/bin/mkyaffs2image
make snod: ignoring dependencies
Target system fs image: out/target/product/generic/system.img
```

Building Out of Tree

If you'd ever like to build code against the AOSP and its Bionic library but don't want to incorporate that into the AOSP, you can use a makefile such as the following to get the job done:[6]

```
# Paths and settings
TARGET_PRODUCT  = generic
ANDROID_ROOT    = /home/karim/android/aosp-2.3.x
BIONIC_LIBC     = $(ANDROID_ROOT)/bionic/libc
PRODUCT_OUT     = $(ANDROID_ROOT)/out/target/product/$(TARGET_PRODUCT)
CROSS_COMPILE   = \
    $(ANDROID_ROOT)/prebuilt/linux-x86/toolchain/arm-eabi-4.4.3/bin/arm-eabi-

# Tool names
AS              = $(CROSS_COMPILE)as
AR              = $(CROSS_COMPILE)ar
CC              = $(CROSS_COMPILE)gcc
CPP             = $(CC) -E
LD              = $(CROSS_COMPILE)ld
NM              = $(CROSS_COMPILE)nm
OBJCOPY         = $(CROSS_COMPILE)objcopy
OBJDUMP         = $(CROSS_COMPILE)objdump
RANLIB          = $(CROSS_COMPILE)ranlib
READELF         = $(CROSS_COMPILE)readelf
SIZE            = $(CROSS_COMPILE)size
STRINGS         = $(CROSS_COMPILE)strings
STRIP           = $(CROSS_COMPILE)strip

export AS AR CC CPP LD NM OBJCOPY OBJDUMP RANLIB READELF \
       SIZE STRINGS STRIP
```

---

6. This makefile is inspired by a blog post (*http://pundiramit.blogspot.com/2011/08/how-to-build-commom-linux-utils-for.html*) by Row Boat developer Amit Pundir and is based on the example makefile provided in Chapter 4 of *Building Embedded Linux Systems, 2nd ed.* (O'Reilly).

```
# Build settings
CFLAGS       = -O2 -Wall -fno-short-enums
HEADER_OPS   = -I$(BIONIC_LIBC)/arch-arm/include \
               -I$(BIONIC_LIBC)/kernel/common \
               -I$(BIONIC_LIBC)/kernel/arch-arm
LDFLAGS      = -nostdlib -Wl,-dynamic-linker,/system/bin/linker \
               $(PRODUCT_OUT)/obj/lib/crtbegin_dynamic.o \
               $(PRODUCT_OUT)/obj/lib/crtend_android.o \
               -L$(PRODUCT_OUT)/obj/lib -lc -ldl

# Installation variables
EXEC_NAME    = example-app
INSTALL      = install
INSTALL_DIR  = $(PRODUCT_OUT)/system/bin

# Files needed for the build
OBJS         = example-app.o

# Make rules
all: example-app

.c.o:
        $(CC) $(CFLAGS) $(HEADER_OPS) -c $<

example-app: ${OBJS}
        $(CC) -o $(EXEC_NAME) ${OBJS} $(LDFLAGS)

install: example-app
        test -d $(INSTALL_DIR) || $(INSTALL) -d -m 755 $(INSTALL_DIR)
        $(INSTALL) -m 755 $(EXEC_NAME) $(INSTALL_DIR)

clean:
        rm -f *.o $(EXEC_NAME) core

distclean:
        rm -f *~
        rm -f *.o $(EXEC_NAME) core
```

In this case, you don't need to care about either *envsetup.sh* or *lunch*. You can just go ahead and type the magic incantation:

$ **make**

Obviously this won't add your binary to any of the images generated by the AOSP. Even the install target here will be of value only if you're mounting the target's filesystem off NFS, and that's valuable only during debugging, which is what this makefile is assumed to be useful for. To an extent, it could also be argued that using such a makefile is actually counterproductive, since it's far more complicated than the equivalent *Android.mk* that would result if this code were added as a module part of the AOSP.

Still, this kind of hack can have its uses. Under certain circumstances, for instance, it might make sense to modify the conventional build system used by a rather large codebase to build that project against the AOSP yet outside of it; the alternative being to copy the project into the AOSP and create *Android.mk* files to reproduce the mechanics of its original conventional build system, which might turn out to be a substantial endeavor in and of itself.

Building Recursively, In-Tree

You can, if you really want to, hack yourself a makefile to build within the AOSP a component that is based on recursive makefiles instead of trying to reproduce the same functionality using *Android.mk* files, as was suggested in the last section. Several of the AOSP forks mentioned in Appendix E, for instance, include the kernel sources at the top level of the AOSP and modify the AOSP's main makefile to invoke the kernel's existing build system.

Here's another example where an *Android.mk* was created by Linaro's Bernhard Rosenkränzer in order to build ffmpeg—which relies on a GNU autotools-like script—using its original build files:

```
include $(CLEAR_VARS)
FFMPEG_TCDIR := $(realpath $(shell dirname $(TARGET_TOOLS_PREFIX)))
FFMPEG_TCPREFIX := $(shell basename $(TARGET_TOOLS_PREFIX))
# FIXME remove -fno-strict-aliasing once the aliasing violations are fixed
FFMPEG_COMPILER_FLAGS = $(subst -I ,-I../../,$(subst -include \
system/core/include/arch/linux-arm/AndroidConfig.h,,$(subst -include \
build/core/combo/include/arch/linux-arm/AndroidConfig.h,, \
$(TARGET_GLOBAL_CFLAGS)))) -fno-strict-aliasing -Wno-error=address \
 -Wno-error=format-security
ifneq ($(strip $(SHOW_COMMANDS)),)
FF_VERBOSE="V=1"
endif

.PHONY: ffmpeg

droidcore: ffmpeg

systemtarball: ffmpeg

REALTOP=$(realpath $(TOP))

ffmpeg: x264 $(PRODUCT_OUT)/obj/STATIC_LIBRARIES/libvpx_intermediates/libvpx.a
mkdir -p $(PRODUCT_OUT)/obj/ffmpeg
cd $(PRODUCT_OUT)/obj/ffmpeg && \
export PATH=$(FFMPEG_TCDIR):$(PATH) && \
$(REALTOP)/external/ffmpeg/configure \
  --arch=arm \
  --target-os=linux \
  --prefix=/system \
```

```
  --bindir=/system/bin \
  --libdir=/system/lib \
  --enable-shared \
  --enable-gpl \
  --disable-avdevice \
  --enable-runtime-cpudetect \
  --disable-libvpx \
  --enable-libx264 \
  --enable-cross-compile \
  --cross-prefix=$(FFMPEG_TCPREFIX) \
  --extra-ldflags="-nostdlib -Wl,-dynamic-linker, \
/system/bin/linker,-z,muldefs$(shell if test $(PRODUCT_SDK_VERSION) -lt 16; \
then echo -n ',-T$(REALTOP)/$(BUILD_SYSTEM)/armelf.x'; fi),-z,nocopyreloc, \
  --no-undefined -L$(REALTOP)/$(TARGET_OUT_STATIC_LIBRARIES) \
-L$(REALTOP)/$(PRODUCT_OUT)/system/lib \
-L$(REALTOP)/$(PRODUCT_OUT)/obj/STATIC_LIBRARIES/libvpx_intermediates -ldl -lc" \
  --extra-cflags="$(FFMPEG_COMPILER_FLAGS) \
-I$(REALTOP)/bionic/libc/include -I$(REALTOP)/bionic/libc/kernel/common \
-I$(REALTOP)/bionic/libc/kernel/arch-arm \
-I$(REALTOP)/bionic/libc/arch-arm/include -I$(REALTOP)/bionic/libm/include \
-I$(REALTOP)/external/libvpx -I$(REALTOP)/external/x264" \
  --extra-libs="-lgcc" && \
$(MAKE) \
TARGET_CRTBEGIN_DYNAMIC_O=$(REALTOP)/$(TARGET_CRTBEGIN_DYNAMIC_O) \
TARGET_CRTEND_O=$(REALTOP)/$(TARGET_CRTEND_O) $(FF_VERBOSE) && \
$(MAKE) install DESTDIR=$(REALTOP)/$(PRODUCT_OUT)
```

Basic AOSP Hacks

You most likely bought this book with one thing in mind: to hack the AOSP to fit your needs. Over the next few pages, we'll start looking into some of the most obvious hacks you'll likely want to try. Of course we're only setting the stage here with the parts that pertain to the build system, which is where you'll likely want to start anyway.

While the following explanations are based on 2.3/Gingerbread, they'll work just the same on 4.2/Jelly Bean, and likely many versions after that one, too. The fact is, these mechanisms have been constant for quite some time. Still, where relevant, changes in 4.2/Jelly Bean are highlighted.

Adding a Device

Adding a custom device is most likely one of the topmost items (if not the topmost) on your list of reasons for reading this book. I'm about to show you how to do just that, so you'll likely want to bookmark this section. Of course I'm actually only showing you the build aspects of the work. There are a lot more steps involved in porting Android to new hardware. Still, adding the new device to the build system will definitely be one of the first things you do. Fortunately, doing that is relatively straightforward.

For the purposes of the current exercise, assume you work for a company called ACME and that you're tasked with delivering its latest gizmo: the CoyotePad, intended to be the best platform for playing all bird games. Let's get started by creating an entry for our new device in *device/*:

```
$ cd ~/android/aosp-2.3.x
$ . build/envsetup.sh
$ mkdir -p device/acme/coyotepad
$ cd device/acme/coyotepad
```

The first thing we'll need in here is an *AndroidProducts.mk* file to describe the various AOSP products that could be built for the CoyotePad:

```
PRODUCT_MAKEFILES := \
    $(LOCAL_DIR)/full_coyotepad.mk
```

While we could describe several products (see *build/target/product/AndroidProducts.mk* for an example), the typical case is to specify just one, as in this case, and it's described in *full_coyotepad.mk*:

```
$(call inherit-product, $(SRC_TARGET_DIR)/product/languages_full.mk)
# If you're using 4.2/Jelly Bean, use full_base.mk instead of full.mk
$(call inherit-product, $(SRC_TARGET_DIR)/product/full.mk)

DEVICE_PACKAGE_OVERLAYS :=

PRODUCT_PACKAGES +=
PRODUCT_COPY_FILES +=

PRODUCT_NAME := full_coyotepad
PRODUCT_DEVICE := coyotepad
PRODUCT_MODEL := Full Android on CoyotePad, meep-meep
```

It's worth taking a closer look at this makefile. First, we're using the `inherit-product` function to tell the build system to pull in other product descriptions as the basis of ours. This allows us to build on other people's work and not have to specify from scratch every bit and piece of the AOSP that we'd like to include. *languages_full.mk* will pull in a vast number of locales, and *full.mk* will make sure we get the same set of modules as if we had built using the `full-eng` combo.

With regard to the other variables:

DEVICE_PACKAGE_OVERLAYS
    Allows us to specify a directory that will form the basis of an overlay that will be applied onto the AOSP's sources, thereby allowing us to substitute default package resources with device-specific resources. You'll find this useful if you'd like to set

custom layouts or colors for Launcher2 or other apps, for instance. We'll look at how to use this in the next section.

PRODUCT_PACKAGES

Allows us to specify packages we'd like to have this product include in addition to those specified in the products we're already inheriting from. If you have custom apps, binaries, or libraries located within *device/acme/coyotepad/*, for instance, you'll want to add them here so that they are included in the final images generated. Notice the use of the += sign. It allows us to append to the existing values in the variable instead of substituting its content.

PRODUCT_COPY_FILES

Allows us to list specific files we'd like to see copied to the target's filesystem and the location where they need to be copied. Each source/destination pair is colon-separated, and pairs are space-separated among themselves. This is useful for configuration files and prebuilt binaries such as firmware images or kernel modules.

PRODUCT_NAME

The TARGET_PRODUCT, which you can set either by selecting a *lunch* combo or passing it as part of the combo parameter to *lunch*, as in:

```
$ lunch full_coyotepad-eng
```

PRODUCT_DEVICE

The name of the actual finished product shipped to the customer. TARGET_DEVICE derives from this variable. PRODUCT_DEVICE has to match an entry in *device/acme/*, since that's where the build looks for the corresponding *BoardConfig.mk*. In this case, the variable is the same as the name of the directory we're already in.

PRODUCT_MODEL

The name of this product as provided in the "Model number" in the "About the phone" section of the settings. This variable actually gets stored as the ro.product.model global property accessible on the device.

Version 4.2/Jelly Bean also includes a PRODUCT_BRAND that is typically set to Android. The value of this variable is then available as the ro.product.brand global property. The latter is used by some parts of the stack that take action based on the device's vendor.

Now that we've described the product, we must also provide some information regarding the board the device is using through a *BoardConfig.mk* file:

```
TARGET_NO_KERNEL := true
TARGET_NO_BOOTLOADER := true
TARGET_CPU_ABI := armeabi
BOARD_USES_GENERIC_AUDIO := true

USE_CAMERA_STUB := true
```

This is a very skinny *BoardConfig.mk* and ensures that we actually build successfully. For a real-life version of that file, have a look at *device/samsung/crespo/BoardConfig Common.mk* in 2.3/Gingerbread, and also at *device/asus/grouper/BoardConfigCom mon.mk* in 4.2/Jelly Bean.

You'll also need to provide a conventional *Android.mk* in order to build all the modules that you might have included in this device's directory:

```
LOCAL_PATH := $(call my-dir)
include $(CLEAR_VARS)

ifneq ($(filter coyotepad,$(TARGET_DEVICE)),)
include $(call all-makefiles-under,$(LOCAL_PATH))
endif
```

It's in fact the preferred modus operandi to put all device-specific apps, binaries, and libraries within the device's directory instead of globally within the rest of the AOSP. If you do add modules here, don't forget to also add them to PRODUCT_PACKAGES as I explained earlier. If you just put them here and provide them valid *Android.mk* files, they'll build, but they won't be in the final images.

If you have several products sharing the same set of packages, you may want to create a *device/acme/common/* directory containing the shared packages. You can see an example of this in 4.2/Jelly Bean's *device/generic/* directory. In that same version, you can also check how *device/samsung/maguro/device.mk* inherits from *device/samsung/tuna/ device.mk* for an example of how one device can be based on another device.

Lastly, let's close the loop by making the device we just added visible to *envsetup.sh* and *lunch*. To do so, you'll need to add a *vendorsetup.sh* in your device's directory:

```
add_lunch_combo full_coyotepad-eng
```

You also need to make sure that it's executable if it's to be operational:

```
$ chmod 755 vendorsetup.sh
```

We can now go back to the AOSP's root and take our brand-new ACME CoyotePad for a ~~run~~chase:

```
$ croot
$ . build/envsetup.sh
$ lunch

You're building on Linux

Lunch menu... pick a combo:
     1. generic-eng
     2. simulator
     3. full_coyotepad-eng
     4. full_passion-userdebug
     5. full_crespo4g-userdebug
```

```
    6. full_crespo-userdebug

Which would you like? [generic-eng] 3


============================================
PLATFORM_VERSION_CODENAME=REL
PLATFORM_VERSION=2.3.4
TARGET_PRODUCT=full_coyotepad
TARGET_BUILD_VARIANT=eng
TARGET_SIMULATOR=false
TARGET_BUILD_TYPE=release
TARGET_BUILD_APPS=
TARGET_ARCH=arm
HOST_ARCH=x86
HOST_OS=linux
HOST_BUILD_TYPE=release
BUILD_ID=GINGERBREAD
============================================

$ make -j16
```

As you can see, the AOSP now recognizes our new device and prints the information correspondingly. When the build is done, we'll also have the same type of output provided in any other AOSP build, except that it will be a product-specific directory:

```
$ ls -al out/target/product/coyotepad/
total 89356
drwxr-xr-x  7 karim karim     4096 2011-09-21 19:20 .
drwxr-xr-x  4 karim karim     4096 2011-09-21 19:08 ..
-rw-r--r--  1 karim karim        7 2011-09-21 19:10 android-info.txt
-rw-r--r--  1 karim karim     4021 2011-09-21 19:41 clean_steps.mk
drwxr-xr-x  3 karim karim     4096 2011-09-21 19:11 data
-rw-r--r--  1 karim karim    20366 2011-09-21 19:20 installed-files.txt
drwxr-xr-x 14 karim karim     4096 2011-09-21 19:20 obj
-rw-r--r--  1 karim karim      327 2011-09-21 19:41 previous_build_config.mk
-rw-r--r--  1 karim karim  2649750 2011-09-21 19:43 ramdisk.img
drwxr-xr-x 11 karim karim     4096 2011-09-21 19:43 root
drwxr-xr-x  5 karim karim     4096 2011-09-21 19:19 symbols
drwxr-xr-x 12 karim karim     4096 2011-09-21 19:19 system
-rw-------  1 karim karim 87280512 2011-09-21 19:20 system.img
-rw-------  1 karim karim  1505856 2011-09-21 19:14 userdata.img
```

Also, have a look at the *build.prop* file in *system/*. It contains various global properties that will be available at runtime on the target and that relate to our configuration and build:

```
# begin build properties
# autogenerated by buildinfo.sh
ro.build.id=GINGERBREAD
ro.build.display.id=full_coyotepad-eng 2.3.4 GINGERBREAD eng.karim.20110921.1908
49 test-keys
ro.build.version.incremental=eng.karim.20110921.190849
ro.build.version.sdk=10
```

```
ro.build.version.codename=REL
ro.build.version.release=2.3.4
ro.build.date=Wed Sep 21 19:10:04 EDT 2011
ro.build.date.utc=1316646604
ro.build.type=eng
ro.build.user=karim
ro.build.host=w520
ro.build.tags=test-keys
ro.product.model=Full Android on CoyotePad, meep-meep
ro.product.brand=generic
ro.product.name=full_coyotepad
ro.product.device=coyotepad
ro.product.board=
ro.product.cpu.abi=armeabi
ro.product.manufacturer=unknown
ro.product.locale.language=en
ro.product.locale.region=US
ro.wifi.channels=
ro.board.platform=
# ro.build.product is obsolete; use ro.product.device
ro.build.product=coyotepad
# Do not try to parse ro.build.description or .fingerprint
ro.build.description=full_coyotepad-eng 2.3.4 GINGERBREAD eng.karim.20110921.190
849 test-keys
ro.build.fingerprint=generic/full_coyotepad/coyotepad:2.3.4/GINGERBREAD/eng.kari
m.20110921.190849:eng/test-keys
# end build properties
...
```

You may want to carefully vet the default properties before using the build on a real device. Some developers have encountered some severe issues due to default values. In both 2.3/Gingerbread and 4.2/Jelly Bean, for instance, `ro.com.android.dataroaming` is set to `true` in some builds. Hence, if you're doing development on a device connected to a live cell network, changing the value to `false` might save you some money.

As you can imagine, there's a lot more to be done here to make sure the AOSP runs on our hardware. But the preceding steps give us the starting point. However, by isolating the board-specific changes in a single directory, this configuration will simplify adding support for the CoyotePad to the next version of the AOSP that gets released. Indeed, it'll just be a matter of copying the corresponding directory to the new AOSP's *device/* directory and adjusting the code therein to use the new APIs.

Adding an App

Adding an app to your board is relatively straightforward. As a starter, try creating a HelloWorld! app with Eclipse and the default SDK; all new Android projects in Eclipse are a HelloWorld! by default. Then copy that app from the Eclipse workspace to its destination:

```
$ cp -a ~/workspace/HelloWorld ~/android/aosp-2.3.x/device/acme/coyotepad/
```

You'll then have to create an *Android.mk* file in *aosp-root/device/acme/coyotepad/HelloWorld/* to build that app:

```
LOCAL_PATH:= $(call my-dir)
include $(CLEAR_VARS)

LOCAL_MODULE_TAGS := optional
LOCAL_SRC_FILES := $(call all-java-files-under, src)
LOCAL_PACKAGE_NAME := HelloWorld

include $(BUILD_PACKAGE)
```

Given that we're tagging this module as `optional`, it won't be included by default in the AOSP build. To include it, you'll need to add it to the `PRODUCT_PACKAGES` listed in the CoyotePad's *full_coyotepad.mk*.

If, instead of adding your app for your board only, you would like to add a default app globally to **all** products generated by the AOSP alongside the existing stock apps, you'll need to put it in *packages/apps/* instead of your board's directory. You'll also need to modify one of the built-in *.mk* files, such as *aosp-root/build/target/product/core.mk*, to have your app built by default. This is not recommended, though, as it's not very portable since it will require you to make this modification to every new AOSP release. As I stated earlier, it's best to keep your custom modifications in *device/acme/coyotepad/* in as much as possible.

Adding an App Overlay

Sometimes you don't actually want to add an app but would rather modify existing ones included by default in the AOSP. That's what app overlays are for. Overlays are a mechanism included in the AOSP to allow device manufacturers to change the resources provided (such as for apps), without actually modifying the original resources included in the AOSP. To use this capability, you must create an overlay tree and tell the build system about it. The easiest location for an overlay is within a device-specific directory such as the one we created in the previous section:

```
$ cd device/acme/coyotepad/
$ mkdir overlay
```

To tell the build system to take this overlay into account, we need to modify our *full_coyotepad.mk* such that:

```
DEVICE_PACKAGE_OVERLAYS := device/acme/coyotepad/overlay
```

At this point, though, our overlay isn't doing much. Let's say we want to modify some of Launcher2's default strings. We could then do something like this:

```
$ mkdir -p overlay/packages/apps/Launcher2/res/values
$ cp aosp-root/packages/apps/Launcher2/res/values/strings.xml \
> overlay/packages/apps/Launcher2/res/values/
```

You can then trim your local *strings.xml* to override only those strings that you need. Most importantly, your device will have a Launcher2 that has your custom strings, but the default Launcher2 will still have its original strings. So if someone relies on the same AOSP sources you're using to build for another product, they'll still get the original strings. You can, of course, replace most resources like this, including images and XML files. So long as you put the files in the same hierarchy as they are found in the AOSP but within *device/acme/coyotepad/overlay/*, they'll be taken into account by the build system.

Overlays can be used only for resources. You can't overlay source code. If you want to customize parts of Android's internals, for instance, you'll still have to make those modifications in situ. There's no way, currently at least, to isolate those changes to your board.

Adding a Native Tool or Daemon

Like the example above of adding an app for your board, you can add your custom native tools and daemons as subdirectories of *device/acme/coyotepad/*. Obviously, you'll need to provide an *Android.mk* in the directory containing the code to build that module:

```
LOCAL_PATH:= $(call my-dir)
include $(CLEAR_VARS)

LOCAL_MODULE := hello-world
LOCAL_MODULE_TAGS := optional
LOCAL_SRC_FILES := hello-world.cpp
LOCAL_SHARED_LIBRARIES := liblog

include $(BUILD_EXECUTABLE)
```

As in the app's case, you'll also need to make sure `hello-world` is part of the CoyotePad's PRODUCT_PACKAGES.

If you intend to add your binary globally for all product builds instead of just locally to your board, you need to know that there are a number of locations in the tree where native tools and daemons are located. Here are the most important ones:

system/core/ and system/
: Custom Android binaries that are meant to be used outside the Android Framework or are standalone pieces.

frameworks/base/cmds/
> Binaries that are tightly coupled to the Android Framework. This is where the Service Manager and *installd* are found, for example.

external/
> Binaries that are generated by an external project that is imported into the AOSP. *strace*, for instance, is here.

Having identified from the list above where the code generating your binary should go, you'll also need to add it as part of one of the global *.mk* files such as *aosp-root/build/target/product/core.mk*. As I said above, however, such global additions are not recommended since they can't be transferred as easily to newer AOSP versions.

Adding a Native Library

Like apps and binaries, you can also add native libraries for your board. Assuming, as above, that the sources to build the library are in a subdirectory of *device/acme/coyotepad/*, you'll need an *Android.mk* to build your library:

```
LOCAL_PATH:= $(call my-dir)
include $(CLEAR_VARS)

LOCAL_MODULE := libmylib
LOCAL_MODULE_TAGS := optional
LOCAL_PRELINK_MODULE := false
LOCAL_SRC_FILES := $(call all-c-files-under,.)

include $(BUILD_SHARED_LIBRARY)
```

Note that LOCAL_PRELINK_MODULE has been removed and is no longer necessary as of 4.0/Ice-Cream Sandwich.

To use this library, you must add it to the libraries listed by the *Android.mk* file of whichever binaries depend on it:

```
LOCAL_SHARED_LIBRARIES := libmylib
```

You'll also likely need to add relevant headers to an *include/* directory located in about the same location as you put your library, so that the code that needs to link against your library can find those headers, such as *device/acme/coyotepad/include/*.

Should you want to make your library apply globally to all AOSP builds, not just your device, you'll need a little bit more information regarding the various locations where libraries are typically found in the tree. First, you should know that, unlike binaries, a lot of libraries are used within a single module but nowhere else. Hence, these libraries will typically be placed within that module's code and not in the usual locations where libraries used systemwide are found. The latter are typically in the following locations:

system/core/
    Libraries used by many parts of the system, including some outside the Android Framework. This is where liblog is, for instance.

frameworks/base/libs/
    Libraries intimately tied to the framework. This is where libbinder is.

frameworks/native/libs/
    In 4.2/Jelly Bean, many libraries that were in *frameworks/base/libs/* in 2.3/Gingerbread have been moved out and into *frameworks/native/libs/*.

external/
    Libraries generated by external projects imported into the AOSP. OpenSSL's libssl is here.

Similarly, instead of using a CoyotePad-specific include directory, you'd use a global directory such as *system/core/include/* or *frameworks/base/include/* or, in 4.2/Jelly Bean, *frameworks/base/include/*. Again, as stated earlier, you should carefully review whether such global additions are truly required, as they'll represent additional work when you try to port for your device to the next version of Android.

Library Prelinking

If you look closely at the example *Android.mk* we provide for the library, you'll notice a LOCAL_PRELINK_MODULE variable. To reduce the time it takes to load libraries, Android versions up to 2.3/Gingerbread used to *prelink* most of their libraries. Prelinking is done by specifying ahead of time the address location where the library will be loaded instead of letting it be figured out at runtime. The file where the addresses are specified in 2.3/Gingerbread is *build/core/prelink-linux-arm.map*, and the tool that does the mapping is called *apriori*. It contains entries such as these:

```
# core system libraries
libdl.so               0xAFF00000 # [<64K]
libc.so                0xAFD00000 # [~2M]
libstdc++.so           0xAFC00000 # [<64K]
libm.so                0xAFB00000 # [~1M]
liblog.so              0xAFA00000 # [<64K]
libcutils.so           0xAF900000 # [~1M]
libthread_db.so        0xAF800000 # [<64K]
libz.so                0xAF700000 # [~1M]
libevent.so            0xAF600000 # [???]
libssl.so              0xAF400000 # [~2M]
...
# assorted system libraries
libsqlite.so           0xA8B00000 # [~2M]
```

```
libexpat.so              0xA8A00000 # [~1M]
libwebcore.so            0xA8300000 # [~7M]
libbinder.so             0xA8200000 # [~1M]
libutils.so              0xA8100000 # [~1M]
libcameraservice.so      0xA8000000 # [~1M]
libhardware.so           0xA7F00000 # [<64K]
libhardware_legacy.so    0xA7E00000 # [~1M]
...
```

If you want to add a custom native library to 2.3/Gingerbread, you need to either add it to the list of libraries in *prelink-linux-arm.map* or set the `LOCAL_PRELINK_MODULE` to `false`. The build will fail if you forget to do one of these.

Library prelinking was dropped starting in 4.0/Ice-Cream Sandwich.

CHAPTER 5
Hardware Primer

Now that you have a good handle on Android's build system, the next step is to incrementally explore how the built images are used on the target. To best accomplish that, we must step back and look at the hardware configurations Android is typically run on. Indeed, while Android can be made to run on a wide variety of embedded systems, it remains deeply rooted in the world of consumer electronics and, most notably, handsets.

We're going to start by going over the typical system architecture of a hardware platform made for running Android. We'll then discuss the architecture of a typical SoC and provide an overview of some of the more notable SoCs out there used to run Android. We'll also cover the difference between virtual and physical address spaces, the typical host-target debug setup, and finish the chapter with a list of evaluation boards that you could use to prototype your embedded Android system and/or use to learn the trade.

Typical System Architecture

As we discussed in Chapter 1, Android should run on any hardware that runs Linux. Android, however, wasn't built in a vacuum. It was originally designed for handsets, and its current architecture still reflects that. Figure 5-1 illustrates the architecture block diagram of a prototypical embedded system made to run Android. Your actual target will likely differ, possibly greatly, from the one I illustrate. But for the sake of discussion, this diagram should be good enough.

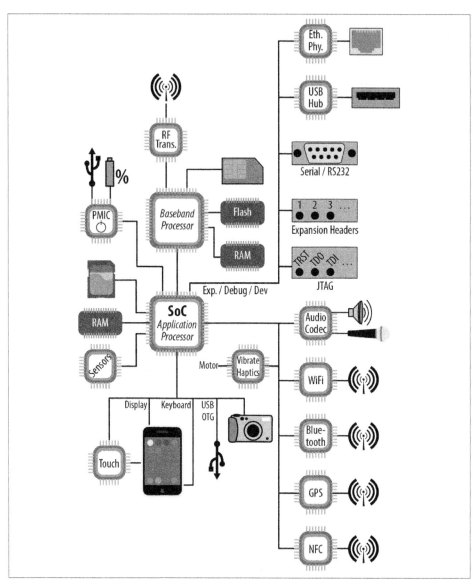

Figure 5-1. Typical system architecture block diagram

The most important thing to note is that at the center of this system lies an SoC. We'll discuss SoCs in greater detail in the next section. Suffice it to say for now that an SoC comprises a CPU and a bunch of peripheral controllers all on the same integrated circuit (IC) die. All other components on the target's board are typically connected in one way or another to the SoC. Android essentially runs on that SoC and therefore controls and/ or accesses everything on the board from that vantage point.

The Baseband Processor

The next component you want to pay attention to is the Baseband Processor. The majority of handsets on the market have separate processing units for running the user-facing software and managing the radio functions. These are typically known as the Application Processor (AP) and the Baseband Processor (BP), respectively.

You might wonder why there are two separate processors instead of just one. The reasons are both legal and technical. First, in the US, the law requires that software-defined radio (SDR) devices be certified by the Federal Communications Commission (FCC). Part of this certification is a requirement that the software controlling the radio may not be modified without authorization. Essentially, this means that under no circumstance should the end user of the device be allowed to change the way the radio operates or which frequencies it uses. In addition, there are hard real-time constraints on the operation of the radio functions. Hence, controlling the radio from the same CPU running the user-facing OS is not an option. There are also benefits in being able to put the AP to sleep while the BP continues operating.

Of course this is but a summary, and there is much more to say on this topic. However, for the purposes of our current discussion, assume that there's no way to have a single processor running both Android and the software that controls the radio. Obviously if your embedded system isn't a handset, or doesn't have radio functions, just assume that the diagram doesn't have a BP or any of the components attached to it.

Nevertheless, it's worth understanding the BP and its interaction with the AP, since the architecture of Android's RIL is tightly coupled to the underlying hardware. At a very simple level, the BP and AP talk to each other over some form of serial bus using AT commands. Notice that the BP has its own flash and RAM. This guarantees that the certified software running on the BP is isolated from the software running on the AP, and that the real-time OS (RTOS) running on the BP is focused on running a single thing: the radio's operation. The BP, for instance, runs software implementing the GSM stack. Notice also that the SIM card and an RF transceiver are connected to the BP. The transceiver takes care of the actual RF transmission and reception with the tower, while the SIM card is used to identify the handset user with the mobile network operator (MNO).

Telephony and wireless radio technologies are a world of their own. There is definitely a lot more to this topic than I could cover here. In fact, I'm barely scratching the surface. Real-life designs are infinitely more subtle than my simplification. Modern AP-BP interaction, for instance, may not actually rely on either a serial line or AT commands, but rather use mapped memory and proprietary handshake protocols. For the sake of the current conversation, though, the simple explanation is again good enough.

If you'd like to get more information on the radio architecture of smartphones, I would suggest reading Harald Welte's "Anatomy of contemporary GSM cellphone hardware" (*http://bit.ly/VPMQWm*) and visiting this xda-developers thread (*http://bit.ly/10hn6ky*).

Core Components

Although many of the components we'll discuss may or may not be present in your embedded system, a handful would most certainly be present in any embedded system, be it Android or another: RAM and storage. There isn't much to be said about RAM, but the storage may come in different incarnations.

Traditionally, most embedded systems would be equipped with either NOR or NAND flash, and a flash filesystem would be used to manage those chips and implement wear-leveling. More recently, however, the trend has been toward using embedded Multi-MediaCard (eMMC) chips. Essentially, these are chips that appear as SD cards and are managed by the Linux kernel as a traditional block device (i.e., the same as a conventional ATA hard drive). Hence, these systems don't have any NOR or NAND flash, just an eMMC chip. Their SoC chips have the required modules to do basic reads and boot directly from a partition on the eMMC.

Also, there may be more than one storage device attached to the system. Android in fact distinguishes between "internal" and "external" storage. "Internal" storage typically designates the onboard eMMC, while "external" storage designates the user-removable SD card attached to the phone or tablet. The former hosts Android itself and is used for booting and regular filesystem operations. The latter stores pictures and other multimedia content. Of course, this distinction is of little use to you if your device isn't a phone or a tablet, but the Android App Development API reflects Android's phone heritage and makes a distinction between those two types of storage.

 Note that on some more recent devices, the "external" storage is nothing more than a FUSE (Filesystem in User SpacE)–mounted filesystem over a specific directory of the system's "internal" storage. Such is the case of all modern Nexus devices, such as the Galaxy Nexus, Nexus 4, 7, and 10.

Another component that you are likely to find in any battery-powered device is a Power Management IC (PMIC). The PMIC's job is to manage all aspects of the battery, including regulating the voltage it provides and controlling its charging. The PMIC is typically connected to the battery and whatever DC power is used to feed the board. On most consumer devices, the external DC power comes from the USB On-the-Go (OTG) connector, which doubles as a plug for the power charger. In the case of nonmobile devices (and even in the case of some mobile devices), the external power isn't provided through USB but through some other type of connector, such as a barrel connector.

The PMIC is connected to the SoC through SPI, I²C, and/or GPIO. It can generate interrupts for such things as low battery or the charger being attached. It can (and increasingly does) also include functionality other than just power management. For instance, it may include a real-time clock (RTC), an audio codec, and a USB transceiver.

Real-World Interaction

Android is of course mainly a user-facing system. As suggested by its Compatibility Definition Document (CDD), a system built with it should allow rich user interaction and comprise quite a few hardware components that allow tying in to the user's immediate physical surroundings. This, in turn, means that there are quite a few hardware components dedicated to this task.

First and foremost, there are the parts tied to direct user interaction, such as the display, touch input, and the keyboard. While phones typically use the SoC's integrated display capabilities directly, devices with larger displays, such as tablets, will typically have a display bridge for low-voltage differential signaling (LVDS)–driven LCD displays. There's also typically a touch controller for handling the onscreen touch sensors and some form of circuitry for handling the use of a keyboard or any physical button on the device.

Second, there are parts that allow the user to have the device interact with the world around it. This includes things such as the camera (or cameras—e.g., some devices have both front- and rear-facing cameras, for video chatting), which is controlled by the SoC, and audio I/O, which is controlled by the audio codec IC. But hardware also includes a variety of components for sensing the physical properties of the device's immediate environment and mechanically interacting with it.

A wide range of sensors, for example, may be found in an Android device, such as an accelerometer, a gyroscope, a thermometer, a barometer, a photometer, a magnetometer, and a proximity sensor. I've illustrated only a "Sensors" IC to simplify the diagram, but there can in fact be many sensor ICs on the board. There are also components for creating vibrations and/or providing haptic feedback to the user. Again, several components may be involved.

Connectivity

One of Android's features is its connectivity, and the hardware used to run it reflects this with controllers, connectors, and antennas for a range of standards such as USB, WiFi, Bluetooth, GPS, and NFC. Again, these tend to increasingly be packaged together instead of being separate ICs.

Most consumer Android devices on the market provide only a USB OTG connector for connecting the device to a computer or plugging in another USB device, such as a camera or a USB stick. A very limited number of devices will also allow the USB OTG connector to be used as a USB host. Even fewer devices provide separate USB host connectors for plugging in peripherals, as you typically would to a USB host such as a PC or a Mac.

Expansion, Development, and Debugging

In addition to the typical components found in the mainstream Android devices I just covered, SoCs can also generally accommodate a slew of other components and peripherals. While most of these won't be found in consumer handsets or tablets, they can definitely be used in other Android-based embedded systems. Some are more or less well supported by the Android stack, while others aren't at all. But that's what got you into embedded development anyway, right? To boldly go where no other sane developer would?

Hence, you'll easily find development boards equipped with components and connectors for Ethernet, USB host, serial (RS-232), JTAG, and expansion headers. The popular BeagleBoard and PandaBoard, for instance, have most of these. JTAG is a hardware-level debugging interface and therefore doesn't need any software support from either the Linux kernel or the Android stack. Expansion headers exposed by development boards will usually allow a peripheral board (i.e., add-on modules connected through the expansion headers) to be connected to some of the SoC's pins, such as I^{2}C or GPIO. It'll then be up to you to make sure you load the appropriate device drivers to enable Linux to talk to the peripherals on the add-on module.

Serial port connectivity is provided by the Linux kernel's TTY layer. So long as your kernel has support for console on serial for your SoC (as it typically would if Linux runs on your SoC), this should work practically "out of the box." Serial-port connectivity is crucial for embedded systems, especially during board bringup, since it's a simple yet effective way for the host and target to communicate.

USB host mode will work if you are using Android 3.1 or later. Earlier versions, including Gingerbread, do not have USB host mode support in the Android stack. But that doesn't preclude the underlying Linux kernel from supporting the same set of USB devices it does by default. It only means that the app API for USB host mode, available starting with Android 3.1, won't be available to you.

A similar situation affects Ethernet. While you can connect an Android device using an Ethernet connection to a network, the Android stack doesn't recognize Ethernet as a valid data communication path—only WiFi and packet switching (i.e., your wireless carrier's data connection.) Hence, while some applications will work when the Ethernet connection is properly set, some others won't.

> ### Adding Ethernet Support to Android
>
> Android doesn't currently deal properly with Ethernet by default, but that hasn't stopped those needing Ethernet from supporting it. If you're interested in this type of functionality, have a look at the following work:
>
> - Fabien Brisset and Benjamin Zores have put together a set of patches for 4.0/Ice-Cream Sandwich and 4.1/Jelly Bean to support Ethernet. The patches are on GitHub (*https://github.com/gxben/aosp-ethernet*), and you can find the presentation Benjamin did about this work at the Embedded Linux Conference Europe (*http://www.slideshare.net/gxben/elce-2012-dive-into-android-networking-adding-ethernet-connectivity*) in November 2012.
> - Linaro has created its own set of patches for adding the same functionality. These changes are available here (*http://review.android.git.linaro.org/#change,2599*), here (*http://review.android.git.linaro.org/#change,2598*), and here (*http://review.android.git.linaro.org/#change,2554*).
>
> It's understandable that the AOSP doesn't officially support Ethernet at this point: It's not a technology commonly found in the type of devices where Google is pushing Android. Should Android be aimed at other types of devices in the future, this may change.

What's in a System-on-Chip (SoC)?

Up to this point, we've discussed the SoC as a black box. Let's take a peek inside and see what's in there. Have a look at Figure 5-2 for a representation of the internals of a typical SoC.

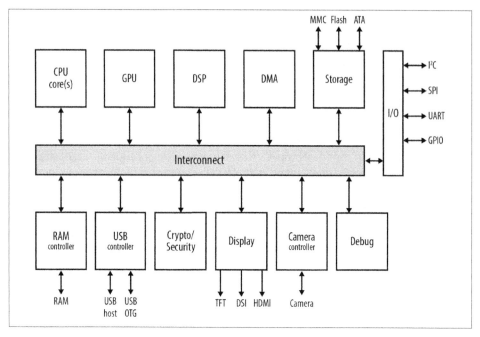

Figure 5-2. A typical System-on-Chip (SoC)

As you can see, there's much more than the CPU cores. An SoC is to some extent its own circuit board, with a bus interconnecting a variety of different components (typically known as the "interconnect fabric"). The number and complexity of each component depends on the SoC and its manufacturer. There's no real standard here, although most SoCs on the market include a similar set of basic components that are essentially interchangeable, even though they come from different manufacturers. And as in the case of the system architecture block diagram covered earlier, many of these components may be grouped together or even further divided into additional modules. This, after all, is a simplified view. Note also that not all components within an SoC operate at the same clock speed. So while the CPU may be listed as operating close to or above the gigahertz mark, for instance, the graphics processing unit (GPU) is likely operating at several hundred megahertz only.

 GPUs typically have a clock speed divided down from the CPU's own speed. If the CPU is clocked at 1GHz, for instance, the GPU may be running at 250MHz. Though they run slower, GPUs are made up of massively parallel computing units. Even if the CPU is dual-core, the GPU may have 16 or 64 cores.

Table 5-1 lists some of the most prominent SoCs used for Android at the time of this writing. As you can see, the market is increasingly offering dual-core Android devices, and quad-core devices are just around the corner. Manufacturers are "out-coring" themselves as fast as they can. That doesn't mean your embedded Android system needs to have that much firepower, but chances are that component pricing will bring the cost of a multicore SoC within your design's reach in the foreseeable future.

Table 5-1. SoC lineup

SoC	Manufacturer	CPU	Speed	GPU
OMAP3	Texas Instruments (TI)	ARM Cortex-A8	600MHz–1.2GHz	PowerVR SGX530
OMAP4	TI	Dual-core ARM Cortex-A9	1–1.8GHz	PowerVR SGX54x
OMAP5	TI	Dual-core ARM Cortex-A15	2GHz	PowerVR SGX544
i.MX51	Freescale	Cortex-A8	800MHz	OpenGL ES 2.0-compatible[a]
i.MX53	Freescale	Cortex-A8	1GHz	OpenGL ES 2.0-compatible
i.MX6	Freescale	Dual- or quad-core Cortex-A9	1GHz	OpenGL ES 2.0-compatible
Tegra 2	Nvidia	Dual-core ARM Cortex-A9	1–1.2GHz	GeForce
Tegra 3	Nvidia	Quad-core ARM Cortex-A9	1.3GHz	GeForce
Snapdragon S2	Qualcomm	Scorpion[b]	800MHz–1.5GHz	Adreno 205
Snapdragon S3	Qualcomm	Dual-core Scorpion	1.2–1.5GHz	Adreno 220
Snapdragon S4	Qualcomm	Dual-core Krait[c]	1–1.7GHz	Adreno 225 or 320
Exynos	Samsung	Single or Dual-core ARM Cortex-A8	1–1.5GHz	PowerVR SGX540 or ARM MALI-400
Exynos 4	Samsung	Quad-core Cortex-A9	1.4–1.6GHz	ARM MALI-400 MP4
Exynos 5	Samsung	Quad-core Cortex-A15	2.0GHz	ARM MALI-T658
Atom	Intel	Single core x86	1.6–2GHz	PowerVR SGX540
MT6575	Mediatek	Cortex-A9	1GHz	PowerVR Series5 SGX
MT6577	Mediatek	Dual-core Cortex-A9	1GHz	PowerVR Series5 SGX

[a] No additional details about the origin of the GPU engine are provided in Freescale's data sheet.
[b] This is specific to Qualcomm and, according to Wikipedia, is similar to an ARM Cortex-A8.
[c] This is specific to Qualcomm and, according to Wikipedia, is similar to an ARM Cortex-A15.

The Linux kernel has supported symmetric multiprocessing for quite some time, so you won't have trouble with its handling of a multicore SoC. The Android stack has only recently started being run on multicore processors, and while it implicitly benefits from Linux's multicore capabilities, the Android stack itself doesn't, at the time of this writing, contain any specific multicore optimizations. Hence, if you have code that must run on multiple CPUs simultaneously, you will need to manually make sure that each thread has its CPU affinity properly set.

Traditionally, Android is used with ARM-based SoCs, as is well reflected by the table above. But as we saw earlier, it has been made to run on a variety of other architectures supported by Linux, such as x86, MIPS, SuperH, and PowerPC. In fact, a number of devices from the likes of Motorola and Lenovo have already shipped with Intel-based chips. Google and Intel collaborated, in fact, to bring x86 support into the upstream AOSP. Most of the tools, documentation, and examples found on the Net remain, however, ARM-centric for the time being.

Another important component in the SoC is the GPU, which is responsible for accelerating the rendering of graphics to the device's display. While most CPU cores for Android SoCs are ARM-based, there's no standard GPU used by all SoC manufacturers. Instead, each manufacturer uses a different GPU, as you can see in Table 5-1. As mentioned earlier, these are clocked at several hundred megahertz (300 to 500) even if the CPU core(s) they're packaged with on the same SoC are clocked at speeds close to or above 1GHz.

Apart from the CPU and the GPU, the role of most of the rest of the components in the SoC can be more modestly described:

RAM controller
    Interfaces with the onboard RAM.

DMA
    Handles the automated transfers of data between the RAM and memory-mapped hardware.

USB controller
    Manages the hardware side of the device's USB connections.

DSP
    Provides hardware acceleration for some signal processing, such as JPEG encoding.

Display
    Enables the SoC to drive various display types.

Camera
    Allows the SoC to interface with a camera.

Storage
    Manages I/O with the various types of storage that can be used with the SoC.

Debug
    Enables the SoC to be connected to hardware debugging tools through various mechanisms, such as JTAG.

The SoC also likely contains some cryptographic and security functionality. This may consist simply of hardware acceleration for common cryptographic functions. It may also include security mechanisms made available by the SoC manufacturer to device manufacturers for locking the device and for preventing unauthorized code from running. Such mechanisms are often used to implement digital rights management (DRM) and can lead to frustration by people wanting to reprogram their devices. Unfortunately, however, consumers aren't the SoC manufacturers' direct customers, and the ethical issues surrounding the use of such technology far exceed our present scope.

Finally, the SoC most likely has capabilities to connect to additional external ICs using a variety of different buses and interfaces. This is how, for instance, most of the components described in the previous section are connected to the SoC through wiring on the PCB. Such buses and interfaces may include I²C, SPI, UART, and GPIO, but may include other mechanisms as well.

The specific capabilities and makeup of each SoC are typically documented by its manufacturer in data sheets it provides to device manufacturers, as well as OS and device-driver developers. Often, SoC manufacturers will provide a set of drivers for the most important components found in the SoC, such as the GPU, for instance. Most SoC vendors tend to, in fact, go much further and provide AOSP trees that are known to work "out of the box" on their own evaluation boards.

Memory Layout and Mapping

To be of any use, the hardware components we just saw must be accessible in some way from software. In general, this is done through device drivers in the Linux kernel. Applications then use the standard interfaces exposed by those drivers to, in effect, talk to the underlying hardware. Figure 5-3 illustrates how this works.

One of the buses connected to any CPU is an address bus. This bus is connected so as to allow the CPU access to the components attached to it using separate address ranges. In fact, most components occupy several, often consecutive, address regions. The addresses accessible by the CPU through its address bus are typically referred to as *physical* addresses, meaning they represent real, physical components connected to the CPU. When the CPU refers to any of these addresses, there are actual electrical signals being applied to the address bus on the printed circuit board (PCB) by the CPU, allowing it to designate a specific IC component.

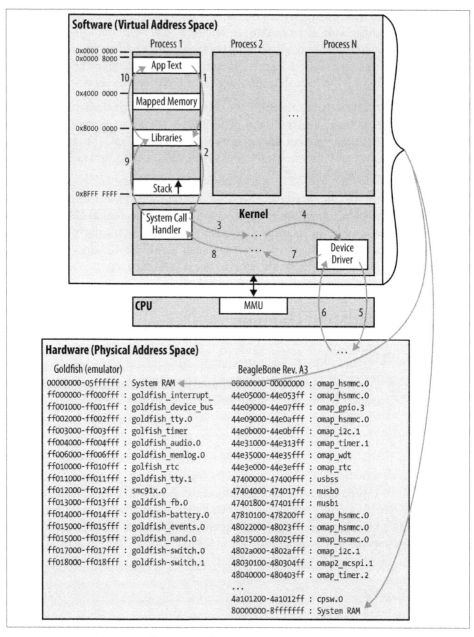

Figure 5-3. Virtual versus physical address spaces

The actual location of each of the components in the physical adress space is typically known as the *physical address map* and is determined by the device's designers as they route the connections from the SoC to the various components included on the PCB.

Two separate boards having identical components can have totally different physical memory maps. What's important is that each device driver know the location of the component or components it needs to talk to. Sometimes, the component the driver communicates with is actually a bus itself. In that case, that component acts as a bridge for additional components connected to it using its own specific bus. Such is the case for components connected to the SoC through I²C, for instance.

If you'd like to look at the physical memory map that your kernel sees at runtime, all you need to do is go to a command line and type *cat /proc/iomem*. That map might not contain all peripherals on your actual board, but it will contain those seen by the kernel. Some ICs or peripherals might not be listed because no driver registered with the kernel recognizes or deals with them.

The mapping between applications and devices works because the CPU manages two entirely separate address spaces through its memory management unit (MMU). Using its MMU, the CPU can present a virtual address space to applications running on it and still use a physical address space to communicate with components connected to it through its address bus.

One of the components residing in the physical address space is the system RAM. As you can see in Figure 5-3, the RAM location in the physical address space can vary greatly. Obviously, RAM is used to hold all active software code and data. However, this code and data is rarely addressed using references to its actual physical location. Instead, the OS collaborates with the MMU to implement a virtual address space wherein each process gets a similar view of the world. Virtual addresses eventually map to actual physical addresses, but the conversion is automatically handled by the MMU based on OS-maintained *page tables*.

It's beyond the scope of this book to explain paging and MMUs' operation in full, but just remember that the address ranges you see in your applications have nothing to do with the actual addresses being put by the CPU on its address bus to access your code and data. Figure 5-3 illustrates the virtual address space where Android processes live —bear in mind that the layout is not proportional. Some objects may be larger or smaller than they appear. The kernel is always seen as occupying an address range starting at 0xC000 0000 as its low address. Android apps, on the other hand, occupy the entire address space below that address.

The actual application "text," that is, the application's code, sits very near the beginning of the virtual address space. It's followed by mapped memory regions. These are virtual addresses that point either to RAM shared with other processes for interprocess communication, or physical address ranges mapped into the process's address space using the corresponding driver's mmap() function.

The mapping of physical address ranges into a process's address space allows that process to directly drive an IC component or another connected device, instead of having to go through the kernel and the device's driver for every operation. This is especially useful for performance-intensive operations such as graphics rendering. However, it's also an effective means of exporting critical device-driver intelligence outside the kernel and, therefore, subtracting it from the kernel's GPL requirements. In fact, it's a very effective way of implementing key driver functionality in Android HAL components.

Finally, libraries start at `0x8000 0000`, and the process's stack grows downward from the process's topmost address. Except where your software uses memory-mapped registers and regions to operate on hardware, the path for calls affecting hardware is usually as follows:

1. Your code calls on a function that interacts with a file descriptor associated with hardware. The immediate code called is actually in one of the system libraries mapped into your process's address space. This function typically has more "sugar-coating" than the raw kernel system call.
2. The library does some processing and eventually calls a matching system call.
3. The system call handler then does further processing and invokes various functions inside the kernel.
4. Eventually some part of the kernel invokes the device driver matching the device associated with the file descriptor held by your application.
5. The device driver interacts with the hardware using whichever method is applicable. The result of this is of course hardware dependent. In some cases, the device driver may be able to read back a status and return it immediately. In other cases, the hardware feedback may occur only at a much later time. In other cases still, there may be no expected feedback.
6. Assuming the hardware does provide some feedback to the driver or generates an interrupt in response to the earlier operation, the call path will start to return from where it came.
7. The call path returns back from the driver to whatever invoked it.
8. The call path returns back to the system call handler.
9. The call path returns back to the system library.
10. The call path returns back to your code.

The only part of the preceding call chain that might involve physical addresses is where the device driver code communicates with its designated hardware. The rest of the calls being made and data being exchanged all happen in virtual address space.

Development Setup

As soon as you have some prototype hardware, and continuing throughout board bringup and development, it's very practical to have your target hooked up to your development workstation. Figure 5-4 illustrates a generic host-target debug setup. Your specific hookup will likely differ, but this setup represents the ideal.

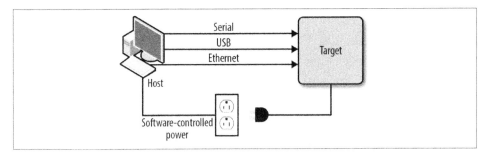

Figure 5-4. Host-target debug setup

Here, the connections between the host and the target can serve a variety of sometimes overlapping purposes. By connecting the target's power to a software-controlled power source managed by the host, the power-on/power-off of the board can be scripted on the host and hence be used to automate the testing of various software versions on the board. There are several power strips on the market that allow you to set up something like this.

The classic way that a target is connected to its host is through a serial connection, typically RS-232. This usually allows you to interact with the board's bootloader, upload and download small files, and generally interact with the target when nothing else works. Obviously this connection is relatively slow, and its purpose is really for basic interaction; transferring large amounts of data is best suited for something like Ethernet.

The Ethernet connection will allow the host to provide a wide range of services to the target, as illustrated in Figure 5-5. To ease the iterative debug process, for instance, it's best to have the target use DHCP to retrieve its IP configuration, use TFTP to load its kernel images, and mount its root filesystem through NFS. If you do that, then any change you make to your project on the host will be deployed to the target via reboot, at worst. At best, you just update a file in the NFS-mounted root filesystem, and all you need to do is restart a command to run its new version. In all cases, you save yourself the trouble of having to manually reprogram the target's storage every time you make a change.

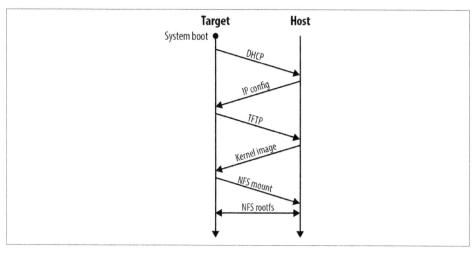

Figure 5-5. Development boot setup

Finally, and especially in the case of Android, USB can be very useful. Indeed, with Android you can rely on USB to connect to the target using ADB very much as an app developer would connect to a consumer phone or tablet for app development. All the typical ADB commands would then be available to you, including shelling into the target, forwarding ports, updating filesystems, etc. Whereas you can configure ADB to run over IP, and therefore over Ethernet, having it available through USB is great because it works "out of the box."

 Setting up ADB over IP is actually relatively simple: It's just that you have fewer command-line parameters to deal with if it's over USB. Most importantly, USB is the case most widely covered by documentation you'll find on the Net. We'll cover this topic in greater detail in the next chapter.

Your specific setup will most likely contain its own quirks, but the configuration shown here should give you a general idea of what you want to aim for. Serial support is usually provided by the bootloader and the kernel. Unless you're bringing up a board based on a whole new CPU, you should already have access to serial-port communication. Ethernet support will require a proper driver for the Ethernet chip used on your board. This may require some work on your part. Finally, USB support will depend on whether the USB hardware on your target is properly supported by the kernel. If you're using a common SoC, this shouldn't be an issue. If you need help setting up a DHCP server, TFTP, or NFS for servicing your target, have a look at O'Reilly's *Building Embedded Linux Systems, 2nd ed* (2008).

Evaluation Boards

If you're still early in your development process or are simply evaluating Android, you'll likely want to rely on an evaluation board. Here are some factors that you may want to consider when selecting one:

SoC
Does it rely on the SoC you're going to use in your final design? Is the SoC of the same family? Or is it a previous iteration of the yet-to-be-released SoC you plan on using from a given manufacturer?

Community
Is there a community around the board, or is the manufacturer the only source of support? How active is this community? Is it built around a single board or a family of boards?

Cost
What's the up-front cost of the board, and what's included for that price? How much do add-ons or extensions cost? What's the price difference between the low-end option and the high-end option, and what are the feature differences?

Features
What functionality is included/exposed by the board? SoCs can increasingly support a wide range of functionality. Yet, the more SoC features the board makes available, the more expensive it tends to be. So does the board you're looking at expose the features you need?

Expandability
The basic features provided by the board may suffice for a certain percentage of what you're trying to accomplish, but does the board allow you to attach additional hardware so you can emulate the final functionality you're aiming for?

Availability
How easy is it to get your hands on the board? Some boards look very nice on paper but have fluctuating supplies.

Licensing
Can you use the board as is for end products? Some manufacturers forbid you from doing that. Do you have access to the bill of materials (BOM) and the schematics? If you want to build a board based on the eval board, these will be critical.

Catalog parts
Is the board using catalog parts? If the board relies on noncatalog parts, then you'll need to go to their manufacturer to get your hands on them. Usually, this situation occurs when the manufacturer wants to sell components only to very-high-volume buyers, making such parts beyond the reach of people doing small projects.

Third-party parts
   Sometimes SoC vendors include third-party parts in their boards. Make sure you apply a similar set of criteria to those components. Keep in mind that, should you use third-party components in your design, you'll be dependent on those suppliers for almost exactly the same kind of support you'd expect from the SoC vendor.

Software support
   How well is Android supported on the board? And by whom? The manufacturer? A third party? Which versions of Android are supported? What's the long-term commitment behind such support?

You'll also most certainly have more criteria for your own project. If you're building your own hardware, however, your starting point will usually be the SoC, as this is a critical decision point involving quite a few stakeholders in your organization, both on the hardware and software sides. And then, your next step will be to go to that SoC's manufacturer site to check the eval board(s) it recommends for that SoC. If you're looking only to get your hands on a decent board that will allow you to experiment with Android, you're likely going to hit your favorite search engine for hours of fun looking at the various evaluation boards out there. Either way, have a look at Table 5-2 for some of the more prominent eval boards as of early 2013.

Table 5-2. Evaluation boards lineup

Board	SoC	Speed	RAM	I/O	Cost
BeagleBone	Sitara AM3358	500MHz (on USB) / 720MHz (on DC)	256MB	USB OTG, USB host, Ethernet, onboard serial, onboard JTAG, expansion headers, microSD	$89
BeagleBoard xM	Davinci DM3730	1GHz	512MB	USB OTG, USB host, Ethernet, serial, JTAG, expansion headers, microSD, DVI-D, LCD header, S-Video, camera header, stereo in/out	$149
iMX53 Quick Start Board	i.MX53	1GHz	1GB	USB OTG, USB host, Ethernet, serial, JTAG, expansion headers, SD, microSD, SATA, VGA, LCD header, stereo in/out	$149
PandaBoard ES	OMAP4 dual-core	1.2GHz	1GB	USB OTG, USB host, Ethernet, WLAN, Bluetooth, serial, JTAG, expansion headers, SD, HDMI, DVI, LCD header, camera header, stereo in/out	$182
AM335x Starter Kit	Sitara AM3358	720MHz	256MB	USB OTG, USB host, Ethernet, WLAN, Bluetooth, onboard serial, onboard JTAG, expansion headers, microSD, capacitive-touch LCD panel, accelerometer, stereo out	$199
Nitrogen6X	i.MX6 quad-core	1GHz	1GB	USB OTG, USB host, serial, JTAG, SATA, SD, CAN, LCD headers	$199
OrigenBoard	Exynos 4210 dual-core	1.2GHz	1GB	USB OTG, USB host, WLAN, Bluetooth, serial, JTAG, SD, HDMI, LCD header, camera header, stereo in/out	$199

Board	SoC	Speed	RAM	I/O	Cost
Origen 4 Quad	Exynos 4 quad-core	1.4GHz	1GB	USB OTG, USB host, Ethernet, SD, JTAG, serial, HDMI, onboard LCD header, audio	$199
DragonBoard APQ8060A	Snapdragon dual-core	1.2GHz	1GB	USB OTG, USB host, Ethernet, WLAN, Bluetooth, GPS, FM radio, accelerometer, gyroscope, compass, magnetometer, pressure sensor, eMMC, SATA HDMI, camera, stereo out, serial, capacitive-touch LCD, JTAG	$499
SABRE	i.MX53	1GHz	1GB	USB OTG, USB host, Ethernet, WLAN, Bluetooth, GPS, ZigBee, accelerometer, light sensor, serial, JTAG, eMMC, SD, SATA, NOR flash, VGA, HDMI, LCD panel, camera, stereo in/out	$999
Snapdragon MDP	Snapdragon S4 dual-core	1.5GHz	1GB	USB OTG, WLAN, Bluetooth, GPS, accelerometer, gyroscope, compass, proximity sensor, temperature sensor, SD, HDMI, LCD panel, camera, stereo out	$999

[a] Most common price at the time of this writing.

Save for the last two entries, all the eval boards listed in Table 5-2 look exactly like what their names imply: a PCB with chips and bare connectors on it. Few of the configurations I listed in Table 5-2 include an LCD panel, though most of these boards can have an LCD touch-panel added to them for anywhere between $100 and $200. The last two eval boards listed actually come in tablet and phone form factors, respectively, with the expected housing and mechanical specifications. If you're trying to build a demo of a final product to show to an end customer, those two systems might be more presentable than a board with wires protruding here and there. They are, as you might have noticed, priced accordingly.

CHAPTER 6
Native User-Space

By this point, you've either already gotten your hands dirty trying a few things here and there or you're very eager to actually play with a live Android system. As with any embedded system you are bringing up, your typical goal would be to get to a minimally functional system and then start adding support for more and more hardware and functionality until your requirements are met.

Obviously, to get a minimally functional Android system, you'll first need to bring the kernel up on your board. As I mentioned earlier, the best way to get yourself an Android-compatible kernel is to talk to your SoC vendor; kernel porting and board bringup being somewhat outside the scope of this book. However, once you've got yourself a minimally functional kernel, the first Android component you'll have to deal with is its native user-space.

As described in Chapter 2, this foundation serves as the hosting environment for all the upper layers of the Android stack, including the Dalvik virtual machine and the services and apps it runs. This is also where a part of Android's hardware support is implemented. Now is therefore a good time to take a closer look at Android's native user environment. If nothing else, it's sufficiently different from what is found in most classic embedded Linux systems to warrant a separate discussion.

Filesystem

In Chapter 4, we discussed how the build system operates and what it generates. Specifically, Table 4-3 provided a detailed list of the images typically created by the build system. Conversely, Figure 6-1 illustrates how these images relate to one another at runtime. Save for a few exceptions that we'll cover later, this filesystem layout is essentially the same in 2.3/Gingerbread and 4.2/Jelly Bean.

To understand how we go from the images generated by the build system to the runtime configuration shown in Figure 6-1, you need to go back to the system startup explanation in Chapter 2 and, more specifically, you need to refer to the boot process illustrated in Figure 2-6. In essence, the kernel mounts the RAM disk image generated by the build system as its root filesystem and launches the init process found in that image. That init process's configuration files will, as we'll see later in this chapter, cause a number of additional images and virtual filesystems to be mounted onto existing directory entries in the root filesystem.

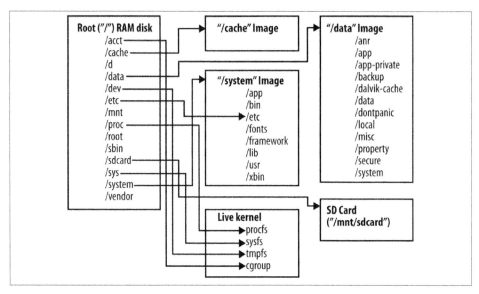

Figure 6-1. Android root filesystem

One of the first questions you might ask is, "Why so many filesystems?" Indeed, why not just a single filesystem image to store everything? The answer lies in the different purposes each image has, along with differences in the nature of the storage devices or technologies being used. The RAM disk image, for example, is meant to be as small as possible, and its sole purpose is to provide the initial skeleton required to get the system going. It's typically stored as a compressed image on some media prior to being loaded into RAM by the kernel and then mounted as a read-only root filesystem.

/cache, */data*, and */system*, on the other hand, are typically mounted from separate partitions on actual storage media. Usually */cache* and */data* are mounted as read-write, while */system* is mounted as read-only.

> ## Using a Single Filesystem
>
> There's nothing preventing you from using a single filesystem for all of Android's build output instead of using separate storage partitions. Texas Instruments' RowBoat distribution, for instance, does exactly that. It generates a single root filesystem image, which is programmed on the target's storage device for use as is. In the case of the BeagleBone or BeagleBoard, for example, the root filesystem in its entirety is programmed into a single partition of the microSD card used for booting and as the device's main storage device.
>
> By consolidating on a single filesystem, however, you're assuming that you can update the entirety of the filesystem in one fell swoop. In sum, it'll be very difficult to create a fail-safe update procedure for your system. In the case of RowBoat's support for the Beagles, this might not be an issue because they are development boards, but in your actual product that has to go in the field, it might well turn out to be a problem.

In Android versions 2.2 and prior, all three directories would typically be mounted from YAFFS2-formatted NAND flash partitions. Since handset manufacturers have slowly been moving toward eMMC instead of NAND flash, YAFFS2 was replaced by ext4 in Google's Android 2.3 lead device, the Samsung Nexus S. Since then, it's been assumed that all Android-based handsets should be using ext4 instead of YAFFS2. Nothing, however, precludes you from using another filesystem type altogether. You just need to modify the build system's makefiles to generate those images and update the parameters used with the *mount* commands as part of *init*'s configuration files, as we'll see shortly.

> ## eMMC versus NOR or NAND Flash
>
> As explained in the book *Building Embedded Linux Systems, 2nd ed.*, Linux's MTD layer is used to manage, manipulate, and access flash devices in Linux; this includes NOR and NAND flash. Various filesystems are then used on top of the MTD layer, such as JFFS2, UBIFS, or YAFFS2, to make the flash device or partition accessible as part of Linux's virtual filesystem switch (VFS.) Those flash filesystems typically implement wear leveling and bad-block management to properly handle the underlying flash devices.
>
> An eMMC device, as explained in Chapter 5, appears as a traditional block device. Essentially, it contains a microcontroller and some RAM that allow it to do the required wear leveling and bad-block management transparently. Therefore, the OS can use a regular disk filesystem such as ext4. While the decision to move toward eMMC is, according to Android developer Brian Swetland (*http://lwn.net/Articles/440826/*), motivated by reduced pin-count on the PCB—and therefore overall cost—there are some additional side benefits to using this type of device.

First, it allows you to use all the traditional commands and methods you're used to with a regular Linux filesystem. The MTD subsystem, while powerful, has always required some getting used to before one could effectively use it. Also, flash filesystems tend to be designed with single-processor systems in mind, while disk filesystems in Linux have had to contend with multiprocessor systems for quite some time. Hence, they're likely a better fit for the coming wave of multicore Android devices.

The SD card always appears as a block device and typically has a VFAT filesystem on it. This should be expected because the user needs to be able to remove it from the Android device and plug it into his regular computer, whatever OS it may be running. */proc*, */sys*, and */acct* are mounted using procfs, sysfs, and cgroupfs, respectively. While */proc* and */sys* are mounted at the same location as in traditional Linux-based systems, cgroups were traditionally mounted as */cgroup* in Linux but are mounted as */acct* in Android. Note also that */dev* is mounted as tmpfs. This means its content is created on the fly and does not reside on any permanent storage. That's fine, because Android relies on Linux's udev mechanism to dynamically create entries in */dev* as devices are plugged in and/or drivers are loaded or initialized.

Procfs, sysfs, tmpfs, and cgroup are all virtual filesystems maintained by the currently running kernel in the system. They don't have any corresponding storage and are, in fact, data structures maintained inside the kernel. Procfs is the traditional way the kernel exports information about itself to user-space. Typically, entries in procfs are seen as text files, or directories containing text files, which can be dumped to the command line for extracting a given piece of information from the kernel. If you're looking for the type of CPU your system is running, for example, you can dump the contents of the */proc/cpuinfo* file.

As the kernel matured and had growing needs, it was eventually agreed that procfs was not necessarily the right mechanism for all interfaces between the kernel and user-space. Enter sysfs, which is very heavily tied to the kernel's device and hardware management. Entries in sysfs can, for instance, be used to get detailed information regarding peripherals, or toggle bits controlling the behavior of certain drivers directly from user-space. Many of Android's power-management features, for example, are controlled via entries in the */sys/power/* directory.

Tmpfs allows you to create a virtual RAM-only filesystem for storing temporary files. As long as power is applied to the RAM, the kernel will allow you to read and write those files. On reboot, however, it's all gone. Cgroupfs is a relatively recent addition to the kernel for managing the control group functionality added in Linux 2.6.24. In sum, cgroups allow you to group certain processes and their children and dictate resource limits and priorities onto those groups. Android uses cgroups to prioritize foreground tasks.

The Root Directory

As we discussed in Chapter 2, the classic structure of Linux root filesystems is specified in the Filesystem Hierarchy Standard (*http://www.pathname.com/fhs/*) (FHS). Android, however, doesn't abide by the FHS, but relies heavily instead on the */system* and */data* directories for hosting most of its key functionality.

Android's root directory is mounted from the *ramdisk.img* generated by the AOSP's build system. Typically, *ramdisk.img* will be stored along with the kernel in the device's main storage device and loaded by the bootloader at system startup. Table 6-1 details the contents of the root directory once mounted.

Table 6-1. Android's root directory

Entry	Type	Description
/acct	dir	cgroup mount-point.
/cache	dir	Temporary location for downloads in progress and other nonessential data.
/d	symlink	Points to */sys/kernel/debug*, the typical mount location for debugfs.[a]
/data	dir	The mount-point for the *data* partition. Usually, the contents of *userdata.img* are mounted here.
/dev	dir	Mounted on tmpfs and contains the device nodes used by Android.
/etc	symlink	Points to */system/etc*.
/mnt	dir	Temporary mount-point.
/proc	dir	The mount-point for procfs.
/root	dir	In traditional Linux systems, the *root* user's home directory. It's generally empty in Android.
/sbin	dir	In Linux, this would hold binaries essential to the system administrator. In Android, it contains only *ueventd* and *adbd*.
/sdcard	dir	The mount-point for the SD card.
/sys	dir	The mount-point for sysfs.
/system	dir	The mount-point for the *system* partition. *system.img* is mounted to this location.
/vendor	symlink	Generally a symbolic link to */system/vendor*. Not all devices actually have a */system/vendor* directory.

Entry	Type	Description
/init	file	The actual *init* binary executed by the kernel at the end of its initialization.
/init.rc	file	*init*'s main configuration file.
/init.<device_name>.rc	file	The board-specific configuration file for *init*.
/ueventd.rc	file	*ueventd*'s main configuration file.
/ueventd.<device_name>.rc	file	The board-specific configuration file for *ueventd*.
/default.prop	file	The default global properties to be set for this system. These are automatically loaded by *init* at startup.

[a] Debugfs is meant as a very flexible, RAM-based filesystem for exporting debugging information from kernel-space to user-space. It's not meant for use in production systems.

As part of 4.2/Jelly Bean, you'll also find some more entries in the root filesystem as listed in Table 6-2.

Table 6-2. Additions to Android's root directory in 4.2/Jelly Bean

Entry	Type	Description
/config	dir	mount-point for configfs.[a]
/storage	dir	Starting with 4.1/Jelly Bean, this directory is used to mount external storage. */storage/sdcard0*, for instance, is typically the fake "external" storage[b] and */storage/sdcard1* is a real SD card.
/charger	file	Native, standalone full-screen application that displays the battery's charge status.
/res	dir	Resources for the *charger* application.

[a] Have at *http://lwn.net/Articles/148973/* for more information on configfs.
[b] "Fake" in the sense that it's essentially a FUSE-mounted "internal" directory made to appear as an external storage device.

/system

As mentioned earlier, */system* contains the immutable components generated by the AOSP build system. To illustrate this further, Figure 6-2 takes the Android architecture diagram presented in Chapter 2 and shows where each part of the stack is found in the filesystem.

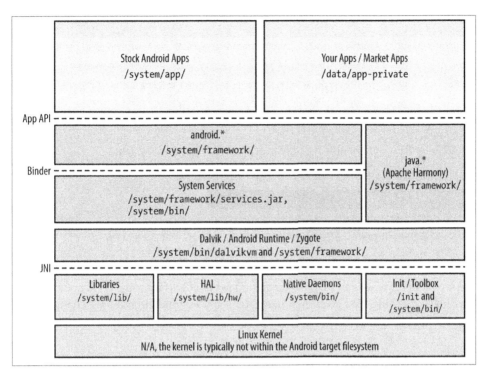

Figure 6-2. Filesystem location of key Android components

As you can see, most of the components are found somewhere within */system* once *system.img* is mounted. Table 6-3 further describes each entry in detail. You can also contrast Figure 6-2 with Figure 3-2 to see where each architecture component is located in the AOSP sources versus the final filesystem.

Table 6-3. /system directory contents

Entry	Type	Description
/app	dir	The stock apps built as part of the AOSP, such as the browser, email app, calendar, etc. All modules built with BUILD_PACKAGE are here.
/bin	dir	All native binaries and daemons built as part of the AOSP. All modules built with BUILD_EXECUTABLE are here. The only exception is *adbd*, which has the LOCAL_MODULE_PATH set to */sbin* and is therefore installed there instead.
/etc	dir	Contains configuration files used by various daemons and utilities, including possibly an *init.<device_name>.sh* script that would be launched by one of *init*'s configuration files at startup.
/fonts	dir	The fonts used by Android.
/framework	dir	Framework *.jar* files.
/lib	dir	The system's native libraries. Essentially this means any module built using BUILD_SHARED_LIBRARY. It's important to note again that Android doesn't use */lib* at all, only this *lib* directory within */system*.

Entry	Type	Description
/modules	dir	An optional directory for storing the dynamically loadable kernel modules required to run the system.
/usr	dir	A miniature /usr akin to the classic /usr directory found in traditional Linux systems.
/xbin	dir	"Extra" binaries generated by some of the packages that are built within the AOSP but aren't essential to the system's operation. This includes things like *strace*, *ssh*, and *sqlite3*.
/build.prop	file	A set of properties generated during the build process of the AOSP. They are loaded by *init* at startup.

In 4.2/Jelly Bean, you'll also find the entries in Table 6-4 in */system*.

Table 6-4. New /system directory entries in 4.2/Jelly Bean

Entry	Type	Description
/media	dir	Files relating to the boot animation and other media.
/tts	dir	Files related to the Text-to-Speech engine.

Generally */system* is mounted read-only because it's called on to change only if the entire Android OS is updated to a newer version. One benefit is that some OTA update scripts do binary patching, and given that this partition is assumed to not have changed since it was shipped, the application of the deltas is guaranteed to be clean.

/data

As discussed earlier, */data* contains all data and apps that can change over time. For example, all the data stored by apps you download from Google Play is found here. The *userdata.img* image generated by the AOSP's build system is mostly empty, so this directory starts off containing little to nothing. As the system starts getting used, however, the content of this directory is naturally populated, and it becomes important to preserve it across reboots. This is why */data* is typically mounted in read-write mode from persistent storage. Table 6-5 shows the contents.

Table 6-5. /data directory contents

Entry	Type	Description
/anr	dir	ANR traces.
/app	dir	Default install location for apps.
/app-private	dir	Install location for apps with *forward locking*.[a] This mechanism has been replaced with an API allowing app developers to check if the running app is a legitimate copy obtained from Google Play. Have a look at the Application Licensing (*https://developer.android.com/google/play/licensing/index.html*) section of the app developers guide for more information on this topic.
/backup	dir	For use by the BackupManager system service.
/dalvik-cache	dir	Holds the cached JIT'ed[b] versions of all dex files.
/data	dir	Contains one subdirectory for each app installed on the system. In effect, this is where each app's "home" directory is located.

Entry	Type	Description
/dontpanic	dir	Last panic output (console and threads)—for use by *dumpstate*.
/local	dir	Shell-writable directory. In other words, any user who can shell into the device, using *adb shell*, for example, can copy anything, including binaries, into this directory and it will be preserved across reboots.
/misc	dir	Miscellaneous data such as for WiFi, Bluetooth, or VPN.
/property	dir	Persistent system properties.
/secure	dir	Used to store user account information if the device uses an encrypted filesystem.
/system	dir	Systemwide data, such as the accounts database and the list of installed packages.
/tombstones	dir	Whenever a native binary crashes, a file whose name is *tombstone_* followed by a sequence number is created here with information about the crash.

[a] When an ISV publishes an app to Google Play, he can set the Copy Protection in the Publishing Options to On or Off. By setting it to Off, the app's *.apk* can be copied off the device, while it can't if it's set to On. In essence, On means the app is installed in */data/app-private* and Off means it's installed in */data/app*.

[b] Remember that Dalvik has a Just-in-Time compiler that converts the byte-code found in *.dex* files to native CPU instructions. This conversion is done once and cached for all future uses.

In 4.2/Jelly Bean, you'll also find the entries described in Table 6-6.

Table 6-6. New /data directory entries in 4.2/Jelly Bean

Entry	Type	Description
/app-asec	dir	Encrypted apps.
/drm	dir	DRM encryption data. Forward-locking control files.
/radio	dir	Radio firmware.
/resource-cache	dir	App resource cache.
/user	dir	User specific data for multiuser systems.

Multiuser support

One of the most important features added to 4.2/Jelly Bean is multiuser support. In fact, some have argued that this addition was a watershed moment, opening Android to new use cases. Though available only in tablet mode, it allows multiple users to share the same device in a coherent fashion. Specifically, it means every user can utilize the device by logging in separately and can have her own set of account credentials and data for each application.

To achieve this, the AOSP's data-storage mechanism has been slightly modified. For instance, */data/data* is now the directory containing the app data for the device's owner (i.e., "administrator"). All other users have their data stored in /data/user/<user_id> instead. Here's the content of /data/user in an emulator running 4.2/Jelly Bean:[1]

```
root@android:/ # ls -l /data/user/
lrwxrwxrwx root     root                    2012-11-30 20:46 0 -> /data/data/
drwxrwx--x system   system                  2012-12-04 23:38 10
root@android:/ # ls -l /data/user/0/
drwxr-x--x u0_a27   u0_a27                  2012-11-30 20:46 com.android.backupconfirm
drwxr-x--x bluetooth bluetooth              2012-11-30 20:46 com.android.bluetooth
drwxr-x--x u0_a17   u0_a17                  2012-12-14 18:01 com.android.browser
drwxr-x--x u0_a43   u0_a43                  2012-11-30 20:46 com.android.calculator2
drwxr-x--x u0_a20   u0_a20                  2012-11-30 20:47 com.android.calendar
drwxr-x--x u0_a33   u0_a33                  2012-11-30 20:46 com.android.certinstaller
drwxr-x--x u0_a0    u0_a0                   2012-11-30 20:47 com.android.contacts
drwxr-x--x u0_a25   u0_a25                  2012-11-30 20:46 com.android.defcontainer
drwxr-x--x u0_a6    u0_a6                   2012-11-30 20:47 com.android.deskclock
...
root@android:/ # ls -l /data/user/10/
drwxr-x--x u10_system u10_system            2012-12-04 23:38 android
drwxr-x--x u10_a27   u10_a27                2012-12-04 23:38 com.android.backupconfirm
drwxr-x--x u10_bluetooth u10_bluetooth      2012-12-04 23:38 com.android.bluetooth
drwxr-x--x u10_a17   u10_a17                2012-12-04 23:38 com.android.browser
drwxr-x--x u10_a43   u10_a43                2012-12-04 23:38 com.android.calculator2
drwxr-x--x u10_a20   u10_a20                2012-12-04 23:38 com.android.calendar
drwxr-x--x u10_a33   u10_a33                2012-12-04 23:38 com.android.certinstaller
drwxr-x--x u10_a0    u10_a0                 2012-12-04 23:38 com.android.contacts
drwxr-x--x u10_a25   u10_a25                2012-12-04 23:38 com.android.defcontainer
drwxr-x--x u10_a6    u10_a6                 2012-12-04 23:38 com.android.deskclock
...
```

Similarly, there are now per-user account credentials for each of the Internet accounts that may be tied to a given user. Prior to 4.2/Jelly Bean, there was a single /data/system/accounts.db to hold all accounts. Now there is one such file for each user:

```
root@android:/ #  ls /data/system/users/ -l
drwx------ system   system                  2013-01-19 01:03 0
-rw------- system   system              155 2012-11-30 20:46 0.xml
drwx------ system   system                  2013-01-19 01:03 10
-rw------- system   system              166 2012-12-04 23:38 10.xml
-rw------- system   system              141 2013-01-19 01:03 userlist.xml
root@android:/ #  ls /data/system/users/0 -l
-rw-rw---- system   system            57344 2012-11-30 20:47 accounts.db
-rw------- system   system             8720 2012-11-30 20:47 accounts.db-journal
-rw------- system   system              534 2013-01-19 01:03 appwidgets.xml
-rw-rw---- system   system              549 2013-01-19 01:03 package-restrictions.xml
-rw------- system   system               97 2013-01-19 01:03 wallpaper_info.xml
root@android:/ #  ls /data/system/users/10 -l
-rw-rw---- system   system            57344 2012-12-04 23:39 accounts.db
```

---

1. The emulator doesn't support multiple users by default. A few hacks must be made to get it to add a fake user.

```
-rw-------  system   system       8720 2012-12-04 23:39 accounts.db-journal
-rw-rw----  system   system        129 2013-01-19 01:03 package-restrictions.xml
```

SD Card

As discussed earlier, consumer devices typically have a microSD card that the user can remove and plug into her computer. The content of this SD card is not critical to the system's operation. In fact, you can relatively safely wipe it out without adverse effects. If a real user is using the device, however, you'll at least want to understand what's on it, because some apps store their information on the SD card, and it might matter to the user. Table 6-7 details some of what you might find in the */sdcard* directory.

Table 6-7. Sample /sdcard directory contents

Entry	Type	Description
/Alarm	dir	Downloaded audio files that can be played as an alarm.
/Android	dir	Contains apps' "External" data and media directories. The former can be used for storing noncritical files and caches, while the latter is for app-specific media.
/DCIM	dir	Pictures and videos taken by the Camera app.
/Download	dir	Files downloaded from the web.
/Movies	dir	Download location for movies.
/Music	dir	The user's music files.
/Notifications	dir	Downloaded audio files that can be selected by the user for playing when notifications occur.
/Pictures	dir	Downloaded pictures available to the user.
/Podcasts	dir	The user's podcasts.
/Ringtones	dir	The downloaded ringtones the user should be able to choose from.

Because */sdcard* is world-writable, the specific contents will depend on the apps running on the device and, of course, what the user decides to manually copy there. Again, just as a reminder, Android's API distinguishes between "internal" and "external" storage, and the SD card is the latter. Also, note that some upgrade procedures use the SD card as the location where the update image is stored during the upgrade.

The Build System and the Filesystem

Chapter 4 covered how the build system generates the various parts of the filesystem. Let's dig a little deeper into how you can control the build system's filesystem generation.

Build templates and file locations

Table 4-2 listed the available build templates. Table 6-8 details the default install location for modules built using each target build template. Note how everything gets installed in one of */system*'s subdirectories.

Table 6-8. Build templates and corresponding output locations

Template	Default Output Location
BUILD_EXECUTABLE	/system/bin
BUILD_JAVA_LIBRARY	/system/framework
BUILD_SHARED_LIBRARY	/system/lib
BUILD_PREBUILT	No default. Make sure you explicitly specify either LOCAL_MODULE_CLASS or LOCAL_MODULE_PATH.
BUILD_MULTI_PREBUILT	Depends on type of module being copied.
BUILD_PACKAGE	/system/app
BUILD_KEY_CHAR_MAP	/system/usr/keychars

Internally, the build system generates a LOCAL_MODULE_PATH for each module built, depending on the module's build template. This is where the compiled output is installed. You can override the default by changing the value of LOCAL_MODULE_PATH within your *Android.mk*. Let's say, for instance, that you have a custom tool for your board that has to be installed in */sbin* instead of */system/bin*. Your *Android.mk* could then look something like this:

```
LOCAL_PATH := $(call my-dir)
include $(CLEAR_VARS)

LOCAL_MODULE_TAGS := optional
LOCAL_SRC_FILES := $(call all-c-files-under, src)
LOCAL_PACKAGE_NAME := calibratebirdradar
LOCAL_MODULE_PATH := $(TARGET_ROOT_OUT_SBIN)

include $(BUILD_PACKAGE)
```

Note that this specifies $(TARGET_ROOT_OUT_SBIN), not */sbin*. This is so the binary gets installed in the proper *out/target/product/PRODUCT_DEVICE/* directory. The TARGET_ROOT_OUT_* macros are defined in *build/core/envsetup.mk*, along with quite a few installation default macros. Here's the relevant snippet for our purposes:

```
TARGET_ROOT_OUT      := $(PRODUCT_OUT)/root
TARGET_ROOT_OUT_BIN  := $(TARGET_ROOT_OUT)/bin
TARGET_ROOT_OUT_SBIN := $(TARGET_ROOT_OUT)/sbin
TARGET_ROOT_OUT_ETC  := $(TARGET_ROOT_OUT)/etc
TARGET_ROOT_OUT_USR  := $(TARGET_ROOT_OUT)/usr
```

Explicitly copying files

In the case of some files, you don't need the build system to build them in any manner; you just need it to copy the files into the filesystem components it generates. That's the purpose of the PRODUCT_COPY_FILES macro that you can use in your product's *.mk*. Here's an updated version of the CoyotePad's *full_coyote.mk* from Chapter 4:

```
$(call inherit-product, $(SRC_TARGET_DIR)/product/languages_full.mk)
$(call inherit-product, $(SRC_TARGET_DIR)/product/full.mk)

DEVICE_PACKAGE_OVERLAYS :=

PRODUCT_PACKAGES +=
PRODUCT_COPY_FILES += \
   device/acme/coyotepad/rfirmware.bin:system/vendor/firmware/rfirmware.bin \
   device/acme/coyotepad/rcalibrate.data:system/vendor/etc/rcalibrate.data

PRODUCT_NAME := full_coyotepad
PRODUCT_DEVICE := coyotepad
PRODUCT_MODEL := Full Android on CoyotePad, meep-meep
```

This will copy *rfirmware.bin* and *rcalibrate.data* from *device/acme/coyotepad/* to the target's */system/vendor/firmware* and */system/vendor/etc* directories, respectively.

Default rights and ownership

One aspect we haven't yet discussed is what and how filesystem rights and ownership are assigned to each directory and file in the Android filesystem. If you're willing to get your hands dirty, I strongly encourage you to take a look at the *system/core/include/private/android_filesystem_config.h* file. It doesn't get a lot of publicity and it's not documented anywhere.[2] It is, however, extremely important, as it provides the list of predefined system users, as well as the rights and ownership assigned to everything in the system. Here's a partial list of the UIDs/GIDs it defines, along with the associated user/group names in 2.3/Gingerbread:

```
#define AID_ROOT            0    /* traditional unix root user */

#define AID_SYSTEM          1000 /* system server */

#define AID_RADIO           1001 /* telephony subsystem, RIL */
#define AID_BLUETOOTH       1002 /* bluetooth subsystem */
#define AID_GRAPHICS        1003 /* graphics devices */
#define AID_INPUT           1004 /* input devices */
...
#define AID_RFU2            1024 /* RFU */
#define AID_NFC             1025 /* nfc subsystem */

#define AID_SHELL           2000 /* adb and debug shell user */
#define AID_CACHE           2001 /* cache access */
#define AID_DIAG            2002 /* access to diagnostic resources */
...
#define AID_MISC            9998 /* access to misc storage */
#define AID_NOBODY          9999
```

2. The file was actually pointed out to me by then–Sony Ericsson engineer Magnus Bäck, who helped review this book, on the android-building mailing list after I inquired about Android's filesystem rights management.

```
#define AID_APP            10000 /* first app user */
...
static const struct android_id_info android_ids[] = {
    { "root",      AID_ROOT, },
    { "system",    AID_SYSTEM, },
    { "radio",     AID_RADIO, },
    { "bluetooth", AID_BLUETOOTH, },
    { "graphics",  AID_GRAPHICS, },
    { "input",     AID_INPUT, },
...
```

If you go to your target's shell and type *ps*, for instance, you'll see something like this:

```
...
root    18048 1     61552  26700 c00a6548 afd0b844 S zygote
system  18090 18048 141756 50224 ffffffff afd0b6fc S system_server
system  18187 18048 75664  21828 ffffffff afd0c51c S com.android.systemui
app_16  18197 18048 78548  19292 ffffffff afd0c51c S com.android.inputmethod.
                                                     latin
radio   18200 18048 86400  19580 ffffffff afd0c51c S com.android.phone
app_19  18201 18048 78636  23472 ffffffff afd0c51c S com.android.launcher
app_1   18234 18048 83904  22232 ffffffff afd0c51c S android.process.acore
app_2   18281 18048 72364  16696 ffffffff afd0c51c S com.android.deskclock
...
```

Notice how the *system_server* runs as the `system` user and how each app is run by a user called `app_N`, with each app having a separate `N`. The kernel itself doesn't provide those names. Instead, Bionic uses the previous definitions to provide PID/GID-to-name conversion. In the case of apps, since each app is installed as a separate user (starting from the base UID/GID for apps, 10000), app user names all start with `app_` and are followed by an integer value matching the actual UID/GID assigned to the app minus 10000. This is slightly different starting with 4.2/Jelly Bean, with multiuser support. Now app names also show user ownership with the form `uM_appN`, where M is the user ID and N is the app ID.

Unlike other aspects of the AOSP's build system, which allow you to isolate most of your board-specific additions within a directory in *device/*, like *device/acme/coyotepad* from our earlier example, there's no substitute for modifying the main *android_filesystem_config.h* if you need to add new default users. The bold lines in the following snippet, for instance, show modifications for adding a `birdradar` user:

```
...
#define AID_RFU2          1024 /* RFU */
#define AID_NFC           1025 /* nfc subsystem */
#define AID_BIRDRADAR     1999 /* Bird radar subsystem */

#define AID_SHELL         2000 /* adb and debug shell user */
#define AID_CACHE         2001 /* cache access */
#define AID_DIAG          2002 /* access to diagnostic resources */
...
```

```
static const struct android_id_info android_ids[] = {
    { "root",      AID_ROOT, },
    { "system",    AID_SYSTEM, },
    { "radio",     AID_RADIO, },
...
    { "media",     AID_MEDIA, },
    { "nfc",       AID_NFC, },
    { "birdradar", AID_BIRDRADAR, },
    { "shell",     AID_SHELL, },
    { "cache",     AID_CACHE, },
...
```

> We're using 1999 instead of 1026 for our new user to avoid as much as possible having to update this integer in future Android releases, should new users be added by Google. In fact, the above snippet is from 2.3/Gingerbread, where the factory IDs stop at 1025. In 4.2/Jelly Bean, the last number used by default by the AOSP is 1028.

Reasons for adding new default users might include the addition of a new, still-unsupported hardware type to the Android stack, or the desire to isolate from the Android stack a custom stack you're running side by side with Android. It could also simply be a matter of isolating a specific daemon using a separate user.

Conversely, here are snippets of the directory and file rights defined in *android_filesystem_config.h*:

```
static struct fs_path_config android_dirs[] = {
    { 00770, AID_SYSTEM, AID_CACHE,  "cache" },
    { 00771, AID_SYSTEM, AID_SYSTEM, "data/app" },
    { 00771, AID_SYSTEM, AID_SYSTEM, "data/app-private" },
    { 00771, AID_SYSTEM, AID_SYSTEM, "data/dalvik-cache" },
    { 00771, AID_SYSTEM, AID_SYSTEM, "data/data" },
...
    { 00750, AID_ROOT,   AID_SHELL,  "sbin" },
    { 00755, AID_ROOT,   AID_SHELL,  "system/bin" },
    { 00755, AID_ROOT,   AID_SHELL,  "system/vendor" },
...
    { 00755, AID_ROOT,   AID_ROOT,   0 },
};
...
static struct fs_path_config android_files[] = {
    { 00440, AID_ROOT,   AID_SHELL,  "system/etc/init.goldfish.rc" },
    { 00550, AID_ROOT,   AID_SHELL,  "system/etc/init.goldfish.sh" },
...
    { 00644, AID_SYSTEM, AID_SYSTEM, "data/app/*" },
    { 00644, AID_SYSTEM, AID_SYSTEM, "data/app-private/*" },
    { 00644, AID_APP,    AID_APP,    "data/data/*" },
...
    { 00755, AID_ROOT,   AID_SHELL,  "system/bin/*" },
```

```
        { 00755, AID_ROOT,      AID_SHELL,   "system/xbin/*" },
        { 00755, AID_ROOT,      AID_SHELL,   "system/vendor/bin/*" },
        { 00750, AID_ROOT,      AID_SHELL,   "sbin/*" },
        { 00755, AID_ROOT,      AID_ROOT,    "bin/*" },
        { 00750, AID_ROOT,      AID_SHELL,   "init*" },
        { 00644, AID_ROOT,      AID_ROOT,    0 },
};
```

If, for any reason, you add a new directory or a file into an unlisted (new) directory in the filesystem, the default ownership and access rights will be dictated by the last entry in the array just shown—the one with a 0 instead of a path within quotes. In other words, a new directory will have 755 access rights and be owned by the AID_ROOT user and group, and a file added to an unlisted directory will have 644 access rights and be owned by the AID_ROOT user and group.

If you'd like to add glibc-linked binaries to your target, as is shown in Appendix A, for instance, you'll likely want to have a /lib directory to host the glibc-libraries; /lib being the default library for traditional C libraries under Linux. However, by default, the libraries in there won't be executable, even if they were on your host as you generated them,[3] and, therefore, any binary linked against glibc will fail to run. To remedy this problem, you'll need to modify the android_files array in *android_filesystem_config.h* to look something like this:

```
...
        { 00750, AID_ROOT,      AID_SHELL,   "sbin/*" },
        { 00755, AID_ROOT,      AID_ROOT,    "bin/*" },
        { 00755, AID_ROOT,      AID_ROOT,    "lib/*" },
        { 00750, AID_ROOT,      AID_SHELL,   "init*" },
        { 00644, AID_ROOT,      AID_ROOT,    0 },
};
```

This is yet another modification that you couldn't isolate into a device-specific directory like *device/acme/coyotepad*.

Note that typically, the */system/vendor* directory is reserved for vendor-specific extensions. In fact, *android_filesystem_config.h* states that all binaries in */system/vendor/bin* should be executable. Hence, if you're going to add a substantial number of files to the filesystem, you might want to look at putting your additions in the */system/vendor* directory. That would be the *clean* way to do it. But, hey, who ever said embedded and clean were synonymous?

---

3. That's because the tools used to generate the filesystem images ignore the rights and ownership set for files on the host. Instead, they rely completely on *android_filesystem_config.h*.

Generally speaking, trying to stay within the boundaries of what's permitted by the AOSP's build system is especially useful if you want to simplify your device support for future Android versions. If you isolate all your device-specific code in a relevant directory in *device/*, adding support for your device in the next AOSP is, theoretically, just a matter of copying your directory into that AOSP's *device/* directory and fixing your code for any API modifications.

While this philosophy makes sense for handsets, embedded systems are often one-offs where previous products get nothing but the most essential updates, if any, and the next product's hardware platform will be the subject of a selection process that might result in the use of a completely different SoC. Hence, abiding by the "rules" in such circumstances might actually be counterproductive, as it'll impose unnecessary limitations and restrictions. I'll keep pointing out the "Android way" and all other possibilities as we move forward, but I'll leave it up to you to decide what's best for your own project.

adb

The filesystem layout we just discussed is only a skeleton for the rest of Android to live in. During board bringup, the first piece of Android software you'll probably want to make sure runs on your device after the kernel is likely going to be *adb*. We already covered its basic operation in Chapter 3. We're now going to delve into its use in much greater detail.

Theory of Operation

While surprisingly simple in use, *adb* is a very powerful tool with uses both for app development and platform development. Whereas several areas of Android build on or replace functionality found in traditional embedded Linux systems, prior to Android there was no project or package that provided functionality similar to *adb* in the Linux world (as far as I know, at least). Hence, *adb* fills an important gap and is a refreshing take on how host-target interactions can be improved and mediated.

adb is actually made up of several components, which themselves connect to several other system components to deliver *adb*'s integrated set of capabilities. Figure 6-3 illustrates *adb*'s interconnections and operation. Interestingly, both *adb*'s host side and target side, save for the *ddms*-related components, are built from a single codebase in *system/core/adb/*, which ensures version coherency among components.

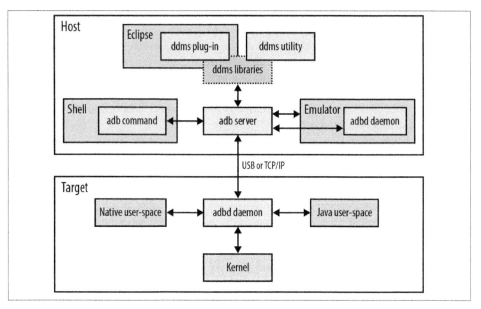

Figure 6-3. ADB and its interconnections

adb acts both as a transparent transport mechanism and as a service provider. Its two most important components are the *adb* server running on the host and the *adbd* daemon running on the target. These two components effectively implement a proxy protocol on which all *adb* services are implemented. They can be linked together either through USB or regular TCP/IP. The command set that *adb* makes available is identical in both cases.

 The names used in Android can be confusing here. Usually, a *server* runs remotely from a client, and some client utility connects to the server through the network. In this case, the *adb* "server" is actually a daemon running in the background on the host, and *adbd* is another, separate daemon running on the target.

The *adb* server is started automatically whenever the *adb* command is invoked on the command line. It monitors connected devices and maintains communication with the remote *adbd* daemons. The latter interface with the native user-space, the Java user-space and the kernel to provide their functionality. We'll discuss some of those interactions in greater detail as we go through *adb*'s functionality below.

On the host side, two major pieces of software initiate connection with the *adb* server: the *adb* command and the ddms (Dalvik Debug Monitor Server) libraries (ddmlib and ddmuilib). The ddms libraries are themselves used by the *ddms* utility, which is a standalone tool, and the ddms plug-in typically added to Eclipse through the installation of Android's ADT plug-in for app developers. The ddms libraries provide primitives both to talk to the *adb* server (ddmlib) and display/manage UI parts (ddmuilib). This is why the user interfaces are identical between parts of the *ddms* utility and the ddms Eclipse plug-in.

Note that the *adb* command and the ddms libraries don't fully expose the *adb* server's capabilities in an equal way. The *adb* server, for instance, can grab the content of the target's framebuffer for the purpose of providing screenshots. This functionality is exposed by the *ddms* utility, but it isn't available on the command line through *adb*.

To provide its services, the *adb* server opens socket 5037 on the host and listens for connections. Anyone can connect to the server as long as he respects the procotol. Have a look at *OVERVIEW.TXT* and *SERVICES.TXT* in *system/core/adb/* if you'd like to implement code that talks directly to the *adb* server. The *adb* server can also interact with an *adbd* daemon running inside an emulator on the host in the same way it would to the same daemon on a remote target.

In addition, *adb* can also interact with the emulator's console. Every emulator instance that starts listens for connections on a different port number; the number is displayed on the upper-left corner of the emulator window and starts from 5554. When you connect to that port number using *telnet*, you are able to issue special commands to control the emulator's behavior, as detailed in Using the Android Emulator (*https://developer.android.com/guide/developing/devices/emulator.html*) in Google's app developers guide. These commands include forwarding ports from the host to the emulator and resizing the emulator's window. To simplify matters, *adb* makes it possible to send the same commands to the emulator without actually having to go through *telnet*.

Main Flags, Parameters, and Environment Variables

As alluded to in Chapter 3 and as we'll see shortly in detail, *adb* provides a lot of commands. However, *adb* can be used to simultaneously interact with several Android devices and AOSP builds. Hence, there are several flags, parameters, and environment variables to gate its behavior, as presented in Table 6-9. If there's only one device connected or emulator instance running, then *adb*'s operation is relatively simple, since it assumes that that single instance is the one you want to execute your commands on.

Table 6-9. adb's flags, parameters, and environment variables

Item	Description
-d	This flag tells *adb* to execute the command passed on the USB-connected device. If you have both an emulator running and an Android device connected through USB to your host, then this will ensure *adb* executes your command on the device, not the emulator. Of course this won't work if you have more than one device connected.

Item	Description
-e	Similarly to -d, this tells *adb* to connect to the emulator instance running, even if there is an Android device connected. Again, it won't work if you have multiple emulator instances running.
-s <serial number>	This tells *adb* to connect to the device designated by the given serial number. Despite it being tedious to have to enter the full serial number of a device to use each *adb* command, this (and ANDROID_SERIAL below) will be the only way to go if you have multiple devices connected or multiple emulators running.
-p <product name or path>	Some of *adb*'s commands require access to the sources that were used to build the target's AOSP. If you're running *adb* from the same shell where you built the AOSP, it will be able to properly find those since the ANDROID_PRODUCT_OUT environment variable will be set. If that's not the case, you'll need to use -p to indicate the path to the product's output directory within an AOSP source tree.
ANDROID_SERIAL	If you constantly have multiple devices connected and want to avoid having to use the -s flag to specify the serial of one specific device that you operate on very frequently, set the ANDROID_SERIAL environment variable to that device's serial number, and *adb* will always connect to that device by default unless you explicitly use -s to override.
ADB_TRACE	If you want to debug or monitor the interaction betweeen the *adb* server on the host and the *adbd* daemon on the target, you can set the ADB_TRACE environment variable to one of or a series of comma-, colon-, semicolon-, or space-separated combinations from the following values: 1, all, adb, sockets, packets, rwx, usb, sync, sysdeps, transport, jdwp.

Basic Local Commands

Let's start with some of *adb*'s basic commands that run locally. First, if you'd like to start the *adb* server manually, you can do so like this:

```
$ adb start-server
* daemon not running. starting it now on port 5037 *
* daemon started successfully *
```

The server will, however, start automatically whenever needed by any other *adb* command you type. So you can usually skip over starting the server manually. There are cases, unfortunately, where you actually have to manually shut the server down—usually you should do this whenever any of your *adb* commands seem to hang:

```
$ adb kill-server
```

If you'd like to know *adb*'s capabilities, you can either start the command without any parameters or type:

```
$ adb help
Android Debug Bridge version 1.0.26

 -d                            - directs command to the only connected USB device
                                 returns an error if more than one USB device is
                                 present.
 -e                            - directs command to the only running emulator.
                                 returns an error if more than one emulator is
                                 running.
```

```
  -s <serial number>           - directs command to the USB device or emulator
                                 with the given serial number. Overrides
                                 ANDROID_SERIAL
...
device commands:
  adb push <local> <remote>    - copy file/dir to device
  adb pull <remote> [<local>]  - copy file/dir from device
  adb sync [ <directory> ]     - copy host->device only if changed
                                 (-l means list but don't copy)
                                 (see 'adb help all')
  adb shell                    - run remote shell interactively
  adb shell <command>          - run remote shell command
  adb emu <command>            - run emulator console command
...
```

The help screen above gave the command's version number as part of the output. But you can ask *adb* to explicitly print its version number:

```
$ adb version
Android Debug Bridge version 1.0.26
```

Like the rest of the AOSP, *adb* is a moving target. Here's the version in 4.2/Jelly Bean:

```
$ adb version
Android Debug Bridge version 1.0.31
```

Device Connection and Status

Let's now take a look at the commands *adb* provides for managing its communications with devices. First, if you want to see which devices are visible to *adb*, you can type:

```
$ adb devices
List of devices attached
emulator-5554   device
0123456789ABCDEF device
emulator-5556   device
```

If you'd like to connect to a remote device whose *adbd* daemon is running on TCP/IP instead of USB, you can use the *connect* command:

```
$ adb connect 192.168.202.79:7878
connected to 192.168.202.79:7878
$ adb devices
List of devices attached
emulator-5554   device
0123456789ABCDEF device
emulator-5556   device
192.168.202.79:7878 device
```

connect's formal description is (5555 being the default port):

```
adb connect <host>[<:port>]
```

To designate that target as the one on which to issue a given command, just use the IP:PORT information displayed by *adb devices* as the serial number. To get a shell, for instance:

```
$ adb -s 192.168.202.79:7878 shell
```

When you're done, you can disconnect from the device; it will then stop appearing in the list of devices seen by the *adb* server:

```
$ adb disconnect 192.168.202.79:7878
```

disconnect's formal description is (if no device is specified then all TCP/IP-connected devices will be disconnected):

```
adb disconnect [<host>[<:port>]]
```

If you'd like *adb* to hang waiting for the device to come online, you can type this:

```
$ adb wait-for-device
```

The shell will then suspend until the device comes online. *adb* will return to the shell when the device is online. This is useful for scripting purposes, as you can make your script wait for a device to be ready before proceeding with other commands.

If you want to inquire about a device's status, type:

```
$ adb -s 0123456789ABCDEF get-state
device
```

States include bootloader, device, offline, and unknown. The `device` value is synonymous with the device being online. `offline` is self-explanatory. `bootloader` means the device is currently in the bootloader. And unknown means *adb* can't recognize the current state of the device.

If, for any reason, you need to explicitly ask about a device's serial number, such as when you're scripting *adb* commands, you can do so:

```
$ adb -d get-serialno
0123456789ABCDEF
```

Finally, if you need to have a shell window open that continuously reports the current device's state, you can do so with this:

```
$ adb -d status-window
```

This will clear the screen and display something like this at the top of the terminal (the state reported beside `State:` being the device's "real-time" state):

```
Android Debug Bridge
State: device
```

To exit, you just type Ctrl-C.

Basic Remote Commands

Up to now, the commands we've seen haven't actually allowed us to do anything on the remote target or get any information about it. So let's start having some fun.

Shell

Obviously, if you're a geek like me, one of the first things you'll want to do is shell into your device for fun and profit. With 2.3/Gingerbread you'll get this:

```
$ adb shell
#
```

4.2/Jelly Bean has a much richer shell, and you'll get this instead:

```
$ adb shell
root@android:/ #
```

In both cases, the command results in the *adbd* daemon spawning a shell on the target to execute the commands you type. All input/output (i.e., stdin, stdout, and stderr) for the commands will then be proxied between the *adb* server running on the host and the *adbd* daemon running on the target.

To exit from the target's shell and return to your host's shell, just type Ctrl-D. You can also launch a specific command by passing it as a parameter to the *shell* command—in this case printing out the CPU information for a BeagleBone:

```
$ adb -d shell cat /proc/cpuinfo
Processor       : ARMv7 Processor rev 2 (v7l)
BogoMIPS        : 718.02
Features        : swp half thumb fastmult vfp edsp thumbee neon vfpv3 tls
CPU implementer : 0x41
CPU architecture: 7
CPU variant     : 0x3
CPU part        : 0xc08
CPU revision    : 2

Hardware        : am335xevm
Revision        : 0000
Serial          : 0000000000000000
```

This is *shell*'s formal description:

```
adb shell [ <command> ]
```

Dumping the logs

If you'd like to dump Android's logger buffer, you can type this:

```
$ adb -d logcat
--------- beginning of /dev/log/main
I/DEBUG   (   59): debuggerd: Mar 27 2012 05:30:39
--------- beginning of /dev/log/system
```

```
I/Vold    (   57): Vold 2.1 (the revenge) firing up
D/Vold    (   57): USB mass storage support is not enabled in the kernel
D/Vold    (   57): usb_configuration switch is not enabled in the kernel
D/Vold    (   57): Volume sdcard state changing -1 (Initializing) -> 0 (No-Media)
D/Vold    (   57): Volume usb state changing -1 (Initializing) -> 0 (No-Media)
D/Vold    (   57): Volume sdcard state changing 0 (No-Media) -> 2 (Pending)
D/Vold    (   57): Volume sdcard state changing 2 (Pending) -> 1 (Idle-Unmounted)
I/Netd    (   58): Netd 1.0 starting
I/        (   61): ServiceManager: 0xad50
W/AudioHardwareInterface(   61): Using stubbed audio hardware. No sound will be
produced.
D/AudioHardwareInterface(   61): setMode(NORMAL)
I/CameraService(   61): CameraService started (pid=61)
I/AudioFlinger(   61): AudioFlinger's thread 0xc638 ready to run
E/dhcpcd  (   65): timed out
D/AndroidRuntime(  224):
D/AndroidRuntime(  224): >>>>>> AndroidRuntime START com.android.internal.os.Zyg
oteInit <<<<<<
D/AndroidRuntime(  224): CheckJNI is ON
D/dalvikvm(  224): creating instr width table
I/SamplingProfilerIntegration(  224): Profiler is disabled.
I/Zygote  (  224): Preloading classes...
...
```

That command is actually an equivalent of this:

```
$ adb -d shell logcat
```

We'll discuss the *logcat* command in greater detail later, but know that you can line up the same parameters after the adb logcat part you typed in as if you were running *logcat* straight from the target's command line. So, for instance, if you want to dump the "radio" buffer instead of the "main" buffer, you can do this:

```
$ adb -d logcat -b radio
I/PHONE   (  394): Network Mode set to 0
I/PHONE   (  394): Cdma Subscription set to 1
I/PHONE   (  394): Creating GSMPhone
D/PHONE   (  394): mDoesRilSendMultipleCallRing=true
D/PHONE   (  394): mCallRingDelay=3000
W/GSM     (  394): Can't open /system/etc/voicemail-conf.xml
W/GSM     (  394): Can't open /system/etc/spn-conf.xml
D/GSM     (  394): [DSAC DEB] registerForPsRestrictedEnabled
D/GSM     (  394): [DSAC DEB] registerForPsRestrictedDisabled
D/GSM     (  394): [GsmDataConnection-1] DataConnection constructor E
D/GSM     (  394): [GsmDataConnection-1] clearSettings
D/GSM     (  394): [GsmDataConnection-1] DataConnection constructor X
D/GSM     (  394): [GsmDataConnection-1] Made GsmDataConnection-1
D/RILJ    (  394): [0000]> RIL_REQUEST_REPORT_STK_SERVICE_IS_RUNNING
D/STK     (  394): StkService: StkService: is running
...
```

adb will also honor the ANDROID_LOG_TAGS environment variable if it's set in the host's shell when you start the command. ANDROID_LOG_TAGS is taken into account by *logcat*, as we'll see later, for filtering the output it prints. This is *logcat*'s formal description:

```
adb logcat [ <parameters> ]
```

---

logcat with ddms Libraries

If you've ever used *ddms* or Android's ADT plug-in, you know they can present the same Android logger information that's printed to the command line by *logcat*. There's a difference in how each retrieves its information, however. While, as I just explained, an *adb logcat* actually just runs the *logcat* command on the target and proxies the output back to the host, *ddms*'s libraries use a different *adb* server mechanism from the one used to proxy shell I/O, the *log service*. This service proxies the content of the relevant */dev/log* buffer directly back to the host, without passing through the target's *logcat*. This is a case where there are in fact two ways to skin a cat.

The protocol between the *adb* server and its client is in fact quite rich, as I alluded to earlier. You'll need to dig into *adb*'s sources to get the full picture, but suffice it to say that the server communicates with the target's *adbd* daemon to provide multiple types of services. The overall ADB functionality exposed through the *adb* command line and *ddms* all rely on those services. However, you can write code that talks directly to the *adb* server to tap into any of the services it provides.

Getting a bug report

Much like the *logcat* target command—for which there's a shortcut with *adb* that doesn't require explicitly telling it to invoke *shell*—*adb* provides a shortcut for *bugreport*. The latter is a target command that dumps the state of the system for bug-reporting purposes. It, in effect, results in the *dumpstate* command to run on the target:

```
$ adb -d bugreport
========================================================
== dumpstate: 2000-01-01 05:05:08
========================================================

Build: beaglebone-eng 2.3.4 GRJ22 eng.karim.20120327.052544 test-keys
Bootloader: unknown
Radio: unknown
Network: (unknown)
Kernel: Linux version 3.1.0-g62911f8-dirty (a0131746@sditapps03) (gcc version 4.
4.3 (GCC) ) #1 Mon Nov 28 22:05:07 IST 2011
Command line: console=ttyO0,115200n8 androidboot.console=ttyO0 mem=256M root=/de
v/mmcblk0p2 rw rootfstype=ext3 rootwait init=/init ip=off

------ MEMORY INFO (/proc/meminfo) ------
MemTotal:          253264 kB
```

```
MemFree:           198308 kB
...
```

You might wonder, why not just do something like this instead, since the *bugreport* command invokes *dumpstate*?

```
$ adb -d shell dumpstate
```

The trouble is that *dumpstate* needs to run as root, and some devices don't allow their shells to run as root. Such is the case of the vast majority of handsets on the market. On those devices, therefore, it wouldn't be possible to type the above command, but it would still be possible to use *bugreport*. Here's what happens on my phone:

```
$ adb -s 4xxxxxxxxxxxxxx shell dumpstate
dumpstate: permission denied
$ adb -s 4xxxxxxxxxxxxxx shell bugreport
========================================================
== dumpstate: 2012-05-04 13:38:05
========================================================

Build: GINGERBREAD.UCKI3
Bootloader: unknown
Radio: unknown
...
```

Essentially, *bugreport* causes *init* to start *dumpsys* in a mode where it opens a Unix domain socket and listens for connections for dumping its output. *bugreport* then connects to that socket and copies the content it reads to its own standard output, which is then proxied through *adb* to your host's shell. Users or technicians can therefore create bug reports for your devices even if you don't give them root access.

Port forwarding

Another very interesting feature of *adb* is that it allows you to forward ports between the host and the target. For instance, this command will forward local port 8080 to the target's port 80:

```
$ adb -d forward tcp:8080 tcp:80
```

Thereafter, any connection you make to your host's port 8080 will be redirected to the target's port 80. If you're running a web server (which runs on port 80 by default) on your Android device, for example, you'll be able to connect your host's web browser to `localhost:8080` to browse your device.

adb's *forward* command can, however, do a lot more than that. It can in fact forward host ports to more than just ports on the target. For instance, you can forward local port 8000 to one of the target's character devices:

```
$ adb -d forward tcp:8000 dev:/dev/ttyUSB0
```

In that case, any read/write operations conducted on port 8000 will result in read/write operations on the remote /dev/ttyUSB0. Table 6-10 lists the connection types supported by *forward* and its formal description is:

    adb forward <local> <remote>

Table 6-10. adb forward's connection types

Connection	Description
tcp:<port>	Regular TCP port. This should be an nonnegative integer value.
localfilesystem:<unix domain socket>	A regular Unix domain socket. This shows up as an entry in the filesystem.
localabstract:<unix domain socket>	An "abstract" Unix domain socket. This is like a Unix Domain socket, but it's a Linux-specific extension. Have a look at the unix man page for more detail: man 7 unix.
localreserved:<unix domain socket>	Android's "reserved" Unix domain sockets. They're all in /dev/socket, and they have very specific uses that we'll cover as we go. These include *dbus*, *installd*, *keystore*, *netd*, *property_service*, *rild*, *rild-debug*, *vold*, and *zygote*.
dev:<character device name>	Actual devices on the target. You must provide the full path to the device in the filesystem.
jdwp:<pid>	Used to specify the PID of a Dalvik process for debugging purposes.

Dalvik debugging

It's worth expanding a bit more on *forward*'s ability to proxy connections to Dalvik processes. Dalvik implements the Java Debug Wire Protocol (JDWP), thereby allowing you to use the regular Java debugger *jdb* to debug your apps. Obviously this is shrink-wrapped into Eclipse for app developers, but if you want to use *jdb* on the command line, *forward*'s ability to redirect Dalvik processes' debug ports to your host becomes essential. Here's an example:

```
$ adb forward tcp:8000 jdwp:376
$ jdb -attach localhost:8000
Set uncaught java.lang.Throwable
Set deferred uncaught java.lang.Throwable
Initializing jdb ...
>
```

To know which PIDs are debuggable through JDWP, you type:

```
$ adb jdwp
271
376
386
389
390
425
473
480
...
```

adb is in fact a crucial component for debugging any Java on the target. When the *adbd* daemon starts on the target, it opens the "abstract" Unix domain socket jdwp-control and awaits connections. Dalvik processes that start **afterward** connect to that socket and therefore make themselves "visible" for debugging. To allow app developers to debug their apps, the ddms Eclipse plug-in goes through ddmlib to talk to the *adb* server to debug the app. Or, as we just saw, you can use *jdb* to debug on the command line.

Note that all of this requires that *adbd* be running on the target before any Dalvik app is started. Only those Dalvik apps that you start **after** *adbd* will be debuggable.

Filesystem Commands

adb also allows you to manipulate and interact with the target's filesystem in a variety of ways. If you want to copy a file to the device, for instance, you can use *push*:

```
$ adb push acme_user_manual.pdf /data/local
```

This will copy the *acme_user_manual.pdf* file to the target's */data/local* directory:

```
$ adb shell ls /data/local
acme_user_manual.pdf
```

You can also copy files from the target to the host:

```
$ adb pull /proc/cpuinfo
$ cat cpuinfo
Processor : ARMv7 Processor rev 2 (v7l)
BogoMIPS : 718.02
...
```

As I explained earlier in this chapter, the target's filesystem parts aren't all mounted with the same rights. */system*, for example, is typically mounted as read-only. If you'd like to remount it in read-write mode, to add or modify a file on it, for instance, you can do so using *remount*. Here's an example:

```
$ adb push acme_utility /system/bin
failed to copy 'acme_utility' to '/system/bin/acme_utility': Read-only file system
$ adb remount
remount succeeded
$ adb push acme_utility /system/bin
$
```

Of course *push*'s functionality is useful only for copying a handful of files. If you're looking to update the entirety of either of the target's */data* or */system* partitions, you can do so using the *sync* command. It will essentially conduct an operation similar to the *rsync* command, making sure that the target's files are synchronized with those on the host. If you run *adb sync* from the same directory where the target's AOSP was built, then it will automatically find the files to sync because the ANDROID_PRODUCT_OUT environment variable will point to the right directory. (Assuming, of course, that you ran *build/envsetup.sh* and *lunch* as required for your target.) Otherwise, you'll need to manually point it to the right output directory like this:

```
$ adb -d -p ~/android/beaglebone/out/target/product/beaglebone/ sync
syncing /system...
push: /home/karim/android/beaglebone/out/target/product/beaglebone/system/xbin/c
rasher -> /system/xbin/crasher
push: /home/karim/android/beaglebone/out/target/product/beaglebone/system/xbin/s
cp -> /system/xbin/scp
push: /home/karim/android/beaglebone/out/target/product/beaglebone/system/xbin/o
pcontrol -> /system/xbin/opcontrol
push: /home/karim/android/beaglebone/out/target/product/beaglebone/system/xbin/t
cpdump -> /system/xbin/tcpdump
push: /home/karim/android/beaglebone/out/target/product/beaglebone/system/xbin/o
profiled -> /system/xbin/oprofiled
push: /home/karim/android/beaglebone/out/target/product/beaglebone/system/xbin/t
imeinfo -> /system/xbin/timeinfo
push: /home/karim/android/beaglebone/out/target/product/beaglebone/system/xbin/c
pueater -> /system/xbin/cpueater
...
491 files pushed. 0 files skipped.
1317 KB/s (81337934 bytes in 60.310s)
syncing /data...
push: /home/karim/android/beaglebone/out/target/product/beaglebone/data/app/gles
2_texture_stream.apk -> /data/app/gles2_texture_stream.apk
push: /home/karim/android/beaglebone/out/target/product/beaglebone/data/app/test
_iterator_host -> /data/app/test_iterator_host
push: /home/karim/android/beaglebone/out/target/product/beaglebone/data/app/test
_iostream_host -> /data/app/test_iostream_host
push: /home/karim/android/beaglebone/out/target/product/beaglebone/data/app/test
_string_host -> /data/app/test_string_host
...
25 files pushed. 0 files skipped.
2804 KB/s (4078615 bytes in 1.420s)
```

You probably want to reboot the target after such an update, as there might be stale file references lingering. Note that *sync* syncs only */system* and */data*. It doesn't sync anything else. In other words, you can't use *sync* to synchronize the contents of the RAM disk mounted as the root filesystem for the target. Even if it allowed you to, it wouldn't be of much use, since the RAM disk lives only in RAM and its contents are not written through to persistent storage.

sync can also be told to sync only the data or the system partitions, instead of both. Simply pass the partition you'd like to sync as a parameter:

```
$ adb -e sync data
syncing /data...
...
```

sync's formal description is:

```
adb sync [ <directory> ]
```

If, instead of copying single files or syncing entire partitions, all you're looking for is to
install new apps, then you should use *install* instead:

```
$ adb install FastBirds.apk
299 KB/s (13290 bytes in 0.043s) pkg: /data/local/tmp/FastBirds.apk
Success
```

Essentially, this will invoke the *pm* (short for "package manager") command on the
target. It will itself interact with the PackageManager system service to get your app
installed. To remove it from the device, you can then use the *uninstall* command:

```
$ adb uninstall com.acme.fastbirds
Success
```

You've likely noted that while *install* relies on the filename, *uninstall* actually needs the
full package name. Each command can actually take a few flags, as explained in
Table 6-11:

```
adb install [-l] [-r] [-s] <file>
adb uninstall [-k] <package>
```

Table 6-11. Flags for install and uninstall

Flag	Description
-l	Tells *install* to ensure that the app is forward-locked. In other words, it disallows the user from copying the *.apk* off the device. In practice, this means that the app is installed in */data/app-private* instead of */data/app*.
-r	Tells *install* to reinstall the app, preserving its data as is.
-s	Tells *install* to install the app on external storage (the SD card) instead of internal storage.
-k	Tells *uninstall* to keep the app's data even though the *.apk* is removed.

State-Altering Commands

For lack of a better name for this category, I've lumped together in this section all the
commands that in one way or another significantly alter the target's behavior. It's not
like the previous commands couldn't or didn't alter the target, it's just that those you'll
find here do so in especially significant ways.

Rebooting

Let's start with one of the more obvious ones:

```
$ adb reboot
```

If you hadn't already guessed, this reboots the target. This actually invokes the re
boot() system call on the target's kernel while passing it the appropriate magic values
to effect a reboot. You can also pass a parameter to *reboot* to tell it to reboot either in
the bootloader or the recovery mode:

```
$ adb reboot bootloader
```

And:

```
$ adb reboot recovery
```

Note, however, that this parameter is passed as is to the kernel. It'll be the job of your board support code in the kernel to deal with this parameter appropriately. If your board-support kernel code doesn't process the string passed to the reboot() function, it's simply ignored, and all that happens is a plain reboot. Another way to reboot into the bootloader is:

```
$ adb reboot-bootloader
```

It's important to highlight that all those reboot commands result in an **immediate** reboot. There is no graceful shutdown of any process or system service. Hence, if you need to do any cleanup, it's best to do so prior to issuing the *reboot* command.

Running as root

By default on a development board, most of *adb*'s commands will work to their full capabilities without a problem, because the *adbd* daemon on the target will likely be running as root. On a production system like a commercial handset, however, it's likely that *adbd* isn't running as root but rather as the shell user, which has far fewer privileges. Hence, commands such as adb shell will also be running only with shell's privileges.

The *adbd* daemon's default privileges will depend on how the AOSP is built and the target that it's running on. If it's running on the emulator, for example, *adbd* will always run as root. In all other cases, *adbd*'s privileges will depend on the TARGET_BUILD_VARIANT chosen to build the AOSP. If it's userdebug or user, *adbd* won't run as root, it'll run as the shell user when started. In the case of userdebug, you can ask it to restart as root by typing:

```
$ adb root
restarting adbd as root
```

If you issue the same command on a user build, you'll get this—in other words, you can't override the default:

```
$ adb root
adbd cannot run as root in production builds
```

If you build with the eng variant, as is likely the case during development, *adbd* will start as root, and here's what happens when you insist:

```
$ adb root
adbd is already running as root
```

The same will happen if the system is already running *adbd* as root because of a previous *adb root* command. All of this behavior is gated by the ro.secure, ro.debuggable, and service.adb.root global properties. The two former are set at build time, while the latter is set by *adb*'s *root* command. Both user and userdebug cause ro.secure to be set to 1, but only userdebug and eng cause ro.debuggable to be set to 1. Obviously those global properties are checked by more than just *adbd*.

Switching connection type

By default, the *adb* server checks for running emulator instances running only on the host and devices physically connected to the host through USB. You can, as we saw earlier, nonetheless connect devices that have their *adbd* daemons listening on a TCP/IP port instead of USB using *adb connect*. What we haven't looked at yet is how to get *adbd* to use TCP/IP instead of USB. Assuming the device is already connected through USB, you can ask it to use TCP/IP instead, like this:

```
$ adb -s 0123456789ABCDEF tcpip 7878
restarting in TCP mode port: 7878
```

Essentially, this will set the service.adb.tcp.port global property on the target to 7878 and restart the *adbd* daemon. Upon restarting, the daemon will then wait for connections on the given port instead of on USB. You can then connect to it like above:

```
$ adb connect 192.168.202.79:7878
connected to 192.168.202.79:7878
```

To switch it back to USB, you can type this:

```
$ adb -s 192.168.172.79:7878 usb
restarting in USB mode
```

Effectively, this command is equivalent to typing:

```
$ adb -s 192.168.172.79:7878 shell
# setprop service.adb.tcp.port 0
# ps
...
root       66    1     3412   164   ffffffff 00008294 S /sbin/adbd
...
# kill 66
```

In both cases, *adbd* is made to exit and is automatically restarted by *init*. It then checks service.adb.tcp.port and starts accordingly. If, for any reason, you don't have a USB connection to your device, you can always manually preset service.adb.tcp.port on the device so that *adbd* always starts on that port number. We'll discuss global property setting later. *connect*'s formal description is:

```
adb tcpip <port>
```

Controlling the emulator

As explained earlier, you can connect to each emulator's console using *telnet*:

```
$ telnet localhost 5554
Trying 127.0.0.1...
Connected to localhost.
```

```
Escape character is '^]'.
Android Console: type 'help' for a list of commands
OK
```
help
```
Android console command help:

    help|h|?         print a list of commands
    event            simulate hardware events
    geo              Geo-location commands
    gsm              GSM related commands
    kill             kill the emulator instance
    network          manage network settings
    power            power related commands
    quit|exit        quit control session
    redir            manage port redirections
    sms              SMS related commands
    avd              manager virtual device state
    window           manage emulator window

try 'help <command>' for command-specific help
OK
```

Google's online manual explains the use of each of these commands in detail. Unfortunately, having to use *telnet* to access each of these commands can be cumbersome, especially if you need to script part of what you need to do. Hence, *adb* allows you to launch these same exact commands like any of its other commands:

```
$ adb -e emu redir add tcp:8080:80
```

This will redirect all connections to the host's port 8080 to the target's port 80. The part of the command line after emu is exactly the same command that you could have typed through the *telnet* session to redirect the port.

Tunneling PPP

One of the external projects included in the AOSP is the standard PPP daemon used in most Linux-based distributions and available at *https://ppp.samba.org/*. You can ask *adb* to set up a PPP connection between the host and the target. This might be for tethering or simply to create a network connection between the host and the target when you have only a USB connection between both. Here's the formal definition of the *ppp* command:

```
adb ppp <adb service name> [ppp opts]
```

Unfortunately, this by itself is insufficient to understand how to use this command. Worse, of all *adb* commands, this one is the most poorly documented. The more common way you're likely to use this command is:[4]

```
adb ppp "shell:pppd nodetach noauth noipdefault /dev/tty" nodetach noauth \
> noipdefault notty <local-ip>:<remote-ip>
```

Essentially, what's happening here is that the host's *pppd* daemon is being started with the following parameters:

```
nodetach noauth noipdefault notty <local-ip>:<remote-ip>
```

And the target's *pppd* is being started with the following parameters:

```
nodetach noauth noipdefault /dev/tty
```

adb then proxies the communication between the two *pppd* daemons and you therefore have a network connection established between the host and the target. You'll likely need to do a little more legwork to figure out exactly what kind of networking connection you want to establish and the specific IP parameters. But with the above, you'll at least have a good starting point. I would encourage you to read *pppd*'s man page on your host for more information on its full capabilities.

I also encourage you to have a look at some of the following articles on the web for more details and examples on the use of this *adb* feature:

- ppp over adb (for linux/unix users) (*http://bit.ly/10qD9jM*)
- device shows up in lsusb + adb but not in ifconfig (*http://bit.ly/WPzHZE*)
- USB Tether for Xperia X10 Mini Pro (*http://bit.ly/13JRrgh*)
- creates a ppp link between my Ubuntu development machine and BeagleBoard running Android connected via USB (*http://bit.ly/10qDhzM*)

Android's Command Line

As I said earlier, one of the first Android-specific tools you're likely to encounter is *adb*, and one of its most common uses is shelling into the target. And since during board bringup you're likely to spend quite some time on the command line before having a functional UI, it's only fitting to now cover Android's command line. In fact, it's possible that you'll likely have to deal directly with Android's command line, probably through a serial console, even before ADB is fully functional: This will be the case if your device doesn't possess USB capabilities or doesn't yet have a functional USB driver or TCP/IP-capable network interface.

---

4. Note that this command is too long to fit in a single line in this book and is therefore line-wrapped. The \ at the end of the first line and the > at the beginning of the second line are there just to show the line-wrapping.

The Shell Up to 2.3/Gingerbread

The standard shell used in Android in versions up to 2.3/Gingerbread is found in *system/core/sh/* in the sources, and the resulting binary is */system/bin/sh* on the target. Unlike many components in the system, this shell is one where Android doesn't reinvent the wheel. Instead, Android uses the NetBSD sh utility with very few tweaks. The AOSP in fact preserves sh's man page as is, so you can do something like this on your host to get more information on how to use the shell:

```
$ man system/core/sh/sh.1
```

This shell is unfortunately a lot more basic than bash or BusyBox's ash. It doesn't, for instance, have tab completion or color-coding of files. If for no other reason, these limitations have been good justification for developers to include BusyBox on their targets, at least during development. For a full comparison of Unix shells, minus Busy-Box, have a look at Arnaud Taddei's Shell Choice, A shell comparison (*http://bit.ly/YHKKGN*). It dates back to 1994, but it's one of the few documents that discusses this topic. There's also Wikipedia's comparison (*http://bit.ly/Yh5VfP*), but it's more shallow.

Comparisons aside, here's an overview of sh's capabilities:

- Output redirection using > and <
- Piping using |
- Running background commands using &
- Scripting using `if/then/fi`, `while/do/done`, `for/do/done`, `continue/break`, and `case/in/pattern/esac`.
- Environment variables
- Parameter expansion (${...})
- Command substitution ($(...))
- Shell patterns (*, ?, !, etc.)

Table 6-12 describes *sh*'s built-in commands.

Table 6-12. sh built-in commands

Command	Description
alias	Substitute one command for another.
bg	Run a suspended task in the background.
command	Run specified command; useful when a script has the same name as a built-in command.
cd	Change directory.
eval	Evaluate an expression.
exec	Replace the running shell with the specified command.
exit	Quit the shell process.

Command	Description
export	Export an environment variable's value for all subsequent commands.
fg	Move background job to the foreground.
getopts	Parse command-line options.
hash	Print out location of commands in shell's cache.
jobid	Print PIDs belonging to job ID.
jobs	List currently running jobs.
pwd	Print working directory.
read	Read a variable from the command line.
readonly	Set an environment variable as read-only.
set	List the environment variables currently set.
setvar	Set an environment variable to a given value.
shift	Shift command-line parameters upward ($1 becomes $2, etc.).
trap	Execute an action when given Unix signals are received.
type	Print the filesystem location of a command or an alias's definition.
ulimit	Print/set the process limits (uses sysctl()).
umask	Set default file creation mode.
unalias	Delete a given alias.
unset	Delete a given environment variable.
wait	Wait for a given job to complete.

If you're using any Android version up to 2.3/Gingerbread, I encourage you to look at sh's man page for more information on how to use each of its features. You'll also be able to benefit from the plethora of online examples and tutorials on Unix shell scripting. None of these aspects is unique to Android or the use of *sh* in an embedded setting.

The Shell Since 4.0/Ice-Cream Sandwich

Starting with 4.0/Ice-Cream Sandwich,[5] Android now relies on the MirBSD Korn Shell (*https://www.mirbsd.org/mksh.htm*). It's found in the *external/mksh/* directory in the host, and the binary is */system/bin/mksh* on the target.

---

5. The change was apparently made in the 3.x series, but the sources for that version were never made available as properly tagged branches, even though newer Android versions include that code.

Even though *mksh* was included in AOSP versions before 4.2/Jelly Bean, it was disabled when building for the emulator. There is a TARGET_SHELL configuration variable in the build system that is set by default to mksh. However, a board config can change the default to whatever is appropriate for that board. Prior to 4.2/Jelly Bean, this variable was set to ash, which is the new name of the executable that replaces the *sh* command described in the previous section.

mksh is a lot more powerful than *sh*. It includes tab completion, for instance, though it doesn't support color-coding of files, and has bash/ksh93/zsh-like extensions. It also has a man page that you can check on the host by typing:

```
$ man system/external/mksh/src/mksh.1
```

Given that *mksh* has a lot more features and built-in commands than *sh*, it would be difficult to give it proper coverage in this book. Instead, I encourage you to look at its man page and its website for more information. It includes, for instance, an implementation for the very useful *history* command, which lists the previous commands you typed on the shell.

Toolbox

Like any other Linux-based system, Android's shell provides only the bare minimum required to have a functional command line. The rest of the functionality comes from individual tools providing specific capabilities that can be started individually from the shell. As we discussed in Chapter 2, the package that provides these tools in Android is called Toolbox, and it's distributed under the BSD license. Toolbox is in *system/core/toolbox/* in the AOSP. The resulting binary and the symbolic links to it reside in */system/bin* on the actual target.

Unfortunately, in addition to not being as feature-rich as BusyBox, Toolbox also severely lacks in documentation. Fortunately, the majority of the commands it provides already exist, albeit in more feature-full form, on the standard Linux desktop. Hence, you can use your development machine's man pages as a primer for using the equivalent Toolbox commands. Beware, as some of the Toolbox variants have slightly different command-line semantics from their standard Linux brethren.

In some cases, this is easy to figure out, as the command will print out its usage if you pass it the wrong type of parameters. However, not all Toolbox commands provide online help. In some cases, you'll even have to dig into Toolbox's sources to figure out exactly how the command's parameters are processed and what the command actually does.

Common Linux commands

Table 6-13 lists the common Linux commands found in Toolbox. If your favorite command isn't in this list, I suggest you check BusyBox—it's likely in there. We'll discuss in Appendix A how to get BusyBox to sit side by side with Toolbox in the same filesystem. If even BusyBox doesn't include the utility you're looking for, then you can compile the full Linux utility for Android, possibly by importing it into the AOSP *external/* directory and deriving an *Android.mk* for it based on its existing build scripts or makefiles.

 For the sake of brevity, I'm omitting the full list of command parameters in Table 6-13 for each command. Have a look at the Linux man pages to get an idea of what they likely are.

Table 6-13. Toolbox's common Linux commands

Command	Description
cat	Dump the contents of a given file to the standard output
chmod	Change the access rights on a file or a directory
chown	Change the ownership of a file or a directory
cmp	Compare two files
date	Print out the current date and time
dd	Copy a file while converting and formatting the content
df	Print the filesystems' disk usage
dmesg	Dump the kernel's log buffer
hd	Dump a file in hexadecimal format
id	Print the current user and group IDs
ifconfig	Configure a networking interface
iftop	Monitor the networking traffic in real-time
insmod	Load a kernel module
ionice	Get/set the I/O priority of a process
ln	Create a symbolic link
kill	Send the TERM signal to a process
ls	List a directory's contents
lsmod	List the currently loaded kernel modules
lsof	List the currently open file descriptors
mkdir	Create a directory
mount	Print the list of mounted filesystems or mount new ones
mv	Rename a file
netstat	Print network statistics

Command	Description
printenv	Print all environment variables exported
ps	Print running processes
reboot	Reboot the system
renice	Change a process's "nice" value
rm	Delete a file
rmdir	Delete a directory
rmmod	Remove a kernel module
route	Print/modify the kernel's routing table
sleep	Sleep for a given number of seconds
sync	Flush the filesystem cache back to persistent storage
top	Monitor processes in real time
umount	Unmount a filesystem
uptime	Print the system's uptime
vmstat	Print out the system's memory use

A few of these are downright annoying in their shortcomings. For example, until 4.0/Ice-Cream, *ls* was unable to print directory listings in alphabetical order or provide color-coding for files, which is standard in most Linux systems. Alphabetical ordering has since been added, but not color-coding. Also, contrary to its typical Linux or BusyBox version, *ifconfig* doesn't actually print out the current network configuration if invoked without any parameters—you have to use *netcfg* instead. Table 6-14 lists additional Linux commands you'll find in 4.2/Jelly Bean.

Table 6-14. Additional common Linux commands found in 4.2/Jelly Bean

Command	Description
cp	Copy files
du	Show file-space usage
grep	Look for strings in files
md5	Like *md5sum* command in Linux, compute files' MD5 checksum
touch	Update a file's timestamp (and create it if it doesn't exist)

Global properties

Chapter 2 explained that one of Android's init features is that it maintains a set of global properties that can be accessed from anywhere in the system. Naturally, Toolbox provides a few tools to interface with these global properties:

```
getprop <key>
setprop <key> <value>
watchprops
```

The first thing you'll likely want to do is list all the properties with their current values:

```
# getprop
[ro.ril.wake_lock_timeout]: [0]
[ro.secure]: [0]
[ro.allow.mock.location]: [1]
[ro.debuggable]: [1]
[persist.service.adb.enable]: [1]
[ro.factorytest]: [0]
[ro.serialno]: []
[ro.bootmode]: [unknown]
[ro.baseband]: [unknown]
[ro.carrier]: [unknown]
[ro.bootloader]: [unknown]
[ro.hardware]: [am335xevm]
[ro.revision]: [0]
[ro.build.id]: [GRJ22]
[ro.build.display.id]: [beaglebone-eng 2.3.4 GRJ22 eng.karim.20120504.160548
  test-keys]
[ro.build.version.incremental]: [eng.karim.20120504.160548]
[ro.build.version.sdk]: [10]
...
```

It should print out over 100, if not a lot more, global properties set for your system. If you just want to print out a single value, you can do this:

```
# getprop ro.hardware
am335xevm
```

You can also set global properties straight from the command line:

```
# setprop acme.birdradar.enable 1
# getprop acme.birdradar.enable
1
```

Once a property has been set, you can change its value again using *setprop*. You can't, however, delete a property that you "created" using *setprop*. The property will, however, disappear at the next reboot unless its name starts with persist. In that case, a file with the property's full name will be created in */data/property* containing the property's value. To delete this property, you would need to delete this file or destroy the data partition.

You can also monitor properties being changed in real-time—assuming the acme.bird radar.enable is set after *watchprop* is started:

```
# watchprops
946709853 acme.birdradar.enable = '1'
```

Input events

Android relies heavily on Linux's input layer to get the user's input events. The devices that expose Linux's input layer are available through entries in */dev/input* which, as we saw in Chapter 2, is the basis of Android's input support. Whenever the user touches or swipes the screen or touches any of the device's buttons, an event is generated. While Android's System Server already handles those events appropriately, you might want to either observe or generate your own events. Toolbox lets you do just that:

```
getevent [-t] [-n] [-s <switchmask>] [-S] [-v [<mask>]] [-p] [-q] [-c <count>]
[-r][<device>]
    -t: show time stamps
    -n: don't print newlines
    -s: print switch states for given bits
    -S: print all switch states
    -v: verbosity mask (errs=1, dev=2, name=4, info=8, vers=16, pos. events=32)
    -p: show possible events (errs, dev, name, pos. events)
    -q: quiet (clear verbosity mask)
    -c: print given number of events then exit
    -r: print rate events are received
sendevent <device> <type> <code> <value>
```

To observe the events, you can do something like this:

```
# getevent
/dev/input/event0: 0003 0000 0000007d
/dev/input/event0: 0003 0001 0000011b
/dev/input/event0: 0001 014a 00000001
/dev/input/event0: 0000 0000 00000000
/dev/input/event0: 0001 014a 00000000
/dev/input/event0: 0000 0000 00000000
/dev/input/event0: 0001 0066 00000001
/dev/input/event0: 0001 0066 00000000
...
```

getevent continuously displays events as they come in until you type Ctrl-C. The output format is event type, event code, and event value. This lets you verify whether your driver is reporting the appropriate information back to Android.

In a similar fashion, if you'd like to monitor Android's handling of events, you can send events of your own:

```
# sendevent /dev/input/event0 1 330 1
```

Note that if you were running *getevent* simultaneously, you would then see this new event:

```
/dev/input/event0: 0001 014a 00000001
```

In other words, while *getevent*'s output is hexadecimal, *sendevent*'s input is decimal.

Controlling services

As we saw in Chapter 2, Android's init starts a number of native daemons for a variety of purposes. Typically, these are described as *services* in init's configuration scripts—init's "services" have nothing to do with either system services or the service components available to app developers. As we'll see shortly, such services can be either started automatically or marked as disabled. Either way, you can start and stop services using the following:

```
start <servicename>
stop <servicename>
```

Neither of these generates any output. There's also unfortunately no way to ask Android for the list of running services. Instead, you're assumed to understand init's configuration scripts enough to know which services you can start and stop. For instance, if you want to stop all the system's Java components, you can do this:

```
# stop zygote
```

Note that this specific command is a pretty drastic measure, as it will stop all apps and kill the System Server. But in some cases it might be exactly what you're looking for. Say you wanted to stop a system service from accessing a given driver because it stopped operating properly, and you want to run some diagnostics on it without the system continuing to use it.

We'll cover Android's init and its handling of services in the next section.

Logging

Another interesting Toolbox feature is its ability to allow you to add your own events to Android's logger:

```
log [-p <prioritychar>] [-t <tag>] <message>
prioritychar should be one of:
v,d,i,w,e
```

For example:

```
# log -p i -t ACME Initiating bird tracking sequence
```

Now, if you check the logs with *logcat*, you see this:

```
# logcat
...
I/ACME    ( 336): Initiating bird tracking sequence
...
```

This can be very useful if you have shell scripts that execute alongside the rest of the Android stack. Also, if you've got custom code using Android's logging capabilities, say within an app or a custom system service, you'll be able to see the relative ordering of the events generated there and those generated from scripts or manually on the command line.

ioctl

As we discussed in Chapter 2, devices appear as entries in */dev*. If you are familiar with Linux's driver model, you know that if a device is controlled by a character device driver, then simply opening that device's entry in /dev and reading/writing from/to it will result in its read()/write() functions getting invoked. So you can do something like this to read from a character device:

```
# cat /dev/birdlocator0
```

Similarly, you can do something like this to write to a character device:

```
# echo "Fire" > /dev/birdlaser0
```

Another very important file operation available on character devices is ioctl(). There is, however, no standard Linux utility for invoking this operation, since it's driver-specific. On embedded systems, however, where those manipulating the system are typically either the driver authors themselves or working with them very closely, it makes sense to have a utility to enable developers to invoke drivers' ioctl() functions. And Toolbox provides just that:

```
ioctl [-l <length>] [-a <argsize>] [-rdh] <device> <ioctlnr>
  -l <length>   Length of io buffer
  -a <argsize>  Size of each argument (1-8)
  -r            Open device in read only mode
  -d            Direct argument (no iobuffer)
  -h            Print help
```

Obviously the use you make of this will be highly driver-specific. You'll need to refer to your driver's documentation and/or sources to know exactly the parameters you need to pass to this command and what effects they'll have.

ioctl() is a very powerful driver operation. Uses can go from benign status reporting to outright hardware destruction. Make sure you know **exactly** what the specific I/O control operation you're about to issue does on the designated device. You probably want to use it only on drivers you wrote.

Wiping the device

In some extreme cases, it's necessary to destroy data on an Android device. This extreme and irreversible operation is made possible using Toolbox's *wipe* command:

```
wipe <system|data|all>

system means '/system'
data means '/data'
```

If you need to destroy all data on a system, you can do this:

```
# wipe data
Wiping /data
Done wiping /data
```

Android's Command Line | 217

I'm sure you understand there's no "undo" here, so be careful with this. You might want to use this as a failsafe in case you have sensitive data or binaries on the device and, for instance, destroy it in case you detect unauthorized access to key system parts.

Other Android-specific commands

Toolbox also includes a few other Android-specific commands, which we'll review briefly, since their uses are either obvious or very limited.

nandread. This utility is for reading the contents of a NAND flash device to a file:

```
nandread [-d <dev>] [-f <file>] [-s <size>] [-vh]
    -d <dev>    Read from <dev>
    -f <file>   Write to <file>
    -s <size>   Number of spare bytes in file (default 64)
    -R          Raw mode
    -S <start>  Start offset (default 0)
    -L <len>    Length (default 0)
    -v          Print info
    -h          Print help
```

newfs_msdos. This command allows you to format a device as a VFAT filesystem:

```
newfs_msdos [ -options ] <device> [<disktype>]
where the options are:
-@ create file system at specified offset
-B get bootstrap from file
-C create image file with specified size
-F FAT type (12, 16, or 32)
-I volume ID
-L volume label
-N don't create file system: just print out parameters
-O OEM string
-S bytes/sector
-a sectors/FAT
-b block size
-c sectors/cluster
-e root directory entries
-f standard format
-h drive heads
-i file system info sector
-k backup boot sector
-m media descriptor
-n number of FATs
-o hidden sectors
-r reserved sectors
-s file system size (sectors)
-u sectors/track
```

newfs_msdos is the tool used by the *vold* daemon to format devices for VFAT; *vold* being itself used by the Mount system service for managing mounted devices.

notify. This command uses the inotify system call an API to monitor directories or files for modifications:

```
notify [-m <eventmask>] [-c <count>] [-p] [-v <verbosity>] <path> [<path> ...]
```

r. In 4.2/Jelly Bean, you'll also find an *r* command. It's shorthand for repeating the previous command you typed on the shell. So, instead of pressing the up arrow and then Enter, you can just type *r*. Here's a simple example:

```
root@android:/ # ls -l /proc/cpuinfo
-r--r--r-- root     root            0 2013-01-19 10:34 cpuinfo
root@android:/ # r
ls -l /proc/cpuinfo
-r--r--r-- root     root            0 2013-01-19 10:34 cpuinfo
```

schedtop. Like *top*, *schedtop* is for continuous, real-time monitoring of the kernel's scheduler. Unlike *top*, which only reports the real-time CPU usage percentage for each process, this command continuously reports on the cumulative execution time of each process:

```
schedtop [-d <delay>] [-bitamun]
        -d refresh every <delay> seconds
        -b batch - continuous prints instead of refresh
        -i hide idle tasks
        -t show threads
        -a use alternate screen
        -m use millisecond precision
        -u use microsecond precision
        -n use nanosecond precision
```

The command description given here stems from my reading of Toolbox's sources. *schedtop* itself doesn't provide any online help, nor is there any documentation on its use.

setconsole. This command lets you switch consoles:

```
setconsole [-d <dev>] [-v <vc>] [-gtncpoh]
    -d <dev>   Use <dev> instead of /dev/tty0
    -v <vc>    Switch to virtual console <vc>
    -g         Switch to graphics mode
    -t         Switch to text mode
    -n         Create and switch to new virtual console
    -c         Close unused virtual consoles
    -p         Print new virtual console
    -o         Print old virtual console
    -h         Print help
```

smd. Of all of Toolbox's commands, this one is the most "mysterious." I had a very hard time finding any useful information about the use of *smd* or actual usage examples. It appears that under certain devices, the Baseband Processor appears as one of */dev/smdN*. This tool then allows you to send AT commands to the Baseband Processor:

    smd [<port>] <commands>

Core Native Utilities and Daemons

As I mentioned in Chapter 2, Android has about 150 utilities spread around its filesystem. In this chapter, we'll cover those used independent of the Java framework and services. Specifically, we'll focus in this section mostly on those in */system/bin*, which we could consider *core* to Android. Some utilities are also found in */system/xbin*, but they aren't essential for the system to operate properly.

We already saw how Toolbox implements a lot of functionality commonly found in standard Linux systems, as well as Android-specific functionality. Similarly, there are two categories of core Android utilities and daemons, some which are derived from external projects and others that are Android specific. Table 6-15 presents a number of core utilities and daemons that are compiled from projects in the *external/* directory.

Table 6-15. Core utilities and daemons from external projects

Utility/Daemon	External Project	Original Location
bluetoothd, sdptool, avinfo, hciconfig, hcitool, l2ping, hciattach and rfcomm.	BlueZ[a]	http://www.bluez.org/
dbus-daemon	D-Bus	http://dbus.freedesktop.org
dnsmasq	Dnsmasq	http://www.thekelleys.org.uk/dnsmasq/
dhcpcd and showlease	dhcpcd	http://roy.marples.name/projects/dhcpcd/
fsck_msdos	NetBSD fsck_msdos	http://cvsweb.netbsd.org/bsdweb.cgi/src/sbin/fsck_msdos/
gdbserver	GNU Debugger	http://www.gnu.org/software/gdb/
gzip	gzip utility	http://www.gzip.org/
iptables	Netfilter	http://www.netfilter.org/
ping	iputils	http://www.skbuff.net/iputils/
pppd	PPP	http://ppp.samba.org/
racoon	IPsec-Tools	http://ipsec-tools.sourceforge.net/
tc	iproute2	http://www.linuxfoundation.org/collaborate/workgroups/networking/iproute2
wpa_supplicant and wpa_cli	WPA Supplicant	http://hostap.epitest.fi/wpa_supplicant/

[a] No longer part of Android starting with 4.2/Jelly Bean.

Not all of these are actually necessary for your system to run. If your embedded system doesn't have WiFi or Bluetooth support, for instance, then there's no need to have either *wpa_supplicant* or any of the BlueZ utilities and daemons. In fact, in those specific cases, the binary isn't built unless the board-specific *.mk* files require it. Remember that BlueZ has been replaced with another stack in 4.2/Jelly Bean.

The following subsections look at the core Android-specific utilities and daemons. Many of these aren't actually meant to be invoked by you directly on the command line but are automatically invoked instead by one part of the system or another. Some, however, are worth mastering.

logcat

Probably one of the commands you'll use most often in Android, *logcat* allows you to dump the Android logger's buffer as we saw earlier while covering *adb*. Here's *logcat*'s full online help:

```
# logcat --help
Usage: logcat [options] [filterspecs]
options include:
  -s              Set default filter to silent.
                  Like specifying filterspec '*:s'
  -f <filename>   Log to file. Default to stdout
  -r [<kbytes>]   Rotate log every kbytes. (16 if unspecified). Requires -f
  -n <count>      Sets max number of rotated logs to <count>, default 4
  -v <format>     Sets the log print format, where <format> is one of:

                  brief process tag thread raw time threadtime long

  -c              clear (flush) the entire log and exit
  -d              dump the log and then exit (don't block)
  -t <count>      print only the most recent <count> lines (implies -d)
  -g              get the size of the log's ring buffer and exit
  -b <buffer>     request alternate ring buffer
                  ('main' (default), 'radio', 'events')
  -B              output the log in binary
filterspecs are a series of
  <tag>[:priority]

where <tag> is a log component tag (or * for all) and priority is:
  V    Verbose
  D    Debug
  I    Info
  W    Warn
  E    Error
  F    Fatal
  S    Silent (supress all output)

'*' means '*:d' and <tag> by itself means <tag>:v
```

```
If not specified on the commandline, filterspec is set from ANDROID_LOG_TAGS.
If no filterspec is found, filter defaults to '*:I'

If not specified with -v, format is set from ANDROID_PRINTF_LOG
or defaults to "brief"
```

You should be able to figure out most of *logcat*'s intricacies using this help and Chapter 2's explanations of the Android logger. You can use the -b flag, for instance, to select which buffer you'd like to dump—main being the default. You can also set the AN DROID_LOG_TAGS environment variable to provide a default output filter. Still, a more confusing aspect of *logcat* is specifically its filtering capabilities. Indeed, the online help seems to indicate that just specifiying a <tag>[:priority] after the command is sufficient to limit the output to that belonging to tag. That doesn't work, though:

```
# logcat ActivityManager
--------- beginning of /dev/log/main
I/DEBUG   (   59): debuggerd: Mar 27 2012 05:30:39
--------- beginning of /dev/log/system
I/Vold    (   57): Vold 2.1 (the revenge) firing up
D/Vold    (   57): USB mass storage support is not enabled in the kernel
D/Vold    (   57): usb_configuration switch is not enabled in the kernel
D/Vold    (   57): Volume sdcard state changing -1 (Initializing) -> 0 (No-Media
)
D/Vold    (   57): Volume usb state changing -1 (Initializing) -> 0 (No-Media)
D/Vold    (   57): Volume sdcard state changing 0 (No-Media) -> 2 (Pending)
D/Vold    (   57): Volume sdcard state changing 2 (Pending) -> 1 (Idle-Unmounted
)
I/Netd    (   58): Netd 1.0 starting
D/AndroidRuntime(   61):
D/AndroidRuntime(   61): >>>>>> AndroidRuntime START com.android.internal.os.Zyg
oteInit <<<<<<
D/AndroidRuntime(   61): CheckJNI is ON
D/dalvikvm(   61): creating instr width table
...
```

Obviously, we're seeing the output from all tags, not just the one matching ActivityManager. The trick is to use the -s flag:

```
# logcat -s ActivityManager
--------- beginning of /dev/log/main
--------- beginning of /dev/log/system
I/ActivityManager(  128): Memory class: 16
I/ActivityManager(  128): Config changed: { scale=1.0 imsi=0/0 loc=md_US touch=1
 keys=1/1/2 nav=1/1 orien=2 layout=268435491 uiMode=0 seq=1}
I/ActivityManager(  128): System now ready
I/ActivityManager(  128): Start proc com.android.systemui for service com.androi
d.systemui/.statusbar.StatusBarService: pid=245 uid=1000 gids={3002, 3001, 3003}
I/ActivityManager(  128): Config changed: { scale=1.0 imsi=0/0 loc=md_US touch=1
 keys=1/1/2 nav=1/1 orien=2 layout=268435491 uiMode=17 seq=2}
I/ActivityManager(  128): Start proc com.android.inputmethod.latin for service c
om.android.inputmethod.latin/.LatinIME: pid=247 uid=10016 gids={}
W/ActivityManager(  128): Unable to start service Intent { act=@0 }: not found
```

```
W/ActivityManager(  128): Unable to start service Intent { act=@0 }: not found
...
```

logcat's online help is unfortunately not very helpful in figuring this out.

netcfg

In addition to Toolbox's *ifconfig*, Android has another utility that lets you manipulate network interfaces:

```
netcfg [<interface> {dhcp|up|down}]
```

Confusingly, *netcfg* and *ifconfig* have overlapping functionality. Both can, for example, bring interfaces up and down. However, *netcfg* can initiate DHCP client requests and print out the current interface's configuration, while *ifconfig* can do neither. *ifconfig*, on the other hand, can set an interface's static IP address and its netmask, while *netcfg* can't do that.

Mostly, *netcfg* is very useful for printing out the interfaces' configurations:

```
# netcfg
lo       UP     127.0.0.1       255.0.0.0       0x00000049
eth0     UP     10.0.2.15       255.255.255.0   0x00001043
tunl0    DOWN   0.0.0.0         0.0.0.0         0x00000080
gre0     DOWN   0.0.0.0         0.0.0.0         0x00000080
```

debuggerd

This daemon is actually started by *init* early during startup. It opens the android:debuggerd abstract Unix domain socket[6] and awaits connections. It remains dormant until a user-space process crashes. It's activated by Bionic's linker, which sets up signal handlers for dealing with crashes and connects to *debuggerd* whenever that happens. *debuggerd* then does two things: creates a *tombstone* file in */data/tombstones* and, if required, allows postmortem debugging to be done through *gdbserver*.

You don't need to do anything special for tombstone files to be generated. They'll be created automatically and will contain information about the crashing process that you might find useful for postmortem analysis. Here's one from the frequently crashing VNC server on my BeagleBone:

```
# cat /data/tombstones/tombstone_06
*** *** *** *** *** *** *** *** *** *** *** *** *** *** *** ***
Build fingerprint: 'TI/beaglebone/beaglebone:2.3.4/GRJ22/eng.karim.20120504.1605
48:eng/test-keys'
pid: 4656, tid: 4656  >>> androidvncserver <<<
signal 11 (SIGSEGV), code 1 (SEGV_MAPERR), fault addr deadbaad
 r0 00000027  r1 deadbaad  r2 a0000000  r3 00000000
```

---

6. Unlike typical Unix domain sockets, which appear as entries in the filesystem, *abstract* sockets are not visible on the filesystem.

```
        r4 00000001  r5 00000000  r6 00069ad8  r7 0005e000
        r8 00069cd8  r9 00000000  10 000003e8  fp 00000001
        ip afd46668  sp beeb4bd0  lr afd191d9  pc afd15ca4  cpsr 60000030
        d0  2e302e302e373220  d1  206f742074636567
        d2  000000000000006f  d3  000000000000006e
        ...
                #00  pc 00015ca4  /system/lib/libc.so
                #01  pc 00013614  /system/lib/libc.so
                #02  pc 000144da  /system/lib/libc.so
                #03  pc 00010290  /system/bin/androidvncserver
                #04  pc 00010296  /system/bin/androidvncserver
                #05  pc 0000fcbe  /system/bin/androidvncserver
                #06  pc 0000bc66  /system/bin/androidvncserver
                #07  pc 0000a87e  /system/bin/androidvncserver
                #08  pc 00014b52  /system/lib/libc.so

        code around pc:
        afd15c84 2c006824 e028d1fb b13368db c064f8df
        afd15c94 44fc2401 4000f8cc 49124798 25002027
        afd15ca4 f7f57008 2106ec7c edd8f7f6 460aa901
        afd15cb4 f04f2006 95015380 95029303 e93ef7f6
        afd15cc4 462aa905 f7f62002 f7f5e94a 2106ec68

        code around lr:
        afd191b8 4a0e4b0d e92d447b 589c41f0 26004680
        afd191c8 686768a5 f9b5e006 b113300c 47c04628
        afd191d8 35544306 37fff117 6824d5f5 d1ef2c00
        afd191e8 e8bd4630 bf0081f0 00028344 ffffff88
        afd191f8 b086b570 f602fb01 9004460c a804a901

        stack:
            beeb4b90  0005e008
            beeb4b94  6f000001
            beeb4b98  6f2e6772
            beeb4b9c  7069616e
            beeb4ba0  afd4270c
            beeb4ba4  afd426b8
            beeb4ba8  00000000
            beeb4bac  afd191d9  /system/lib/libc.so
        ...
```

Also, if you set the debug.db.uid to some UID larger than that of the crashing process (just use a large integer value such as 32767 [2^15 - 1]), *debuggerd* will then use the ptrace() system call to attach to the dying process and allow you to start *gdbserver* to take control of it. Here's the output printed out by *debuggerd* to the log when I do that on my BeagleBone:

```
I/DEBUG   (   59): *********************************************************
I/DEBUG   (   59): * Process 4656 has been suspended while crashing.  To
I/DEBUG   (   59): * attach gdbserver for a gdb connection on port 5039:
I/DEBUG   (   59): *
I/DEBUG   (   59): *     adb shell gdbserver :5039 --attach 4656 &
```

```
I/DEBUG   (   59): *
I/DEBUG   (   59): * Press HOME key to let the process continue crashing.
I/DEBUG   (   59): ***********************************************************
```

Once *gdbserver* is attached to the dying process, you can then use one of the *arm-eabi-gdb* debuggers that are part of the AOSP's *prebuilt/* directory to attach to the *gdbserver* running on the target and proceed with debugging the dying process.

Other Android-specific core utilities and daemons

There are also a few other core utilities and daemons you should know about, though you're unlikely to use these very often.

check_prereq. This allows you to check whether the currently running build is older than a given timestamp:

```
# check_prereq 1336847591
current build time: [1336162137]  new build time: [1336847591]
```

This is mainly useful for upgrading purposes, allowing you to invoke this command from *adb* to check whether your current builder is older or newer than the one running on your device. The build time is stored in the *build.prop* file found in the *system/* partition in the ro.build.date.utc global property.

linker. This is Bionic's dynamic linker. You never need to invoke this manually. It is automatically loaded whenever a Bionic-linked binary is executed, and its job is to load all the libraries required by that binary. The *readelf* utility part of the GNU toolchain provides some more insight as to what occurs during this process:

```
$ arm-eabi-readelf -a logcat
ELF Header:
  Magic:   7f 45 4c 46 01 01 01 00 00 00 00 00 00 00 00 00
  Class:                             ELF32
  Data:                              2's complement, little endian
  Version:                           1 (current)
  OS/ABI:                            UNIX - System V
  ABI Version:                       0
  Type:                              EXEC (Executable file)
  Machine:                           ARM
  Version:                           0x1
  Entry point address:               0x8ed0
  Start of program headers:          52 (bytes into file)
  Start of section headers:          13020 (bytes into file)
  Flags:                             0x5000000, Version5 EABI
  Size of this header:               52 (bytes)
  Size of program headers:           32 (bytes)
...
Program Headers:
  Type           Offset   VirtAddr   PhysAddr   FileSiz MemSiz  Flg Align
  PHDR           0x000034 0x00008034 0x00008034 0x000e0 0x000e0 R   0x4
```

```
    INTERP         0x000114 0x00008114 0x00008114 0x00013 0x00013 R   0x1
      [Requesting program interpreter: /system/bin/linker] ❶
    LOAD           0x000000 0x00008000 0x00008000 0x02470 0x02470 R E 0x1000
    LOAD           0x003000 0x0000b000 0x0000b000 0x001cc 0x00608 RW  0x1000
    DYNAMIC        0x003020 0x0000b020 0x0000b020 0x000c8 0x000c8 RW  0x4
    GNU_STACK      0x000000 0x00000000 0x00000000 0x00000 0x00000 RW  0
    EXIDX          0x002410 0x0000a410 0x0000a410 0x00060 0x00060 R   0x4
    ...
    Dynamic section at offset 0x3020 contains 25 entries:
      Tag        Type                         Name/Value
    0x00000003 (PLTGOT)                     0xb0fc
    0x00000002 (PLTRELSZ)                   376 (bytes)
    ...
    0x00000001 (NEEDED)                     Shared library: [liblog.so] ❷
    0x00000001 (NEEDED)                     Shared library: [libc.so]
    0x00000001 (NEEDED)                     Shared library: [libstdc++.so]
    0x00000001 (NEEDED)                     Shared library: [libm.so]
    ...
```

❶   This is the linker required by the binary.

❷   These are the libraries that must be loaded by the linker.

There's of course a lot more output to *readelf* than the above, but this shows you that *logcat*'s "program interpreter" is */system/bin/linker* and that it needs the following libraries: *liblog.so*, *libc.so*, *libstdc++.so*, and *libm.so*.

logwrapper. This command allows you to run another command and redirect its stdout and stderr to the Android logger:

```
logwrapper [-x] <binary> [ <args> ...]
```

The log tag used in this case is the same string as the `binary`'s name. Using the `-x` option causes *logwrapper* to generate a segmentation fault (`SIGSEGV`) when `binary` terminates, with the fault address being the status returned by the `wait()` system call on the existing binary.

run-as. Allows you to run a binary as if it were executed with the rights associated with an app package:

```
run-as <package-name> <command> [<args>]
```

The `command` will run from the directory associated with `package-name` in */data/data* with that app's UID/GID.

sdcard utility. This utility uses Linux's Filesystem in User SpacE (FUSE) to emulate in any directory on the filesystem the rights and permissions you'd find on any FAT-formatted SD card:

    sdcard <path> <uid> <gid>

In other words, files and directories in the designated directory will all be executable, as you'd expect in FAT. The directory provided as path will be mounted to */mnt/sdcard*. And while *sdcard* must be issued as root, it'll run as uid/gid. This is useful for devices that don't actually have a removable SD card. In those cases, the "external" storage is emulated on the "internal" storage using the *sdcard* command.

Extra Native Utilities and Daemons

Android also packs a certain number of extra utilities and daemons that aren't essential to the system's operation. Most of these are in */system/xbin*, and they may, in some circumstances, be useful to you. Tables 6-16 and 6-17 list those utilities and daemons.

Table 6-16. Extra utilities and daemons from external projects

Utility/Daemon	External Project	Original Location
dbus-monitor and *dbus-send*	D-Bus	http://dbus.freedesktop.org
ssh and *scp*	Dropbear	http://matt.ucc.asn.au/dropbear/
nc	Netcat	http://nc110.sourceforge.net/
skia_text	skia 2D graphics library	http://code.google.com/p/skia/
sqlite3	SQLite	http://www.sqlite.org/
strace	strace utility	http://sourceforge.net/projects/strace/
tcpdump	tcpdump utility	http://www.tcpdump.org/
netperf and *netserver*	netperf	http://www.netperf.org/netperf/
oprofiled and *opcontrol*	OProfile	http://oprofile.sourceforge.net/

Table 6-17. Extra Android-specific utilities and daemons

Utility/Daemon	Description
cpueater and *daemonize*	*cpueater* does a while(1) loop, eating as much CPU as possible, and *daemonize* allows you to run it as a daemon in the background.
crasher	This utility is packaged with *debuggerd* and essentially simulates a crashing process.
directiotest	Provided with a block device's mount, does write/readback tests on the block device to test it.
latencytop	Provides per-process latency information.
librank	Prints memory usage information for each object mapped into any process's memory. This includes libraries and memory-mapped devices and regions.
procmem	Prints memory usage information for each section of a running PID.
procrank	Ranks processes by memory used.

Utility/Daemon	Description
schedtest	Tests the scheduler to see how reliable it is at promptly waking up tasks that request 1ms sleeps.
showmap	Prints out a process's memory map.
showslab	Prints out information on the slab allocator.
su	Allows the root user to change his UID/GID.
timeinfo	Reports realtime, uptime, awake percentage, and sleep percentage to the standard output.

Framework Utilities and Daemons

In addition to the utilities and daemons just covered, Android contains quite a number of others that are tightly tied to the system services and Android Framework, such as *servicemanager*, *installd*, and *dumpsys*. We'll discuss those in the next chapter.

Init

One of the most important tasks in the system is initializing the user-space environment once the kernel has finished initializing device drivers and its own internal structures. As we discussed in Chapter 2, this is the *init* process's job once it's started by the kernel. And, as we discussed then, Android has its own custom init, with its own specific features. Now that we've covered a good part of what's available in the native user-space once the system is up, let's take a closer look at the process that's responsible for starting it all.

Theory of Operation

Figure 6-4 illustrates how *init* integrates with the rest of the Android components. After getting started by the kernel, it essentially reads its configuration files, prints out a boot logo or text to the screen, opens a socket for its property service, and starts all the daemons and services that bring up the entire Android user-space. There's of course more to each of these steps.

> ### Android init versus "Normal" init
>
> In a typical Linux system, *init*'s role would be limited to starting daemons, but, if only because of its property service, Android's *init* is special. Like any Linux *init*, however, Android's *init* isn't expected to ever die. *init* is, as we discussed earlier, the first process started by the kernel and, as such, its PID is always 1. Should it ever die, the kernel would panic.

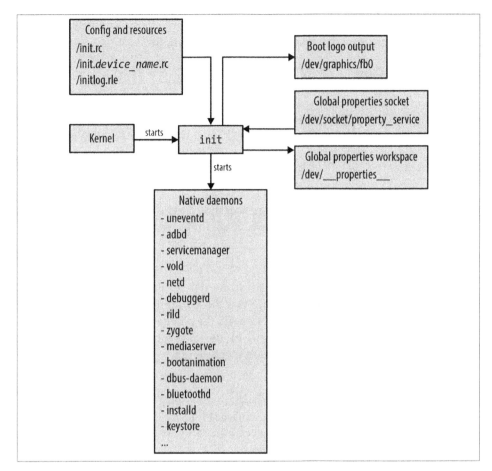

Figure 6-4. Android's init

One of the first things *init* does is check whether it was invoked as *ueventd*. As I mentioned in Chapter 2, *init* includes an implementation of the udev hotplug events handler. Because this code is compiled within *init*'s own code, *init* checks the command-line that was used to invoke it, and if it was invoked through the */sbin/ueventd* symbolic link to */init*, then *init* immediately runs as *ueventd*.

The next thing *init* does is create and mount */dev*, */proc*, and */sys*. These directories and their entries are crucial to many of the things *init* does next. *init* then reads the */init.rc* and */init.<device_name>.rc* files, parses their content into its internal structures, and proceeds to initialize the system based on a mix of its configuration files and built-in rules. We'll discuss this in much greater detail in the next subsection.

Once all initialization is done, *init* then enters an infinite loop in which it restarts any services that might have exited and that need restarting, and then polls file descriptors it handles, such as the property service's socket, for any input that needs to be processed. This is how *setprop* property setting requests are serviced, for instance.

Configuration Files

The main way to control *init*'s behavior is through its configuration files. Given that Android has its own *init*, there is much to say about those configuration files. Let's go over their location and semantics. Then we'll cover the main *init.rc* file and board-specific configuration files.

Location

The main location for all things *init* is the root directory (/). This is where you'll find the actual *init* binary itself and its two configurations files: *init.rc* and *init.<device_name>.rc*. The first file's name is fixed in stone, while the second file's name depends on the hardware.

In essence, the *<device_name>* is extracted from */proc/cpuinfo*. Earlier in this chapter, we used *adb shell* to dump the content of that file for the BeagleBone. In that dump, you'll notice a line that starts with Hardware. It's the content of that line that is parsed by *init* to retrieve the *<device_name>*. In the case of the BeagleBone, this is am335xevm, and in the case of the emulator, it's goldfish.

The string displayed beside Hardware is converted to lowercase before the final *init.<device_name>.rc* is fetched from disk. Hence, though the emulator reports Goldfish as being the hardware in */proc/cpuinfo*, the file being fetched is */init.goldfish.rc*.

One very important thing to highlight is that *init* reads both files before it executes any of the instructions. There is therefore little incentive for adding board-specific modifications to the main *init.rc* file instead of the board-specific *.rc* file. Also, while the *.rc* files typically have their execute permission enabled, *init* itself doesn't really check for that.

Semantics

init's *.rc* files contain a series of declarations that fall in one of two types: *actions* and *services*. Each declarative section starts with a keyword identifying the type of declaration, on for an action and service for a service, and is followed by a number of lines with more details on the declaration:

```
on <trigger>
   <command>
   <command>
```

```
    <command>
...
service <name> <pathname> [ <argument> ]*
    <option>
    <option>
    <option>
...
```

init's "services" have nothing to do with system services or the *service* component used by app developers.

Interestingly, there's a *readme.txt* within *init*'s sources in the AOSP. You'll find it in *system/core/init/*. Some of the things it describes are likely to have been initial design goals but aren't actually in the current *init*, such as the device-added and device-removed triggers. Overall, though, it remains a good reference.

The configuration files can, of course, declare many actions and services. Typically, actions and services are left-aligned, and the commands or options that follow are indented as shown above. Action and service declarations are similar in scope in that a given declaration ends whenever the next on or service keyword appears. Only an action, however, results in the execution of commands. Service declarations serve only to describe services; they don't actually start anything. The services are typically started or stopped when an action is triggered.

There are two types of action triggers: predefined triggers and triggers activated on property-value changes. *init* defines a fixed set of predefined triggers that are run in a specific order. Property-activated triggers, however, are activated whenever a given property takes on a certain value specified in the *init.rc* file. Here's the list of predefined triggers that can be used in an *init* configuration file:

- early-init
- init
- early-fs
- fs
- post-fs
- early-boot
- boot

The meaning of each of these triggers and the commands they consist of will become clearer in the next section, as we look at the main *init.rc* file. For the time being, here's the order in which predefined triggers and built-in actions are executed by *init* after having parsed its configuration files:

1. Run `early-init` commands.
2. *coldboot*: Check that *ueventd* has populated */dev*.
3. Initialize property service's internal data structures.
4. Set up handler for keychords.
5. Initialize the console and display startup text or image.
6. Set up initial properties such as `ro.serialno`, `ro.baseband`, and `ro.carrier`.
7. Run `init` commands.
8. Run `early-fs` commands.
9. Run `fs` commands.
10. Run `post-fs` commands.
11. Start the property service.
12. Prepare to receive `SIGCHLD` signals.
13. Make sure that the property service socket and `SIGCHLD` handler are ready.
14. Run `early-boot` commands.
15. Run `boot` commands.
16. Run all property-triggered commands based on current property values.

Property-based triggers. Actions can also be taken based on property value changes:

```
on property:<name>=<value>
```

Essentially, this allows you to run a set of commands when the property called `name` is set to `value`. A very good example of this is the default *init.rc*'s starting or stopping the *adbd* daemon based on the toggling of the "USB debugging" checkbox in Settings:

```
on property:persist.service.adb.enable=1
    start adbd

on property:persist.service.adb.enable=0
    stop adbd
```

Action commands. After having declared a new action using the on keyword, what's important is what commands are actually executed as part of this action. *init* includes a slew of commands as part of its lexicon. While many of these bear a strong resemblance to their command-line equivalents and you should be able to recognize their use, some are Android-specific. Table 6-18 lists *init*'s commands.

Table 6-18. init's commands in 2.3/Gingerbread

Command	Description
chdir <directory>	Same as *cd* command.
chmod <octal-mode> <path>	Change path's access permissions.
chown <owner> <group> <path>	Change path's ownership.
chroot <directory>	Set process's root directory.
class_start <serviceclass>	Start all services that belong to serviceclass.
class_stop <serviceclass>	Stop all services that belong to serviceclass and disable them.
copy <path> <destination>	Copy a file to destination.
domainname <name>	Set the system's domain name.
exec <path> [<argument>]*	Forks and executes a program. It's suggested to use an init service instead, as this operation is blocking.
export <name> <value>	Set environment variable name to value.
ifup <interface>	Start interface up.
import <filename>	Import an additional *init* config file to the one currently parsed.
insmod <path>	Insert a kernel module.
hostname <name>	Set the system's hostname.
loglevel <level>	Set the current log level.
mkdir <path> [mode] [owner] [group]	Create the path directory with the appropriate permission and ownership.
mount <type> <device> <dir> [<mountoption>]*	Mount device to dir.
restart <service>	Stop and then start service.
setkey <table> <index> <value>	Set a keyboard entry value.
setprop <name> <value>	Set property name to value.
setrlimit <resource> <cur> <max>	Set the resource's rlimit.
start <service>	Start service.
stop <service>	Stop service.
symlink <target> <path>	Create a symbolic link.
sysclktz <mins_west_of_gmt>	Set time zone.
trigger <event>	Start action called event.
wait <path>	Wait until a file appears in the filesystem.
write <path> <string> [<string>]*	Open a file and write strings to it.

Even though many of *init*'s commands resemble command-line equivalents from Toolbox or elsewhere, it's important to note that only those listed in Table 6-18 are recognized. *init* will **not** attempt to issue commands to the command line. Commands that aren't recognized are simply ignored.

4.2/Jelly Bean also has a few additional commands that are recognized by *init*, as you can see in Table 6-19.

Table 6-19. New init commands in 4.2/Jelly Bean

Command	Description
class_reset <serviceclass>	Like class_stop but doesn't disable the services.
load_persist_props	Load persistent properties.
mount_all <path>	Mount all the partitions based on the information found in the *path* file.
restorecon <path>	Restore SELinux context.
rm <path>	Delete file.
rmdir <path>	Delete directory.
setcon <string>	Set security context (SELinux.)
setenforce <value>	Enable or disable security enforcement (SELinux.)
setsebool	Set SELinux Boolean.

As you can see, a number of commands have been added to support SELinux. For more information about SEAndroid, which is an extension of the SELinux work, have a look at the project website (*http://selinuxproject.org/page/SEAndroid*).

Service declarations. *init* refers only to service names and cannot recognize pathnames to files in order to run processes. Therefore, any process that has to be run from a file must first be assigned to a service. As we saw earlier, services are declared this way:

    service <name> <pathname> [<argument>]*

What's important to highlight here is that once this line is parsed, the service will be known by *init* as `name`. The actual name of the binary that is pointed to by `pathname` will itself not be recognized. One of the best examples of that is the Zygote (note that the line is wrapped to fit the page's width in this book):

    service zygote /system/bin/app_process -Xzygote /system/bin --zygote --start
    -system-server

The actual binary being run here is *app_process*. Yet that's not the service being referred to by the rest of the main *init.rc* file. Instead, you'll find references to `zygote`:

    onrestart restart zygote

Service options. Much like actions, the service declaration is often followed by a number of lines that provide more information on the options to use for the service and how to run it. Table 6-20 details those options.

Table 6-20. init's service options

Option	Description
class <name>	This service belongs the class called name, the default class being default.
console	Service requires and runs on console.
critical	If this service crashes five times, reboot into recovery mode.
disabled	Don't automatically start this service. It'll need to be manually started using *start*.
group <groupname> [<group name>]*	Run this service under the given group(s).
ioprio <rt|be|idle> <ioprio 0-7>	Set the service's I/O scheduler and priority.[a]
keycodes <keycode> [<key code>]*	Start the service whenever the given keycodes are activated.
oneshot	Service runs only once. Service is set as disabled on exit.
onrestart <command>	If the service restarts, run command.
seclabel <string>	Set the service's SELinux label; available starting in 4.1/Jelly Bean.
setenv <name> <value>	Set the name environment variable before starting this service.
socket <name> <type> <perm> [<user> [<group>]]	Create a Unix domain socket and pass its file descriptor to the process as it starts.
user <username>	Run this service as username.

[a] Have a look at the man page for ioprio_set() for more information.

Obviously the use of some of these is more obvious than others. Running a service under a certain user or as part of some group should be straightforward. Running a service based on a certain set of key combinations may be less obvious, though. Here's an example of how this is used by the board-specific *.rc* file for the Nexus S (a.k.a. "Crespo") in 2.3/Gingerbread:

```
# bugreport is triggered by holding down volume down, volume up and power
service bugreport /system/bin/dumpstate -d -v -o /sdcard/bugreports/bugreport
    disabled
    oneshot
    keycodes 114 115 116
```

Main init.rc

As we discussed earlier, *init* reads two *.rc* files to figure out its configuration. One of those is provided by default for all boards within the AOSP, and you'll find two versions of that file in Appendix D: one for 2.3/Gingerbread and the other for 4.2/Jelly Bean. I **very strongly** encourage you to read through that appendix, as *init.rc* is the cornerstone of a lot of the system's behavior. If nothing else, have a look at the comments (i.e., lines starting with #). Both default files are in fact commented well enough that you should be able to make sense of their content fairly easily using the earlier tables as guides for specifics.

Some of the operations conducted by *init.rc* are subtle but have profound repercussions on various pieces of Android. It's wise to bookmark the version of the file that's relevant to you and come back to it every so often when you're trying to figure out one thing or another about the system. And while default *init.rc* files are typically an easy read, understanding what specific parts are doing often requires a very solid grasp of the rest of the system and the order in which *init* executes actions.

Always keep in mind that the specific order of actions, commands, and services found in the default *init.rc* file is crucial to the system's operation. You could try to craft your own *init.rc* from scratch, but you'd rapidly find out that a lot of things in the system will break if the steps in the default aren't preserved. Some of the services, for instance, will simply not operate properly unless the appropriate options are used to start them. You are much better off tweaking the default *init.rc* provided with your AOSP or, better yet, adding your own board-specific *.rc* file if you need board-specific actions or services to be started.

Note that not all predefined actions are necessarily in use in your AOSP's default *init.rc*. Neither `early-fs` nor `early-boot` are actually used in 2.3/Gingerbread's, for example. You can therefore use these in your board-specific *.rc* file if you need to preempt commands run in the `fs` or `boot` actions.

Board-specific .rc files

If you need to add board-specific configuration instructions for *init*, the best way is to use an *init.<device_name>.rc* tailored to your system. What it does specifically is up to you. However, I suggest you take a look at the board-specific *.rc* files that are already part of your AOSP. Here are the files from 2.3/Gingerbread, for example:

- *system/core/rootdir/etc/init.goldfish.rc*
- *device/htc/passion/init.mahimahi.rc*
- *device/samsung/crespo4g/init.herring.rc*
- *device/samsung/crespo/init.herring.rc*

Here are the ones in 4.2/Jelly Bean:

- *system/core/rootdir/etc/init.goldfish.rc*
- *build/target/board/vbox_x86/init.vbox_x86.rc*
- *device/asus/tilapia/init.tilapia.rc*
- *device/asus/grouper/init.grouper.rc*
- *device/samsung/tuna/init.tuna.rc*
- *device/ti/panda/init.omap4pandaboard.rc*
- *device/lge/mako/init.mako.rc*

As you'd expect, these files typically contain hardware-specific commands. Very often, for instance, they'll include specific mount instructions for the board. Here's an example from the Crespo-specific file in 2.3/Gingerbread:

```
on fs
    mkdir /efs 0775 radio radio
    mount yaffs2 mtd@efs /efs nosuid nodev
        chmod 770 /efs/bluetooth
        chmod 770 /efs/imei
    mount ext4 /dev/block/platform/s3c-sdhci.0/by-name/system /system wait ro
    mount ext4 /dev/block/platform/s3c-sdhci.0/by-name/userdata /data wait noati
me nosuid nodev
```

As you can see, this mounts */system* and */data* from ext4 partitions found in the onboard eMMC. Another example is the snippet from an earlier section that showed how the *bugreport* command was activated when a certain key combination was pressed on the device.

Again, as I had mentioned earlier, *init* reads both its main *init.rc* and the board-specific *.rc* file before executing any of the actions therein. Hence, by declaring a boot action or an fs action in your board-specific file, the commands therein will be queued up for running right after the commands found in the same action in the main config file. They will, therefore, still run within that action. Hence, commands found in boot actions will run after commands found in fs actions, regardless of which file either set of commands are declared in.

Here's, for example, an *init.coyotepad.rc*:

```
on property:acme.birdradar.enable=1
    start birdradar

service birdradar /system/vendor/bin/bradard -d /system/vendor/etc/rcalibrate.data
    user birdradar
    group birdradar
    disabled
```

This states that the birdradar service should be started whenever the acme.birdradar.enable property is set to 1. In the earlier explanation about Toolbox, we used the *setprop* command on the command line to set the property to 1. Had the above *init.coy otepad.rc* been part of the system at startup, that previous *setprop* command would have therefore resulted in *bradard* being started.

> ### What about init.<device_name>.sh?
>
> In some cases, it makes sense to have a shell script run in addition to the commands run by *init*'s configuration files. The emulator, for instance, relies on a *init.goldfish.sh* found in */system/etc*. Despite the name of the file, *init* itself doesn't recognize such scripts and has no code that looks for them. Instead, board-specific *.rc* files can be made to run shell scripts like they'd run any other service. Here's how *init.goldfish.rc* gets *init.goldfish.sh* to be executed:
>
>   service goldfish-setup /system/etc/init.goldfish.sh
>   oneshot
>
> In this specific case, the shell script runs commands that are available on the shell but aren't part of *init*'s lexicon. And that is in fact a very good reason for having a shell script such as this if you need one.

Global Properties

Though I've mentioned global properties a number of times already, we've yet to take a deeper look at that aspect of Android. As I hinted at earlier, global properties are an important part of Android's overall architecture. As a somewhat distant cousin of the infamous Windows Registry, Android's global properties very often serve as a trivial way of sharing important yet relatively stable values globally among all parts of the stack.

Theory of operation

As I mentioned earlier, *init* maintains a property service as part of its other responsibilities. As you can see in Figure 6-4, there are two ways that this property service is exposed to the rest of the system:

/dev/socket/property_service
    This is a Unix domain socket that processes can open to talk to the property service and have it set and/or change the value of global properties.

/dev/__properties__
    This is an "invisible" file (i.e., you won't see it in */dev* if you look for it) that is created within the tmpfs-mounted */dev* and that is memory-mapped into the address space

of all services started by *init*. It's through this mapped region that descendants of *init* (i.e., all user-space processes in the system) can read global properties.

/dev/__properties__'s Invisibility

You won't find */dev/__properties__* in the filesystem because of the way *init* handles the file. Here's what it actually does to the file during initialization:

1. Creates */dev/__properties__* in read-write mode.
2. Sets its size to a desired global properties *workspace* size.
3. Memory-maps the file into *init*'s address space.
4. Closes the file descriptor.
5. Opens the file as read-only.
6. Deletes the file from the filesystem.

By deleting the file as a last step, anyone looking into */dev* won't actually see the file. However, since the file was memory-mapped while it was still open in read-write mode, *init*'s property service is able to continue writing to the memory-mapped file. Also, since it was opened in read-only mode before it was deleted, *init* also has a file descriptor it can pass to its children, so they can in turn memory-map the file, which will remain read-only for them.

As explained in the sidebar, the property service essentially maintains a RAM-based *workspace* where it stores all global properties. Because of the way it's set up, only the property service can write to this workspace, though any process can read from it. Hence we have a single-writer/multiple-readers configuration. This design allows the property service to apply permission checks on the write requests submitted to it through the */dev/socket/property_service* Unix domain socket. The specific permissions required to set certain global properties are hardcoded. Here's the relevant snippet from 2.3/ Gingerbread's *system/core/init/property_service.c*:

```
/* White list of permissions for setting property services. */
struct {
    const char *prefix;
    unsigned int uid;
    unsigned int gid;
} property_perms[] = {
    { "net.rmnet0.",   AID_RADIO,   0 },
    { "net.gprs.",     AID_RADIO,   0 },
    { "net.ppp",       AID_RADIO,   0 },
    { "ril.",          AID_RADIO,   0 },
    { "gsm.",          AID_RADIO,   0 },
    { "persist.radio", AID_RADIO,   0 },
```

```
        { "net.dns",          AID_RADIO,    0 },
        { "net.",             AID_SYSTEM,   0 },
        { "dev.",             AID_SYSTEM,   0 },
        { "runtime.",         AID_SYSTEM,   0 },
        { "hw.",              AID_SYSTEM,   0 },
        { "sys.",             AID_SYSTEM,   0 },
        { "service.",         AID_SYSTEM,   0 },
        { "wlan.",            AID_SYSTEM,   0 },
        { "dhcp.",            AID_SYSTEM,   0 },
        { "dhcp.",            AID_DHCP,     0 },
        { "vpn.",             AID_SYSTEM,   0 },
        { "vpn.",             AID_VPN,      0 },
        { "debug.",           AID_SHELL,    0 },
        { "log.",             AID_SHELL,    0 },
        { "service.adb.root", AID_SHELL,    0 },
        { "persist.sys.",     AID_SYSTEM,   0 },
        { "persist.service.", AID_SYSTEM,   0 },
        { "persist.security.",AID_SYSTEM,   0 },
        { NULL, 0, 0 }
    };
```

To understand the meaning of each AID_* UID, please refer to the discussion about the *android_filesystem_config.h* file in "The Build System and the Filesystem" on page 185 where user IDs and other core filesystem properties are defined. For instance, the above says that only processes running as the system user can change properties that start with sys. or hw., while only processes running as the radio user—the *rild*, for instance—can change properties that start with ril. or gsm.

Note that processes running as root can change any property they wish. Note also that in the case of properties whose names starts with ro., these three characters are stripped from the name before permissions are checked with the above array. Such properties can be set only once, however. Trying to change the value of an existing property whose name starts with ro. will fail. Furthermore, if a permission isn't explicitly granted by the above array for a given property (or property set) to the user under which a process is running, that process won't be allowed to set that property. Here's an attempt to set acme.birdradar.enable from a non-root shell for example:

```
$ setprop acme.birdradar.enable 1
[ 1992.292414] init: sys_prop: permission denied uid:2000  name:acme.birdradar
.enable
```

As we discussed in the Toolbox section, you can use *getprop*, *setprop*, and *watchprops* to interact with the property service from the command line. You can also interact with the property service from within the code you build as part of the AOSP. If you're coding in Java, have a look at the *frameworks/base/core/java/android/os/SystemProperties.java* class. To use this class, you would need to import android.os.SystemProperties. If you're coding in C, have a look at *system/core/include/cutils/properties.h*. To use the functions in this header, you need to include <cutils/properties.h>.

Global properties aren't accessible through the regular app development API exposed by the SDK.

Nomenclature and sets

As you likely noticed from all previous discussions on global properties, they seem to follow a certain naming convention where each part of the name is separated by a period character (.), with each part of the name following the period, further narrowing the subcategory to which the property belongs. Beyond that, there are few conventions. Of course, the permissions array we saw earlier somewhat dictates a base set of root categories. And quite a few properties are created as part of the build system, as we'll see shortly. There are also a few *special* properties worth keeping in mind. Still, each device has its own specific set of global properties. There is, therefore, no definitive dictionary or official list of global properties that are to be expected across Android devices.

There's nothing stopping you from creating your own set of global properties specifically for your embedded system. Up to now, I've used the acme.birdradar.enable property to illustrate some of the examples. I could very much have a whole slew of acme.* properties, each used for a separate purpose in my system. You can also modify some of the existing global properties as needed for your purpose. Make sure you fully investigate how a specific global property you modify is used by the rest of Android, as some of these properties are read or set by vastly different parts of the stack. A good *grep* across the entire codebase for the property should rapidly help you isolate its users.

You should use *getprop* after the initial boot of your system to get your device's base list of properties. Also, you can look at the default list of properties loaded at startup from property files. We'll take a look at those in the next section.

There are, as I said, some special properties, as well as some properties that are processed differently based on their prefixes:

ro.*
> Properties that start with this prefix are meant to be read-only. Hence, they can be set only once in the system's lifetime. The only way to change their value is to change the source of the information to which they are set and reboot the system. Such is the case for ro.hardware and ro.build.id, for example.

persist.*
> Properties marked with this prefix are committed to persistent storage each time they are set. Such is the case for persist.service.adb.enable, which is used to start/stop *adbd*.

ctl.*
> There's a `ctl.start` and a `ctl.stop`, and setting them doesn't actually result in any property being saved to the global set of properties. Instead, when the property service receives a request to set either of these, it starts/stops the service whose name is provided as the value for the property. The Surface Flinger, for instance, does this as part of its startup:
>
> ```
> property_set("ctl.start", "bootanim");
> ```
>
> This effectively results in the `bootanim` service being started by *init*. The `bootanim` service and its options are described in the *init.rc* file we covered earlier. Toolbox's *start* and *stop* also rely on `ctl.*` to start/stop services.

net.change
> Whenever a `net.*` property is changed, `net.change` is set to the name of that property. Hence, `net.change` always contains the name of the last `net.*` property that was modified.

Storage

There isn't a single location in which global properties are stored or from which they're set. Instead, different pieces of the system are responsible for setting different sets of global properties, and several system parts are involved in creating the final set of global properties found in any single Android device.

The build system. Two property files are generated by the build system:

/system/build.prop
> This one contains information about the build itself, such as the version of Android and the date it was built.

/default.prop
> This one contains default values for certain key properties, such as the `persist.service.adb.enable` property that we saw earlier.

Both of these files are found in the target's root filesystem for the initial boot and serve as the base set of properties for the system. You can find them in the *root/* and *system/* subdirectories of *out/target/product/PRODUCT_DEVICE/*.

The files contain one-liner name-value pairs. They're read and parsed by the property service started early during *init*'s own startup. Most of the content of these files is generated by the core AOSP build code in *build/core/*. Still, as in the following snippet from Crespo's makefiles in 2.3/Gingerbread, some of it is device specific:

```
PRODUCT_PROPERTY_OVERRIDES += \
    wifi.interface=eth0 \
    wifi.supplicant_scan_interval=15 \
    dalvik.vm.heapsize=32m
```

Additional property files. In addition to the files generated by the build system, you can add your own target-specific */system/default.prop* and device-specific */data/local.prop*, both of which will be read by the property service alongside the files generated by the build system we just discussed.

.rc files. As we saw earlier, both the *init.rc* file and *init.<device_name>.rc* can set global properties. *init.rc* in fact sets quite a few crucial global properties.

Code. Some parts of the code also set properties. The Connectivity Service, for instance, does this:

```
SystemProperties.set("net.hostname", name);
```

To confuse things even further, some parts of the code attempt to read global properties and apply defaults if the value isn't found. The following is from *frameworks/base/core/jni/AndroidRuntime.cpp*:

```
property_get("dalvik.vm.heapsize", heapsizeOptsBuf+4, "16m");
```

In this case, the caller attempts to get `dalvik.vm.heapsize`, and if it isn't found, the value `16m` is used as the default.

/data/property. All the storage methods we've seen thus far require manual intervention to either make changes to the AOSP before it's built or to edit files on the device. Obviously, the system needs to be able to automatically update values at runtime and have them available at the next reboot. That's the role of the entries in the */data/property* directory. Indeed, any property that starts with `persist.` is stored as an individual file in that directory. Each of the files there contains the value assigned to the property. Hence, the */data/property/persist.service.adb.enable* file contains the value of `persist.service.adb.enable`.

Properties found in */data/property* are read by the property service at startup and restored. As I mentioned earlier when discussing Toolbox's *setprop*, the only way to destroy a persistent stored property is to delete its file from */data/property*.

ueventd

As discussed earlier, *init* includes functionality to handle kernel hotplug events. When the */init* binary is invoked through the */sbin/ueventd* symbolic link, it immediately switches its identity from running as the regular *init* to running as *ueventd*. Figure 6-5 illustrates *ueventd*'s operation.

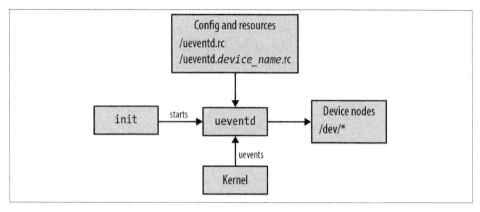

Figure 6-5. Android's ueventd

ueventd is one of the very first services started by the default *init.rc*. It proceeds to read its main configuration files, */ueventd.rc* and */ueventd.<device_name>.rc*,[7] replays all kernel uevents (i.e., hotplug events), and then waits, listening for all future uevents. Kernel uevents are delivered to *ueventd* through a netlink socket, a common way for certain kernel functionalities to communicate with user-space tools and daemons.

Based on the events *ueventd* receives and its configuration files, it automatically creates device node entries in */dev*. And since the latter is mounted as a tmpfs filesystem, and therefore lives only in RAM, these entries are re-created from scratch, based on *ueventd*'s configuration files, at every reboot. The key to *ueventd*'s operation, therefore, is its configuration files.

Unlike *init*, *ueventd*'s configuration files have a rather simple format. Essentially, every device entry is described with a one-liner such as this:

```
/dev/<node>          <mode>    <user>      <group>
```

When a uevent corresponding to node occurs, *ueventd* creates */dev/node* with access permissions set to mode and assigns the entry to user/group. Permissions and ownership are very important, since key daemons and services must have access to relevant */dev* entries in order to operate properly. The System Server, for instance, runs as the sys tem user.

Here's a snippet from the default *ueventd.rc* from 2.3/Gingerbread, for example:

```
/dev/null             0666    root        root
/dev/zero             0666    root        root
/dev/full             0666    root        root
/dev/ptmx             0666    root        root
/dev/tty              0666    root        root
```

---

7. This file's naming is similar to that of the */init.<device_name>.rc* we saw earlier.

```
...
# these should not be world writable
/dev/diag                    0660    radio       radio
/dev/diag_arm9               0660    radio       radio
/dev/android_adb             0660    adb         adb
/dev/android_adb_enable      0660    adb         adb
/dev/ttyMSM0                 0600    bluetooth   bluetooth
/dev/ttyHS0                  0600    bluetooth   bluetooth
/dev/uinput                  0660    system      bluetooth
/dev/alarm                   0664    system      radio
/dev/tty0                    0660    root        system
/dev/graphics/*              0660    root        graphics
/dev/msm_hw3dm               0660    system      graphics
/dev/input/*                 0660    root        input
/dev/eac                     0660    root        audio
...
```

As with *init*, you should put your board-specific node entries in *ueventd.<device_name>.rc*. Here's a device entry from *ueventd.coyotepad.rc*, for example:

```
/dev/bradar                  0660    system      birdradar
```

Note that some uevents might require *ueventd* to load firmware files on behalf of the kernel. There's no configuration option available for that in *ueventd*'s configuration files. Instead, make sure those firmware files are in either */etc/firmware* or */vendor/firmware*. In the case of the CoyotePad, for instance, we put *rfirmware.bin* in */system/vendor/firmware* using PRODUCT_COPY_FILES.

Boot Logo

Not counting whatever the device's bootloader might display at startup, Android devices' screens typically go through four stages during boot:

Kernel boot screen
Usually, an Android device won't show the kernel boot messages to its LCD screen during boot. Instead, the kernel might either maintain the screen black until *init* starts, or it might display a static logo, built as part of the kernel image, to the framebuffer. Any such display is beyond the scope of this book.

Init boot logo
This is a text string or an image displayed very early by *init* while it initializes the console. This section's purpose is to discuss what *init* displays here.

Boot animation
This is a series of animated images, possibly a loop, that displays during the Surface Flinger's start up. We'll discuss the boot animation when we cover the Java userspace later.

Home screen
> This is the starting screen of the Launcher, which is activated at the complete end of the boot sequence. You'll need to dig into the Launcher's sources if you'd like to customize what it displays.

If you refer back to the earlier explanation in "Configuration Files" on page 230 of the execution order enforced by *init* on predefined actions and built-in commands, you'll notice that the fifth step is initializing the console and display startup text or image. During this step, *init* attempts to load a logo image from the */initlogo.rle* file and display it to the screen. If it doesn't find such a file, it displays the familiar text string that is displayed by the emulator as it starts, as shown in Figure 6-6.

Figure 6-6. init's boot logo

If you'd like to change that string, have a look at the `console_init_action()` in *system/core/init/init.c*. If you'd like to have a graphic logo to display instead of just text, you'll need to create a proper *initlogo.rle*. Let's see how that's done.

First, you'll need to figure out the screen size of your device. For instance, the emulator's default resolution when started from the AOSP's command line after build is 320 by 480 pixels. Assuming you have a PNG of that size, you first need to convert it to the format recognized by *init*. Two tools on the host are required to do that: *convert*, which is part of the ImageMagick (*http://www.imagemagick.org*) package, and *rgb2565*, which is built as part of the AOSP:[8]

```
$ cd device/acme/coyotepad
$ convert -depth 8 acmelogo.png rgb:acmelogo.raw
$ rgb2565 -rle < acmelogo.raw > acmelogo.rle
153600 pixels
```

This will take the *acmelogo.png* and convert it into an *acmelogo.rle*, which you can then copy by modifying the CoyotePad's *full_coyote.mk* to add this snippet:

```
PRODUCT_COPY_FILES += \
        device/acme/coyotepad/acmelogo.rle:root/initlogo.rle
```

After you rebuild the AOSP, update your device, and restart it, you'll see the logo instead of the previous text string, as illustrated in Figure 6-7.

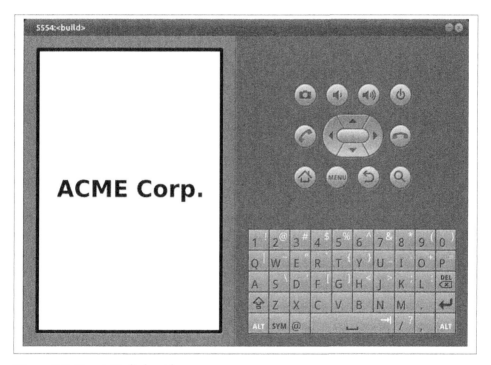

Figure 6-7. CoyotePad's boot logo

Generally, the LCD screen will then remain unchanged until the Surface Flinger starts and launches the boot animation while the rest of the system services are starting.

---

8. Remember that you'll need to run *build/envsetup.sh* and *lunch* before the paths are properly set to use host tools built as part of the AOSP.

CHAPTER 7
Android Framework

Ultimately, your goal is to get your embedded system to run the Android environment users and developers are accustomed to, not simply the native user-space we just covered. That includes not only the full set of system services and the packages that provide the standard APIs used by app developers, but also some less visible components, such as a set of native daemons that support the system services and the Hardware Abstraction Layer. This chapter will cover how the Android Framework operates on top of the native user-space and will discuss how to interact with and customize it.

Note that unlike the previously discussed components of Android, whose behavior can be modified in a number of ways, most of the Android Framework has to be used as is. You can't, for instance, pick and choose which system services to run, as they aren't started from a script or based on a configuration file. Instead, modifying the Framework typically requires diving into its sources and/or adding your own code to customize its behavior.

Such customization work therefore requires becoming intimately familiar with Android's sources and is inherently version dependent. Still, we'll try to cover enough of the essentials to enable you to start navigating Android's internals on your own. Nevertheless, expect this to be the start of a long-term endeavor, as Android's sources are fairly big, and new releases come out at a very rapid pace.

> ### What Exactly Is the "Android Framework"?
>
> If you refer back to Figure 2-1, the Android Framework includes the android.* packages, the System Services, the Android Runtime, and some of the native daemons. Sourcewise, the Android Framework is typically composed of all the code located in the *frameworks/* directory of the AOSP.
>
> At a certain level, I'm using "Android Framework" here to designate practically everything "Android" that runs on top of native user-space. So my explanations here do

sometimes go beyond just *frameworks/*. Namely, I will discuss such things as Dalvik and the HAL, which are intrinsic to the Android Framework.

Kick-Starting the Framework

We closed the last chapter on the *init* command, and how it can be configured and used. I only briefly hinted, however, at how the Android Framework is started by way of the Zygote when describing the default *init.rc*. There is of course much to say on this topic, as we'll see shortly. Much of what I described in the last chapter can be easily compared to components that exist in the embedded Linux world; however, very little of what follows has any such equivalent. Indeed, the Android developers' contribution to the world of mobile is the stack they built on top of a BSD/ASL-licensed embedded Linux equivalent.

> ### Building the AOSP Without the Framework
>
> As odd as it may seem, there are cases where you actually may want to build the AOSP without all the fancy, Java-based system services and apps that Android is most widely known for. Whether it be to run "Android" on a "headless" system or simply because you're in the midst of a board bringup and would like a minimal build of the AOSP to get just the basic tools and environment of the native user-space, there's an AOSP build for you: Tiny Android.
>
> To make the AOSP generate Tiny Android, you just need to go to the AOSP's source directory and type this:
>
> ```
> $ BUILD_TINY_ANDROID=true make -j16
> ```
>
> This will get you a set of output images with the minimal set of Android components for a functional native Android user-space to run with a kernel. Mainly, you'll get Toolbox, Bionic, *init*, *adbd*, *logcat*, *sh*, and a few other key binaries and libraries. No part of the Android Framework, such as the system services or any of the apps, will be included in those images.
>
> It's questionable whether this is "Android" anymore, but in some cases it's exactly what you're looking for. Whether you want to refer to the end result as "Android" is really up to you. Hey, apparently beauty is in the eye of the beholder.

Core Building Blocks

The Framework's operation relies on a handful of key building blocks: the Service Manager, the Android Runtime, the Zygote, and Dalvik. Without these, none of the components that make up what we know to be Android work. We've already covered most of these and their role in the system's startup in Chapter 2. I encourage you to go back to that chapter for an in-depth discussion, but let's still recap the highlights here, especially now that we've just looked at *init* and its scripts. You may, in fact, want a finger on the pages from Appendix D about the main *init.rc* file as you read the following explanations.

One of the first services started by *init* is the *servicemanager*. As I explained earlier, this is the "Yellow Pages" or the directory of all system services running. Obviously, at the time it starts no system services have started, but it needs to be available very early on so that system services that do start can register with it and therefore become visible to the rest of the system.

If the *servicemanager* isn't running, none of the system services will be able to advertise themselves, and the Framework simply will not work. Hence, the *servicemanager* is not an optional component, and its ordering in the *init.rc* file isn't subject to customization. You must leave it exactly where it is in the main *init.rc* file with the options that are specified for it by default.

The next core component to get started is the Zygote. Here's the relevant line from *init.rc*:

```
service zygote /system/bin/app_process -Xzygote /system/bin --zygote --start-sys
tem-server
```

There is a lot happening in that simple line. First, note that what's actually getting run is this *app_process* command. Here's its formal parameter list:

```
Usage: app_process [java-options] cmd-dir start-class-name [options]
```

app_process is a little-known command that packs a punch. It lets you start a new Dalvik VM for running Android code straight from the command line. This doesn't mean you can use it to start regular Android apps from the command line; in fact you can't use it for that purpose, but you'll soon learn about a command that does: *am*. However, some key system components and tools must be started from the command line without a reference to any existing Dalvik VM instance. The Zygote is one of these, as it's the first Dalvik process to run; *am* and *pm* are two more, which we'll cover later.

To do its magic, *app_process* relies on the Android Runtime. Packaged as a shared library, *libandroid_runtime.so*, the Android Runtime is capable of starting and managing a Dalvik VM for the purpose of running Android-type code. Among other things, it preloads this VM with a number of libraries that are typically used by any code that relies on the Android APIs. This includes all the native calls, which are required by any of the Android Framework's Java code. These are registered with the VM so it can find them whenever a Java-coded Android Framework package calls on one of its native functions.

The Runtime also includes functions for facilitating operations typically done for all Android-type applications running on Dalvik. You can, in fact, consider Dalvik to be a very raw, low-level VM that doesn't assume you're running Android-type code on top of it. To run Android-type code on top of Dalvik, the Runtime starts Dalvik with parameters specifically tailored for its use to run Java code that relies on the Android Java APIs—either those publicly documented in the developer documentation and made available through the SDK, or internal APIs available only as part of building internal Android code within the AOSP.

Furthermore, the Runtime relies on many native user-space functionalities. For instance, it takes into account some of the init-maintained global properties in order to gate the starting of the Dalvik VM, and it uses Android's logging functions to log the progress of the Dalvik VM's initialization. In addition to setting up the parameters used to start the Dalvik VM used to run Java code, the Runtime also initializes some key aspects of the Java and Android environment before calling the code's main( ) method. Most importantly, it provides a default exception handler for all threads running on the just-instantiated VM.

Note that the Runtime doesn't preload classes: That's something the Zygote does when it sets up the system for running Android apps. And since each use of the *app_process* command results in starting a separate VM, all non-Zygote instances of Dalvik will load classes on demand, not before your code starts running.

Dalvik's Global Properties

In addition to the global properties maintained by *init* that we discussed in the last chapter, Dalvik continues to provide the property system found in Java through java.lang.System. As such, if you're browsing some of the system services' sources, you might notice calls to System.getProperty( ) or System.setProperty( ). Note that those calls and the underlying set of properties are completely independent from *init*'s global properties.

The Package Manager Service, for instance, reads the java.boot.class.path at startup. Yet, if you use *getprop* on the command line, you won't find this property as part of the list of properties returned by *init*. Instead, such variables are maintained within each Dalvik instance for retrieval and/or use by running Java code. The specific java.boot.class.path, for instance, is set in *dalvik/vm/Properties.c* using the BOOT CLASSPATH variable set in *init.rc*.

You can find out more about Java System Properties in Java's official documentation (*http://docs.oracle.com/javase/tutorial/essential/environment/sysprop.html*). Note that the semantics of the variable names used by *init*'s global properties are very similar to those used by Java System Properties.

Once it's started, a Java class launched using *app_process* can start using "regular" Android APIs and talk to existing system services. If it's built as part of the AOSP, it can use many of the android.* packages available to it at build time. The *am* and *pm* commands, for instance, do exactly that. It follows that you, too, could write your own command-line tool completely in Java, using the Android API, and have it start separately from the rest of the Framework. In other words, it would be started and would run independently of the Zygote and everything that the Zygote causes to start as part of its own initialization.

But this still won't let you write a regular Android app that is started by *app_process*. Android apps can be started only by the Activity Manager using intents, and the Activity Manager is itself started as part of the rest of the system services once the Zygote itself is started. Which brings the discussion back to the startup of the Zygote.

For the Zygote to start properly and have it start the System Server, you must leave its corresponding *app_process* line intact in *init.rc*, in its default location. There's nothing that you can configure about the Zygote's startup. You can, however, influence the way the Android Runtime starts any of its Dalvik VMs by modifying some of the system's global properties. Have a look at the AndroidRuntime::startVm(JavaVM** pJavaVM, JNIEnv** pEnv) function in *frameworks/base/core/jni/AndroidRuntime.cpp* in either 2.3/Gingerbread or 4.2/Jelly Bean to see which global properties are read by the Android Runtime as it prepares to start a new VM. Note that any use of these properties to influence the setup of Dalvik VMs is likely to be version specific.

Once the Zygote's VM is started, the com.android.internal.os.ZygoteInit class's main() function is called, and it will preload the entire set of Android packages, proceed to start the System Server, and then start looping around and listening for connections from the Activity Manager asking it to fork and start new Android apps. Again, there is nothing to be customized here unless you can see something relevant to you in the list of parameters used to start the System Server in the startSystemServer() function in *frameworks/base/core/java/com/android/internal/os/ZygoteInit.java*. My recommendation is to leave this as is unless you have a very strong understanding of Android's internals.

Disabling the Zygote

While you can't configure what the Zygote does at startup, you can nevertheless disable its startup entirely by adding the disabled option to its section in *init.rc*. Here's how this is done in 2.3/Gingerbread:

```
service zygote /system/bin/app_process -Xzygote /system/bin --zygote
--start-system-server
    socket zygote stream 666
    onrestart write /sys/android_power/request_state wake
    onrestart write /sys/power/state on
    onrestart restart media
```

```
    onrestart restart netd
disabled
```

This will effectively prevent *init* from starting the Zygote at boot time, so none of the Android Framework's parts will start, including the System Server. This may be very useful if you're in the process of debugging critical system errors or developing one of the HAL modules, and you must manually set up debugging tools, load files, or monitor system behavior **before** key system services start up.

You can then start the Zygote, and the rest of the system:

```
# start zygote
```

System Services

As we saw in the last section, the System Server is started as part of the Zygote's startup, and we'll continue delving into that part of the process in this section. However, and as was discussed in Chapter 2, there are also system services started from processes other than the System Server, and we'll discuss those in this section.

Starting with 4.0/Ice-Cream Sandwich, the very first system service to get started is the Surface Flinger. Up to 2.3/Gingerbread, it had been started as part of the System Server, but with 4.0/Ice-Cream Sandwich, it's started right before the Zygote and runs independently from the System Server and the rest of the system services. Here's the relevant snippet that precedes the Zygote's entry in *init.rc* in 4.2/Jelly Bean:

```
service surfaceflinger /system/bin/surfaceflinger
    class main
    user system
    group graphics drmrpc
    onrestart restart zygote
```

The Surface Flinger's sources are in *frameworks/base/services/surfaceflinger/* in 2.3/Gingerbread and *frameworks/native/services/surfaceflinger/* in 4.2/Jelly Bean. Its role is to composite the drawing surfaces used by apps into the final image displayed to the user. As such, it's one of Android's most fundamental building blocks.

In Android 4.0, because the Surface Flinger is started before the Zygote, the system's boot animation comes up much faster than in earlier versions. We'll discuss the boot animation later in this chapter.

To start the System Server, the Zygote forks and runs the `com.android.server.SystemServer` class' `main()` function. The latter loads the *libandroid_servers.so* library, which contains the JNI parts required by some of the system services and then invokes native code in *frameworks/base/cmds/system_server/library/system_init.cpp*, which starts C-coded system services that run in the `system_server` process. In 2.3/Gingerbread, this includes the Surface Flinger and the Sensor Service. In 4.2/Jelly Bean, however, the Surface Flinger is started separately, as we just saw, and the only C-coded system service started by `system_server` is the Sensor Service.

The System Server then goes back to Java and starts initializing the critical system services such as the Power Manager, Activity Manager, and Package Manager. It then continues to initialize all the system services it hosts and registers them with the Service Manager. This is all done in code in *frameworks/base/services/java/com/android/server/SystemServer.java*. None of this is configurable. It's all hardcoded into *SystemServer.java*, and there are no flags or parameters you can pass to tell the System Server not to start some of the system services. If you want to disable any, you'll have to go in by hand and comment out the corresponding code.

The system services are interdependent, and almost all of Android's parts, including the Android API, assume that all the system services built into the AOSP are available at all times. As I mentioned in Chapter 2, as a whole, the system services form an object-oriented OS built on top of Linux—and the parts of that OS weren't built with modularity in mind. So if you take one of the system services away, it's fair to assume that some of Android's parts will start breaking under your feet.

That doesn't mean it can't be done, though. As part of a presentation titled "Headless Android" (*http://www.opersys.com/blog/headless-android-1*) at the 2012 Android Builders Summit, I showed how I successfully disabled the Surface Flinger, the Window Manager, and a couple of other key system services, to run the full Android stack on a headless system. As I warned in that presentation, that work was very much a proof of concept and would require a lot more effort to be production ready.[1]

So, by all means, feel free to tinker around, but you've been warned that if you're going to play this deep in Android's guts, you'd better saddle up.

What's /system/bin/system_server?

You might notice while browsing your target's root filesystem that there's a binary called *system_server* in */system/bin*. That binary, however, has nothing to do with the startup of the System Server or with any of the system services. It's unclear what purpose, if any, this binary has. It's very likely that this is a legacy utility from Android's early days.

This factoid is often a source of confusion, because a quick look at the list of binaries and the output of *ps* may lead you to believe that the `system_server` process is in fact

---

1. Interestingly, a new `ro.config.headless` global property has been added to the official AOSP releases since 4.1/Jelly Bean. That property appears to allow the execution of the stack without a user interface.

> started by the *system_server* command. I was in fact very skeptical of my own reading of the sources on that matter and posted a question about it to the android-building mailing list. The ensuing response (*https://groups.google.com/forum/?fromgroups=#!topic/android-platform/x2ToX7x5Yzw*) seems to confirm my reading of the sources, however.

In addition to the Surface Flinger and the system services started by the System Server, another set of system services stems from the starting of *mediaserver*. Here's the relevant snippet from 2.3/Gingerbread's *init.rc* (4.2/Jelly Bean's is practically identical):

```
service media /system/bin/mediaserver
    user media
    group system audio camera graphics inet net_bt net_bt_admin net_raw
    ioprio rt 4
```

The *mediaserver*, whose sources are in *frameworks/base/media* in 2.3/Gingerbread and *frameworks/av/media* in 4.2/Jelly Bean, starts the following system services: Audio Flinger, Media Player Service, Camera Service, and Audio Policy Service. Again, none of this is configurable, and it's recommended that you leave the relevant *init.rc* portions untouched unless you fully understand the implications of your modifications. For instance, if you try to remove the startup of the *mediaplayer* service from *init.rc* or use the `disabled` option to prevent it from starting, you will notice messages such as these in *logcat*'s output:

```
...
I/ServiceManager(   56): Waiting for service media.audio_policy...
I/ServiceManager(   56): Waiting for service media.audio_policy...
I/ServiceManager(   56): Waiting for service media.audio_policy...
W/AudioSystem(   56): AudioPolicyService not published, waiting...
I/ServiceManager(   56): Waiting for service media.audio_policy...
I/ServiceManager(   56): Waiting for service media.audio_policy...
...
```

And the system will hang and continue to print out those messages until the *mediaserver* is started.

Note that the *mediaserver* is one of the only init services that uses the `ioprio` option. Presumably—and there's unfortunately no official documentation to confirm this—this is used to make sure that media playback has an appropriate priority to avoid choppy playback.

There is finally one odd player in this game, the Phone app, which provides the Phone system service. Generally speaking, apps are the wrong place to put system services because apps are lifecycle managed and can therefore be stopped and restarted at will. System services, on the other hand, are supposed to live from boot to reboot and cannot therefore be stopped midstream without affecting the rest of the system. The Phone app is different, however, because its manifest file has the `android:persistent` property of the `application` XML element set to `true`. This indicates to the system that this app should not be lifecycle managed, which therefore enables it to house a system service. It will also lead to this app being automatically started as part of the initialization of the Activity Manager.

Again, there's nothing typically configurable about the Phone app's startup. You can, however, relatively easily remove the Phone app from the list of apps built into the AOSP. The result, however, will be that any part of the system depending on that system service will fail to function correctly. Again, you might as well leave it in. If you want to remove the dialer icon from the home screen, then what you actually want to remove is the Contacts app. As counterintuitive as it may sound, the typical phone dialer Android users are accustomed to isn't part of the Phone app; it's part of the Contacts app.

 Another example of an app that houses a system service is the NFC app found in *packages/apps/Nfc/*.

The Phone app way of providing a system service is very interesting, because it opens the door for us to emulate its example and to add system services as apps within our own *device/acme/coyotepad/* directory—without having to modify the sources of the default system services in *frameworks/base/services/*.

Boot Animation

As I explained when discussing the boot logo in the previous chapter, Android's LCD goes through four stages during boot. One of those is a boot animation. Here's the corresponding entry in 2.3/Gingerbread's *init.rc* (the one in 4.2/Jelly Bean is practically identical):

```
service bootanim /system/bin/bootanimation
    user graphics
    group graphics
    disabled
    oneshot
```

Notice that this service is marked as `disabled`. Hence, *init* won't actually start this right away. Instead, it must be explicitly started somewhere else. In this case, it's the Surface Flinger that actually starts the boot animation *after* it has finished its own initialization by setting the `ctl.start` global property. Here's code from the `SurfaceFlinger::ready ToRun()` function in 2.3/Gingerbread's *frameworks/base/services/surfaceflinger/SurfaceFlinger.cpp*:

```
// start boot animation
property_set("ctl.start", "bootanim");
```

The code in 4.2/Jelly Bean's *frameworks/native/services/surfaceflinger/SurfaceFlinger.cpp* does effectively the same thing:

```
...
void SurfaceFlinger::startBootAnim() {
    // start boot animation
    property_set("service.bootanim.exit", "0");
    property_set("ctl.start", "bootanim");
}
...
status_t SurfaceFlinger::readyToRun()
{
...

    // start boot animation
    startBootAnim();

    return NO_ERROR;
}
...
```

And given that the Surface Flinger is one of the first system services started—if not the first—the boot animation ends up continuously displaying while critical parts of the system are initializing. Typically, it will stop only when the phone's home screen finally comes to the fore. We'll take a look at some of the things happening during the boot animation shortly.

As you can see in the previous *init.rc* snippet, the `bootanim` service corresponds to the *bootanimation* binary. The latter's sources are in *frameworks/base/cmds/bootanimation/*, and if you dig into them you'll notice that this utility talks directly through Binder to the Surface Flinger in order to render its animation; hence the need for the Surface Flinger to be live before the animation can start. Figure 7-1 illustrates the default Android boot animation displayed by *bootanimation*, with the moving light reflection projected on the Android logo.

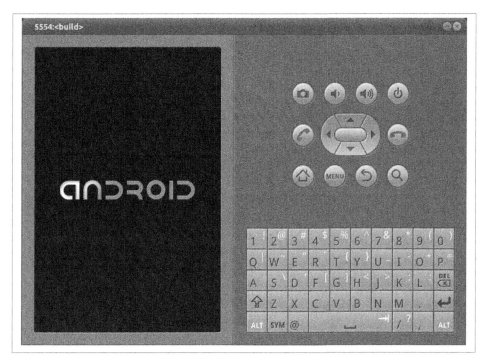

Figure 7-1. Default boot animation

bootanimation actually has two modes of operation. In one mode it creates the default Android logo boot animation using the images in *frameworks/base/core/res/assets/images/*. It's likely best not to try modifying the boot animation by touching these files. Instead, by providing either */data/local/bootanimation.zip* or */system/media/bootanimation.zip*, you will force *bootanimation* to enter its other mode of operation, where it uses the content of one of those ZIP files to render a boot animation. It's worth taking some time to see how that can be done, even though a book is not the ideal medium for illustrating a running animation.

bootanimation.zip

The *bootanimation.zip* is a regular, **uncompressed** ZIP file with at least a *desc.txt* file at the top-level directory inside and a bunch of directories containing PNG files. The latter are animated in sequence according to the rules in the *desc.txt* file. Note that *bootanimation* doesn't support anything but PNG files. Here are the semantics of the *desc.txt* file:

```
<width> <height> <fps>
p <count> <pause> <path>
p <count> <pause> <path>
```

Note that the file's format is very simplistic and doesn't allow for any fluff. So stick to the above semantics as is. The first line indicates the width, height, and frame rate (frames per second) for the animation. Each subsequent line is a *part* of the animation. For each part, you must provide the number of times this part is played (count), the number of frames to pause after each time the part is played (pause), and the directory where that part of the animation is located (path). Parts are played in the order they appear in the *desc.txt*.

Each animation part, and therefore the associated directory, is made of several PNG files, with filenames being a string representing the sequential number of that frame in the full sequence. Files could, for instance, be named *001.png*, *002.png*, *003.png*, etc. If the count is set to zero, the part will loop playing until the system has finished booting and the Launcher starts. Typically, initial parts are likely to have a count of 1, while the last part usually has a count of 0, so it continues playing until the boot is done.

The best way to create your own boot animation is to look at the existing *bootanimation.zip* files that have been created by others. If you look for that filename in your favorite search engine, you should find a few examples relatively easily. Have a look, for example, at some of the latest boot animations created for the CyanogenMod (http://bit.ly/WapOZ9) aftermarket Android distribution.

Again, make sure the ZIP file you created isn't compressed. Otherwise it won't work. Have a look at the *zip* command's man page—especially the -0 flag.

Disabling the boot animation

You can also outright disable the boot animation if you don't want it. Just use the *setprop* command in *init.rc* to set the debug.sf.nobootanimation to 1:

    setprop debug.sf.nobootanimation 1

In this case, the screen will go black at some point after the boot logo has been displayed, and stay black until the Launcher app displays the home screen.

Dex Optimization

One of the system services started during the boot animation is the Package Manager. We haven't covered its functionality in detail, but suffice it to say that the Package Manager manages all the *.apks* in the system. Among other things, it'll deal with the installation and removal of *.apks* and help the Activity Manager resolve intents.

One of the Package Manager's responsibilities is also to make sure that JIT-ready versions of any DEX byte-code are available prior to the corresponding Java code ever executing. To achieve this, the Package Manager's constructor (the Package Manager system service is implemented as a Java class) goes through all *.apk* and *.jar* files in the system and requests that *installd* run the *dexopt* command on them.

This process should happen on the first boot only. Subsequently, the */data/dalvik-cache* directory will contain JIT-ready versions of all *.dex* files, and the boot sequence should be substantially faster. If you look into *logcat*'s output at first boot, you'll actually see entries like these:

```
D/dalvikvm(    32): DexOpt: --- BEGIN 'core.jar' (bootstrap=1) ---
D/dalvikvm(    62): Ignoring duplicate verify attempt on Ljava/lang/Object;
D/dalvikvm(    62): Ignoring duplicate verify attempt on Ljava/lang/Class;
D/dalvikvm(    62): DexOpt: load 349ms, verify+opt 4153ms
D/dalvikvm(    32): DexOpt: --- END 'core.jar' (success) ---
D/dalvikvm(    32): DEX prep '/system/framework/core.jar': unzip in 405ms, rewrite 5337ms
D/dalvikvm(    32): DexOpt: --- BEGIN 'bouncycastle.jar' (bootstrap=1) ---
D/dalvikvm(    63): DexOpt: load 54ms, verify+opt 779ms
D/dalvikvm(    32): DexOpt: --- END 'bouncycastle.jar' (success) ---
D/dalvikvm(    32): DEX prep '/system/framework/bouncycastle.jar': unzip in 48ms, rewrite 1023ms
D/dalvikvm(    32): DexOpt: --- BEGIN 'ext.jar' (bootstrap=1) ---
D/dalvikvm(    64): DexOpt: load 129ms, verify+opt 1497ms
D/dalvikvm(    32): DexOpt: --- END 'ext.jar' (success) ---
D/dalvikvm(    32): DEX prep '/system/framework/ext.jar': unzip in 91ms, rewrite 1923ms
...
D/installd(    35): DexInv: --- BEGIN '/system/framework/am.jar' ---
D/dalvikvm(    95): DexOpt: load 15ms, verify+opt 58ms
D/installd(    35): DexInv: --- END '/system/framework/am.jar' (success) ---
D/installd(    35): DexInv: --- BEGIN '/system/framework/input.jar' ---
D/dalvikvm(    96): DexOpt: load 5ms, verify+opt 28ms
D/installd(    35): DexInv: --- END '/system/framework/input.jar' (success) ---
D/installd(    35): DexInv: --- BEGIN '/system/framework/pm.jar' ---
D/dalvikvm(    97): DexOpt: load 12ms, verify+opt 64ms
D/installd(    35): DexInv: --- END '/system/framework/pm.jar' (success) ---
...
D/installd(    35): DexInv: --- BEGIN '/system/app/ApplicationsProvider.apk' ---
D/dalvikvm(   249): DexOpt: load 31ms, verify+opt 110ms
D/installd(    35): DexInv: --- END '/system/app/ApplicationsProvider.apk' (success) ---
D/installd(    35): DexInv: --- BEGIN '/system/app/UserDictionaryProvider.apk' ---
D/dalvikvm(   253): DexOpt: load 19ms, verify+opt 52ms
D/installd(    35): DexInv: --- END '/system/app/UserDictionaryProvider.apk' (success) ---
D/installd(    35): DexInv: --- BEGIN '/system/app/Settings.apk' ---
D/dalvikvm(   254): DexOpt: load 155ms, verify+opt 894ms
D/installd(    35): DexInv: --- END '/system/app/Settings.apk' (success) ---
D/installd(    35): DexInv: --- BEGIN '/system/app/Launcher2.apk' ---
D/dalvikvm(   256): DexOpt: load 178ms, verify+opt 581ms
D/installd(    35): DexInv: --- END '/system/app/Launcher2.apk' (success) ---
```

At first, the Package Manager Service isn't yet running, so we can see Dalvik running *dexopt* directly for some *.jar* files instead of being run by *installd*, as happens when the Package Manager Service requests it. Once the Package Manager is started, it then runs the rest of this optimization process in the following order:

1. *.jar* files listed in the BOOTCLASSPATH variable in *init.rc*
2. *.jar* files listed as libraries in */system/etc/permission/platform.xml*
3. *.apk* and *.jar* files found in */system/framework*
4. *.apk* files found in */system/app*
5. *.apk* files found in */vendor/app*
6. *.apk* files found in */data/app*
7. *.apk* files found in */data/app-private*

Obviously this process takes some time. On my quad-core CORE i7, it takes the emulator image of a freshly compiled 2.3/Gingerbread AOSP 75 seconds for its first full boot (i.e., up to the home screen) and 24 seconds for subsequent boots. In a production system, such as a phone, boot times like this can be unacceptable.

You'll therefore be happy to hear that you can actually stop this optimization process from happening at boot time and do it at build time instead. You just need to set the WITH_DEXPREOPT build flag to true when building the AOSP:

```
$ make WITH_DEXPREOPT=true -j16
```

You can also set this variable in your device's *BoardConfig.mk* instead, and avoid having to add it to the *make* command every time. In the case of the emulator build, this wasn't done by default in 2.3/Gingerbread but is in 4.2/Jelly Bean.

The build will of course take more time, but the first boot will be significantly faster. On the same workstation mentioned previously, it takes 30 minutes to build 2.3/Gingerbread instead of 20 with the WITH_DEXPREOPT flag. However, the emulator image comes up in 40 seconds instead of 75 on a first boot. When the option is used, the */data/dalvik-cache* directory ends up being empty on the target after the first boot. Instead, at build time, *.odex* files are placed side by side in the same filesystem path as their corresponding *.jar* and *.apk* files.

Apps Startup

As the startup of the system services nears its end, apps start to get activated, including the home screen. As I explained in Chapter 2, the Activity Manager ends its initialization by sending an intent of type Intent.CATEGORY_HOME, which causes the Launcher app to start and the home screen to appear. That's only part of the story, though. The startup of the system services will in fact cause quite a few apps to start. Here's a portion of the output of the *ps* command on a freshly booted 2.3/Gingerbread emulator image:

```
# ps
...
root       32    1    60832  16240 c009b74c afd0b844 S zygote
media      33    1    17976  1056  ffffffff afd0b6fc S /system/bin/mediaserver
bluetooth  34    1    1256   220   c009b74c afd0c59c S /system/bin/dbus-daemon
root       35    1    812    232   c02181f4 afd0b45c S /system/bin/installd
keystore   36    1    1744   212   c01b52b4 afd0c0cc S /system/bin/keystore
root       38    1    824    268   c00b8fec afd0c51c S /system/bin/qemud
shell      40    1    732    200   c0158eb0 afd0b45c S /system/bin/sh
root       41    1    3364   168   ffffffff 00008294 S /sbin/adbd
system     61    32   124096 26352 ffffffff afd0b6fc S system_server
app_19     113   32   80336  17400 ffffffff afd0c51c S com.android.inputmethod.
                                                       latin
radio      121   32   87112  17972 ffffffff afd0c51c S com.android.phone
system     122   32   73160  18452 ffffffff afd0c51c S com.android.systemui
app_26     132   32   76608  20812 ffffffff afd0c51c S com.android.launcher
app_1      169   32   85368  20584 ffffffff afd0c51c S android.process.acore
app_12     234   32   70752  15748 ffffffff afd0c51c S com.android.quicksearchbox
app_8      242   32   73108  16908 ffffffff afd0c51c S android.process.media
app_10     266   32   70928  16572 ffffffff afd0c51c S com.android.providers.
                                                       calendar
app_29     300   32   72764  17484 ffffffff afd0c51c S com.android.email
app_18     315   32   70272  15428 ffffffff afd0c51c S com.android.music
app_22     323   32   69712  15220 ffffffff afd0c51c S com.android.protips
app_3      335   32   71432  16756 ffffffff afd0c51c S com.cooliris.media
...
```

All the processes that have a Java-style process name[2] are actually apps that were automatically started with no user intervention whatsoever at system startup. Various system mechanisms cause these apps to start given the content of their respective manifest files. And this is a welcome change, since controlling apps' activation requires a lot less internals work than is required for controlling many other aspects of the startup, as we've seen above. Instead, it's all about creating carefully crafted apps for packaging with the AOSP. Sure, there's the case where you'll want to modify a stock app to make it behave or start differently, but at least we're into the app world, where functionality is more loosely coupled and documentation more readily accessible.

Which leads us to discussing the triggers that cause stock apps to be activated.

Input methods

One of the earliest types of apps to start are input methods. The Input Method Manager Service's constructor goes around and activates all app services that have an intent filter

---

2. These are dot-separated names, such as com.android.launcher for the Launcher app, for example.

for `android.view.InputMethod`. This is how, for example, the LatinIME app, which runs as the `com.android.inputmethod.latin` process, is activated.

As you can see by reading the Creating an Input Method (*http://android-developers.blogspot.com/2009/04/creating-input-method.html*) blog post on the Android Developers Blog, input methods are actually carefully crafted services.

Persistent apps

Apps that have the `android:persistent="true"` attribute in the `<application>` element of their manifest file will be automatically spawned at startup by the Activity Manager. In fact, should such an app ever die, it will also be automatically restarted by the Activity Manager.

As I explained earlier, unlike regular apps, apps that are marked as persistent are not lifecycle managed by the Activity Manager. Instead, they are kept alive throughout the lifetime of the system. This allows using such apps to implement special functionality. The status bar and the phone app, for example, running as the `com.android.systemui` and `com.android.phone` processes, are persistent apps.

While the app development documentation does explain the role of `android:persistent`, the use of that attribute is reserved for apps that are built within the AOSP.

Home screen

Typically there's only one home screen app, and it reacts to the `Intent.CATEGORY_HOME` intent, which is sent by the Activity Manager at the end of the system services' startup. There's a sample home app in *development/samples/Home/*, but the real home app activated is in *packages/apps/Launcher2/*. Here's the Launcher's main activity and its intent filter in 2.3/Gingerbread (4.2/Jelly Bean's is basically the same):

```
<activity
    android:name="com.android.launcher2.Launcher"
    android:launchMode="singleTask"
    android:clearTaskOnLaunch="true"
    android:stateNotNeeded="true"
    android:theme="@style/Theme"
    android:screenOrientation="nosensor"
    android:windowSoftInputMode="stateUnspecified|adjustPan">
    <intent-filter>
        <action android:name="android.intent.action.MAIN" />
        <category android:name="android.intent.category.HOME" />
        <category android:name="android.intent.category.DEFAULT" />
        <category android:name="android.intent.category.MONKEY"/>
    </intent-filter>
</activity>
```

Obviously, if you want to start a custom app to be the home screen instead of Launcher2, you'll need to remove the latter and add your own that reacts to that same intent. If more than one app reacts to that intent, users will get a dialog asking them which of the home screens they want to use.

Note that this intent isn't sent just at startup. Depending on the state of the system, the Activity Manager will send this intent whenever it needs to bring the home screen to the foreground.

BOOT_COMPLETED intent

The Activity Manager also broadcasts the Intent.BOOT_COMPLETED intent at startup. This is an intent commonly used by apps to be notified that the system has finished booting. A number of stock apps in the AOSP actually rely on this intent, such as Media provider, Calendar provider, Mms app, and Email app. Here's the broadcast receiver used by the Media Provider in 2.3/Gingerbread, along with its intent filter (4.2/Jelly Bean's is very similar):

```xml
<receiver android:name="MediaScannerReceiver">
    <intent-filter>
        <action android:name="android.intent.action.BOOT_COMPLETED" />
    </intent-filter>
    <intent-filter>
        <action android:name="android.intent.action.MEDIA_MOUNTED" />
        <data android:scheme="file" />
    </intent-filter>
    <intent-filter>
        <action android:name="android.intent.action.MEDIA_SCANNER_SCAN_
          FILE" />
        <data android:scheme="file" />
    </intent-filter>
</receiver>
```

In order to receive this intent, apps must explicitly request permission to do so:

```xml
<uses-permission android:name="android.permission.RECEIVE_BOOT_COMPLETED" />
```

APPWIDGET_UPDATE intent

In addition to apps, the App Widget Service, which is itself a system service, registers itself to receive the Intent.BOOT_COMPLETED. It uses the receipt of that intent as a trigger to activate all app widgets in the system by sending Intent.APPWIDGET_UPDATE. Hence, if you've developed an app widget as part of your app, your code will be activated at this point. Have a look at the App Widgets (*http://bit.ly/VQ0k4u*) section of the Android developer documentation for more information on how to write your own app widget.

Several stock AOSP apps have app widgets, such as Quick Search Box, Music, Protips, and Media. Here's the Quick Search Box's app widget declaration in its manifest file, for example:

```xml
<receiver android:name=".SearchWidgetProvider"
        android:label="@string/app_name">
    <intent-filter>
        <action android:name="android.appwidget.action.APPWIDGET_
            UPDATE" />
    </intent-filter>
    <meta-data android:name="android.appwidget.provider" android:
      resource="@xml/search_widget_info" />
</receiver>
```

Utilities and Commands

Once the Framework and the basic set of apps is up and running, there are quite a few commands that you can use to query or interact with system services and the Framework. Much like the commands covered in Chapter 6, these can be used on the command line once you shell into the device. But these commands have no meaning, and therefore no effect, unless the Framework is running. Of course you'll find many of these useful, even crucial, as you're bringing up Android on new devices and/or debugging parts of the Framework. And as with the commands in the native user-space, the tools available for interacting with the Framework vary greatly in terms of documentation and capabilities. Yet they provide the essential capabilities required to bring Android up on new hardware or to troubleshoot it on existing products. Let's take a look at the command set available to you for interacting with the Android Framework.

Many of the commands here are located in the *frameworks/base/cmds/* directory of the AOSP sources, though in 4.2/Jelly Bean, some of those commands have been moved to *frameworks/native/cmds/*. I encourage you to refer to those sources when using some of these commands, as their effects aren't always obvious just by looking at their online help, when it exists.

General-Purpose Utilities

In contrast with some utilities we'll see later, a certain number of utilities are useful for interacting with various parts of the Framework. Some of these are very powerful.

service

The *service* command allows us to interact with any system service registered with the Service Manager:

```
# service -h
Usage: service [-h|-?]
       service list
       service check SERVICE
       service call SERVICE CODE [i32 INT | s16 STR] ...
```

```
Options:
    i32: Write the integer INT into the send parcel.
    s16: Write the UTF-16 string STR into the send parcel.
```

As you can see, it can be used for querying but can also be used for invoking methods from system services. Here's how it can be used to query the list of existing system services in 2.3/Gingerbread:

```
# service list
Found 50 services:
0 phone: [com.android.internal.telephony.ITelephony]
1 iphonesubinfo: [com.android.internal.telephony.IPhoneSubInfo]
2 simphonebook: [com.android.internal.telephony.IIccPhoneBook]
3 isms: [com.android.internal.telephony.ISms]
4 diskstats: []
5 appwidget: [com.android.internal.appwidget.IAppWidgetService]
6 backup: [android.app.backup.IBackupManager]
7 uimode: [android.app.IUiModeManager]
8 usb: [android.hardware.usb.IUsbManager]
9 audio: [android.media.IAudioService]
10 wallpaper: [android.app.IWallpaperManager]
11 dropbox: [com.android.internal.os.IDropBoxManagerService]
12 search: [android.app.ISearchManager]
13 location: [android.location.ILocationManager]
14 devicestoragemonitor: []
15 notification: [android.app.INotificationManager]
16 mount: [IMountService]
17 accessibility: [android.view.accessibility.IAccessibilityManager]
...
```

The interface names provided in between square brackets allow you to browse the AOSP sources to find the matching *.aidl* file that defines the interface.

You can also check if a given service exists:

```
# service check power
Service power: found
```

Most interestingly, you can use *service call* to directly invoke system services' Binder-exposed methods. In order to do that, you first need to understand that service's interface. Here's the IStatusBarService interface definition from 2.3/Gingerbread's *frameworks/base/core/java/com/android/internal/statusbar/IStatusBarService.aidl* (4.2/Jelly Bean's interface name is the same, though setIcon()'s prototype has changed):

```
...
interface IStatusBarService
{
    void expand();
    void collapse();
    void disable(int what, IBinder token, String pkg);
    void setIcon(String slot, String iconPackage, int iconId, int iconLevel);
...
```

Note that *service call* actually needs a method code, not a method's name. To find the codes matching the method names defined in the interface, you'll need to look up the code generated by the *aidl* tool based on the interface definition. Here's the relevant snippet from the *IStatusBarService.java* file generated in *out/target/common/obj/ JAVA_LIBRARIES/framework_intermediates/src/core/java/com/android/internal/ statusbar/*:

```
...
static final int TRANSACTION_expand = (android.os.IBinder.FIRST_CALL_
TRANSACTION + 0);
static final int TRANSACTION_collapse = (android.os.IBinder.FIRST_CALL_
TRANSACTION + 1);
static final int TRANSACTION_disable = (android.os.IBinder.FIRST_CALL_
TRANSACTION + 2);
static final int TRANSACTION_setIcon = (android.os.IBinder.FIRST_CALL_
TRANSACTION + 3);
...
```

Also, note that *frameworks/base/core/java/android/os/IBinder.java* has the following definition for FIRST_CALL_TRANSACTION:

```
int FIRST_CALL_TRANSACTION  = 0x00000001;
```

Hence, expand( )'s code is 1 and collapse( )'s code is 2. Therefore, this command will cause the status bar to expand:

```
# service call statusbar 1
```

While this command will cause the status bar to collapse:

```
# service call statusbar 2
```

This is a very simple case where the action is rather obvious and the methods invoked don't take any parameters. In other cases, you'll need to look more closely at the system service's API and understand the parameters expected. In addition, keep in mind that system services' interfaces aren't necessarily exposed through *.aidl* files. In some cases, such as for the Activity Manager, the interface definition is hardcoded directly into a regular Java file instead of being autogenerated. And in the case of C-based system services, the Binder marshaling and unmarshaling is all done straight in C code. Hence, try using *grep* on the AOSP's *frameworks/* directory in addition to *out/target/ common/* to find all instances of FIRST_CALL_TRANSACTION.

dumpsys

Another interesting thing to do is to query system services' internal state. Indeed, every system service implements a dump( ) method internally that can be queried using the *dumpsys* command:

```
dumpsys [ <service> ]
```

By default, if no system service name is provided as a parameter, *dumpsys* will first print
out the list of system services and will then dump their status:

```
# dumpsys
Currently running services:
  SurfaceFlinger
  accessibility
  account
  activity
  alarm
  appwidget
  audio
  backup
  battery
  batteryinfo
  clipboard
  connectivity
  content
  cpuinfo
  device_policy
  devicestoragemonitor
  diskstats
  dropbox
  entropy
  hardware
...
-------------------------------------------------------------------------------
DUMP OF SERVICE SurfaceFlinger:
+ Layer 0x1e5788
      z=   21000, pos=(   0,   0), size=( 320, 480), needsBlending=0, needsDith
ering=0, invalidate=0, alpha=0xff, flags=0x00000000, tr=[1.00, 0.00][0.00, 1.00]
      name=com.android.internal.service.wallpaper.ImageWallpaper
      client=0x1ed2a8, identity=3
      [ head= 1, available= 2, queued= 0 ] reallocMask=00000000, identity=3, sta
tus=0
      format= 4, [320x480:320] [320x480:320], freezeLock=0x0, bypass=0, dq-q-tim
e=2034 us
  Region transparentRegion (this=0x1e5918, count=1)
    [  0,   0,   0,   0]
  Region transparentRegionScreen (this=0x1e57bc, count=1)
    [  0,   0,   0,   0]
  Region visibleRegionScreen (this=0x1e5798, count=1)
    [  0,  25, 320, 480]
+ Layer 0x268b70
      z=   21005, pos=(   0,   0), size=( 320, 480), needsBlending=1, needsDith
ering=1, invalidate=0, alpha=0xff, flags=0x00000000, tr=[1.00, 0.00][0.00, 1.00]
...
-------------------------------------------------------------------------------
DUMP OF SERVICE accessibility:
-------------------------------------------------------------------------------
DUMP OF SERVICE account:
Accounts: 0
```

```
Active Sessions: 0

RegisteredServicesCache: 1 services
  ServiceInfo: AuthenticatorDescription {type=com.android.exchange}, ComponentIn
fo{com.android.email/com.android.email.service.EasAuthenticatorService}, uid 100
29
-------------------------------------------------------------------------------
DUMP OF SERVICE activity:
Providers in Current Activity Manager State:
  Published content providers (by class):
  * ContentProviderRecord{4060d0e0 com.android.deskclock.AlarmProvider}
...
```

Obviously the output is very verbose and, most importantly, requires understanding the corresponding system service's internals. If you're implementing your own system service, however, being able to query its state at runtime can be crucial. Of course, if you're not interested in dumping the state of all system services, you just need to provide the name of the specific service you'd like to get information about as a parameter to *dumpsys*:

```
# dumpsys power
Power Manager State:
  mIsPowered=true mPowerState=1 mScreenOffTime=46793204 ms
  mPartialCount=1
  mWakeLockState=SCREEN_ON_BIT
  mUserState=
  mPowerState=SCREEN_ON_BIT
  mLocks.gather=SCREEN_ON_BIT
  mNextTimeout=94351 now=46880555 -46786s from now
  mDimScreen=true mStayOnConditions=1
  mScreenOffReason=0 mUserState=0
  mBroadcastQueue={-1,-1,-1}
  mBroadcastWhy={0,0,0}
  mPokey=0 mPokeAwakeonSet=false
...
```

dumpstate

In some cases, what you're trying to do is get a snapshot of the entire system, not just the system services. This is what *dumpstate* takes care of. In fact, you might recall our discussion of this command when we covered *adb*'s *bugreport* in Chapter 6, since *dumpstate* is what provides *bugreport* with its information. Here's *dumpstate*'s detailed help in 2.3/Gingerbread:

```
# dumpstate -h
usage: dumpstate [-d] [-o file] [-s] [-z]
  -d: append date to filename (requires -o)
  -o: write to file (instead of stdout)
  -s: write output to control socket (for init)
  -z: gzip output (requires -o)
```

In 4.2/Jelly Bean, *dumpstate*'s capabilities have expanded:

```
root@android:/ # dumpstate -h
usage: dumpstate [-b soundfile] [-e soundfile] [-o file [-d] [-p] [-z]] [-s] [-q]
  -o: write to file (instead of stdout)
  -d: append date to filename (requires -o)
  -z: gzip output (requires -o)
  -p: capture screenshot to filename.png (requires -o)
  -s: write output to control socket (for init)
  -b: play sound file instead of vibrate, at beginning of job
  -e: play sound file instead of vibrate, at end of job
  -q: disable vibrate
```

If you invoke it without any parameters, it goes ahead and queries several parts of the sytem to provide you with a complete snapshot of the system's status:

```
# dumpstate
========================================================
== dumpstate: 2012-10-10 03:15:26
========================================================

Build: generic-eng 2.3.4 GINGERBREAD eng.karim.20120913.141233 test-keys
Bootloader: unknown
Radio: unknown
Network: Android
Kernel: Linux version 2.6.29-00261-g0097074-dirty (digit@digit.mtv.corp.google.c
om) (gcc version 4.4.0 (GCC) ) #20 Wed Mar 31 09:54:02 PDT 2010
Command line: qemu=1 console=ttyS0 android.checkjni=1 android.qemud=ttyS1 androi
d.ndns=1

------ MEMORY INFO (/proc/meminfo) ------
MemTotal:           94096 kB
MemFree:             1296 kB
Buffers:                0 kB
Cached:             32424 kB
...
------ CPU INFO (top -n 1 -d 1 -m 30 -t) ------


User 2%, System 11%, IOW 33%, IRQ 0%
User 3 + Nice 0 + Sys 15 + Idle 67 + IOW 42 + IRQ 0 + SIRQ 0 = 127

  PID   TID CPU% S     VSS   RSS PCY UID      Thread             Proc
  339   339  13% R    976K  440K  fg shell    top                top
  121   121   0% S  86100K 18484K  fg radio    m.android.phone com.android.phone
    3     3   0% S      0K    0K  fg root     ksoftirqd/0
    4     4   0% S      0K    0K  fg root     events/0
...
------ PROCRANK (procrank) ------
  PID       Vss      Rss      Pss      Uss  cmdline
   61    25676K   25076K   10581K    8552K  system_server
  124    21412K   21412K    6851K    4908K  com.android.launcher
```

```
     122    19268K    19268K    5698K    4388K com.android.systemui
     121    18484K    18484K    4744K    3568K com.android.phone
     295    18176K    18176K    4337K    3132K com.android.email
     115    17836K    17836K    4118K    2960K com.android.inputmethod.latin
...
------ VIRTUAL MEMORY STATS (/proc/vmstat) ------
nr_free_pages 553
nr_inactive_anon 6708
nr_active_anon 6068
nr_inactive_file 3449
nr_active_file 2062
...
------ VMALLOC INFO (/proc/vmallocinfo) ------
0xc684c000-0xc684e000     8192 __arm_ioremap_pfn+0x68/0x2fc ioremap
0xc6850000-0xc6852000     8192 __arm_ioremap_pfn+0x68/0x2fc ioremap
0xc6854000-0xc6856000     8192 __arm_ioremap_pfn+0x68/0x2fc ioremap
0xc6880000-0xc68a1000   135168 binder_mmap+0xb4/0x200 ioremap
...
------ SLAB INFO (/proc/slabinfo) ------
slabinfo - version: 2.1
# name            <active_objs> <num_objs> <objsize> <objperslab> <pagesperslab>
 : tunables <limit> <batchcount> <sharedfactor> : slabdata <active_slabs> <num_s
labs> <sharedavail>
rpc_buffers            8     8   2048    2    1 : tunables   24   12    0 : sla
bdata       4     4     0
rpc_tasks              8    24    160   24    1 : tunables  120   60    0 : sla
bdata       1     1     0
rpc_inode_cache        0     0    416    9    1 : tunables   54   27    0 : sla
bdata       0     0     0
bridge_fdb_cache       0     0     64   59    1 : tunables  120   60    0 : sla
bdata       0     0     0
...
------ ZONEINFO (/proc/zoneinfo) ------
Node 0, zone   Normal
  pages free     550
        min      312
        low      390
        high     468
        scanned  0 (aa: 0 ia: 0 af: 26 if: 0)
...
------ SYSTEM LOG (logcat -v time -d *:v) ------
10-10 01:38:02.762 I/DEBUG   (   30): debuggerd: Feb 26 2012 21:06:53
10-10 01:38:02.882 I/Netd    (   29): Netd 1.0 starting
10-10 01:38:02.932 D/qemud   (   38): entering main loop
10-10 01:38:02.972 I/Vold    (   28): Vold 2.1 (the revenge) firing up
10-10 01:38:02.972 D/Vold    (   28): USB mass storage support is not enabled in
the kernel
...
------ VM TRACES JUST NOW (/data/anr/traces.txt.bugreport: 2012-10-10 03:15:26)
------
```

```
----- pid 61 at 2012-10-10 03:15:26 -----
Cmd line: system_server

DALVIK THREADS:
(mutexes: tll=0 tsl=0 tscl=0 ghl=0 hwl=0 hwll=0)
"main" prio=5 tid=1 NATIVE
  | group="main" sCount=1 dsCount=0 obj=0x4001f1a8 self=0xce48
  | sysTid=61 nice=0 sched=0/0 cgrp=default handle=-1345006528
  | schedstat=( 1116789165 392598071 782 )
  at com.android.server.SystemServer.init1(Native Method)
  at com.android.server.SystemServer.main(SystemServer.java:625)
...
------ EVENT LOG (logcat -b events -v time -d *:v) ------
10-10 01:38:03.642 I/boot_progress_start(   32): 5126
10-10 01:38:04.221 I/boot_progress_preload_start(   32): 5706
10-10 01:38:04.251 I/dvm_gc_info(   32): [8825198673194415294,-90644969689662529
97,-4012584086963399109,0]
10-10 01:38:04.281 I/dvm_gc_info(   32): [8825198673194406507,-92148046065296736
57,-4012584086963329465,0]
10-10 01:38:04.331 I/dvm_gc_info(   32): [8825198673194406993,-91348657131437777
12,-4012584086963259824,0]
10-10 01:38:04.371 I/dvm_gc_info(   32): [8825198673194415172,-91399322627244589
19,-4012584086963149223,0]
...
------ RADIO LOG (logcat -b radio -v time -d *:v) ------
10-10 01:58:04.988 D/AT     (   31): AT< +CSQ: 7,99
10-10 01:58:04.988 D/AT     (   31): AT< OK
10-10 01:58:04.988 D/RILJ   (  121): [0114]< SIGNAL_STRENGTH {7, 99, 0, 0, 0
, 0, 0}
10-10 01:58:24.998 D/RILJ   (  121): [0115]> SIGNAL_STRENGTH
10-10 01:58:25.008 D/RIL    (   31): onRequest: SIGNAL_STRENGTH
...
------ NETWORK INTERFACES (netcfg) ------
*** exec(netcfg): Permission denied
*** netcfg: Exit code 255
[netcfg: 0.1s elapsed]

------ NETWORK ROUTES (/proc/net/route) ------
Iface   Destination     Gateway         Flags   RefCnt  Use     Metric  Mask
        MTU     Window  IRTT
eth0    0002000A        00000000        0001    0       0       0       00FFFFFF
        0       0       0

eth0    00000000        0202000A        0003    0       0       0       00000000
        0       0       0


------ ARP CACHE (/proc/net/arp) ------
IP address      HW type     Flags    HW address          Mask    Device
10.0.2.2        0x1         0x2      52:54:00:12:35:02   *       eth0

------ SYSTEM PROPERTIES ------
```

```
[dalvik.vm.heapsize]: [16m]
[dalvik.vm.stack-trace-file]: [/data/anr/traces.txt]
[dev.bootcomplete]: [1]
[gsm.current.phone-type]: [1]
[gsm.defaultpdpcontext.active]: [true]
...
------ KERNEL LOG (dmesg) ------
Initializing cgroup subsys cpu
Linux version 2.6.29-00261-g0097074-dirty (digit@digit.mtv.corp.google.com) (gcc
 version 4.4.0 (GCC) ) #20 Wed Mar 31 09:54:02 PDT 2010
CPU: ARM926EJ-S [41069265] revision 5 (ARMv5TEJ), cr=00093177
CPU: VIVT data cache, VIVT instruction cache
Machine: Goldfish
Memory policy: ECC disabled, Data cache writeback
On node 0 totalpages: 24576
...
------ KERNEL WAKELOCKS (/proc/wakelocks) ------
name      count    expire_count   wake_count       active_since       total_time
sleep_time         max_time       last_change
"alarm" 106     0       0       0       1632946980     0        41697763
5822030632794
"KeyEvents"     27      0       0       0       123592046      0        94064309
        27084159991
"event0-61"     26      0       0       0       48780811       0        12891126
        27083608920
"radio-interface"       3       0       0       0       3472899963     0
1459986280      25362482435
...
------ KERNEL CPUFREQ (/sys/devices/system/cpu/cpu0/cpufreq/stats/time_in_state)
------
*** /sys/devices/system/cpu/cpu0/cpufreq/stats/time_in_state: No such file or di
rectory

------ VOLD DUMP (vdc dump) ------
000 Dumping loop status
000 Dumping DM status
000 Dumping mounted filesystems
000 rootfs / rootfs ro 0 0
...
------ SECURE CONTAINERS (vdc asec list) ------
200 asec operation succeeded
[vdc: 0.1s elapsed]

------ PROCESSES (ps -P) ------
USER     PID   PPID  VSIZE  RSS     PCY  WCHAN    PC        NAME
root     1     0     268    180     fg   c009b74c 0000875c  S /init
root     2     0     0      0       fg   c004e72c 00000000  S kthreadd
root     3     2     0      0       fg   c003fdc8 00000000  S ksoftirqd/0
root     4     2     0      0       fg   c004b2c4 00000000  S events/0
root     5     2     0      0       fg   c004b2c4 00000000  S khelper
root     6     2     0      0       fg   c004b2c4 00000000  S suspend
...
```

```
------ PROCESSES AND THREADS (ps -t -p -P) ------
USER     PID   PPID  VSIZE   RSS    PRIO  NICE  RTPRI  SCHED  PCY  WCHAN      PC
    NAME
root     1     0     268     180    20    0     0      0      fg   c009b74c  0000875c
  S /init
root     2     0     0       0      15    -5    0      0      fg   c004e72c  00000000
  S kthreadd
root     3     2     0       0      15    -5    0      0      fg   c003fdc8  00000000
  S ksoftirqd/0
root     4     2     0       0      15    -5    0      0      fg   c004b2c4  00000000
  S events/0
root     5     2     0       0      15    -5    0      0      fg   c004b2c4  00000000
  S khelper
root     6     2     0       0      15    -5    0      0      fg   c004b2c4  00000000
  S suspend
...
------ LIBRANK (librank) ------
RSStot       VSS      RSS     PSS     USS    Name/PID
16658K                                       /dev/ashmem/dalvik-heap
             6980K    6980K   3218K   2896K  system_server [61]
             5208K    5208K   1371K   1048K  com.android.launcher [124]
             5272K    5272K   1343K   1012K  com.android.phone [121]
...
------ BINDER FAILED TRANSACTION LOG (/sys/kernel/debug/binder/failed_transactio
n_log) ------
*** /sys/kernel/debug/binder/failed_transaction_log: No such file or directory

------ BINDER TRANSACTION LOG (/sys/kernel/debug/binder/transaction_log) ------
*** /sys/kernel/debug/binder/transaction_log: No such file or directory

------ BINDER TRANSACTIONS (/sys/kernel/debug/binder/transactions) ------
*** /sys/kernel/debug/binder/transactions: No such file or directory

------ BINDER STATS (/sys/kernel/debug/binder/stats) ------
*** /sys/kernel/debug/binder/stats: No such file or directory

------ BINDER STATE (/sys/kernel/debug/binder/state) ------
*** /sys/kernel/debug/binder/state: No such file or directory

------ FILESYSTEMS & FREE SPACE (df) ------
Filesystem            1K-blocks     Used  Available  Use%  Mounted on
tmpfs                 47048         32    47016      0%    /dev
tmpfs                 47048         0     47048      0%    /mnt/asec
tmpfs                 47048         0     47048      0%    /mnt/obb
/dev/block/mtdblock0  65536         65536 0          100%  /system
/dev/block/mtdblock1  65536         25292 40244      39%   /data
/dev/block/mtdblock2  65536         1156  64380      2%    /cache
[df: 0.1s elapsed]

------ PACKAGE SETTINGS (/data/system/packages.xml: 2012-10-10 01:38:16) ------
<?xml version='1.0' encoding='utf-8' standalone='yes' ?>
<packages>
```

```
<last-platform-version internal="10" external="0" />
...
------ PACKAGE UID ERRORS (/data/system/uiderrors.txt: 2012-09-24 21:06:14) ----
--
2012-09-24 21:06: No settings file; creating initial state

------ LAST KMSG (/proc/last_kmsg) ------
*** /proc/last_kmsg: No such file or directory

------ LAST RADIO LOG (parse_radio_log /proc/last_radio_log) ------
*** exec(parse_radio_log): Permission denied
*** parse_radio_log: Exit code 255
[parse_radio_log: 0.1s elapsed]

------ LAST PANIC CONSOLE (/data/dontpanic/apanic_console) ------
*** /data/dontpanic/apanic_console: No such file or directory

------ LAST PANIC THREADS (/data/dontpanic/apanic_threads) ------
*** /data/dontpanic/apanic_threads: No such file or directory


------ BLOCKED PROCESS WAIT-CHANNELS ------
------ BACKLIGHTS ------
LCD brightness=*** /sys/class/leds/lcd-backlight/brightness: No such file or dir
ectory
Button brightness=*** /sys/class/leds/button-backlight/brightness: No such file
 or directory
Keyboard brightness=*** /sys/class/leds/keyboard-backlight/brightness: No such f
ile or directory
ALS mode=*** /sys/class/leds/lcd-backlight/als: No such file or directory
LCD driver registers:
*** /sys/class/leds/lcd-backlight/registers: No such file or directory


========================================================
== Android Framework Services
========================================================
------ DUMPSYS (dumpsys) ------
Currently running services:
  SurfaceFlinger
...
```

In most cases, as you can see, *dumpstate* is in fact invoking other commands such as *logcat*, *dumpsys*, and *ps* to retrieve its information. As you can also see, the command is very verbose.

rawbu

In some cases, you may want to back up and later restore the contents of */data*. You can use the *rawbu* command to do that:

```
# rawbu help
Usage: rawbu COMMAND [options] [backup-file-path]
```

```
commands are:
  help            Show this help text.
  backup          Perform a backup of /data.
  restore         Perform a restore of /data.
options include:
  -h              Show this help text.
  -a              Backup all files.

The rawbu command allows you to perform low-level
backup and restore of the /data partition.  This is
where all user data is kept, allowing for a fairly
complete restore of a device's state.  Note that
because this is low-level, it will only work across
builds of the same (or very similar) device software.
```

Here's how it can be used to create a backup:

```
# rawbu backup /sdcard/backup.dat
Stopping system...
Backing up /data to /sdcard/backup.dat...
Saving dir /data/local...
Saving dir /data/local/tmp...
Saving dir /data/app-private...
Saving dir /data/app...
Saving dir /data/property...
Saving file /data/property/persist.sys.localevar...
Saving file /data/property/persist.sys.country...
Saving file /data/property/persist.sys.language...
Saving file /data/property/persist.sys.timezone...
...
Backup complete!  Restarting system...
```

The first thing the command does is stop the Zygote, thereby stopping all system services. It then proceeds to copy everything from /data and finishes by restarting the Zygote. Once data is backed up, you can restore it later:

```
# rawbu restore /sdcard/backup.dat
Stopping system...
Wiping contents of /data...
warning -- rmdir() error on '/data/system': Directory not empty
warning -- rmdir() error on '/data/system': Directory not empty
Restoring from /sdcard/backup.dat to /data...
Restoring dir /data/local...
Restoring dir /data/local/tmp...
Restoring dir /data/app-private...
Restoring dir /data/app...
...
Restore complete!  Restarting system, cross your fingers...
```

Obviously, as the command's output implies, this is a fragile operation and you should be aware that results will vary.

Service-Specific Utilities

As we saw earlier, there are dozens of system services. Typically, using these system services requires writing code that interacts with their Binder-exposed API in some way, shape, or form. In some cases, however, the AOSP includes command-line utilities for directly interacting with certain system services. Some of these utilities are very powerful and allow us to tap into Android's functionality straight from the command line. This opens the door for using many of the following utilities as part of scripts either in production or during development.

Circumventing Android's Permission System

The system services' APIs are typically protected by Android's permission system, which requires apps' manifest files to declare upfront which permissions they require. Generally, a system service will check whether its caller has the appropriate permissions before going ahead and servicing the caller's request. Part of this checking will require checking the caller's PID and using the Package Manager's services to verify the originating *.apk*'s rights.

There is one case, however, that circumvents all safeguards: when the caller is running as root. Indeed, if you look at the permission-checking code of the Activity Manager, which is used by the other system services to check for permissions, you will see this snippet in *frameworks/base/services/java/com/android/server/am/ActivityManagerService.java* in 2.3/Gingerbread:

```
    int checkComponentPermission(String permission, int pid, int uid,
...
        // Root, system server and our own process get to do everything.
        if (uid == 0 || uid == Process.SYSTEM_UID || pid == MY_PID ||
           !Process.supportsProcesses()) {
           return PackageManager.PERMISSION_GRANTED;
        }
...
```

In 4.2/Jelly Bean, you'll find this instead:

```
    int checkComponentPermission(String permission, int pid, int uid,
...
        if (pid == MY_PID) {
            return PackageManager.PERMISSION_GRANTED;
        }

        return ActivityManager.checkComponentPermission(permission, uid,
                owningUid, exported);
    }
```

With `ActivityManager.checkComponentPermission()` being defined as the following in *frameworks/base/core/java/android/app/ActivityManager.java*:

```
        public static int checkComponentPermission(String permission, int uid,
            int owningUid, boolean exported) {
        // Root, system server get to do everything.
        if (uid == 0 || uid == Process.SYSTEM_UID) {
            return PackageManager.PERMISSION_GRANTED;
        }
...
```

Hence, in both versions of the AOSP, any of the commands you see here that talk to a system service will typically be granted a green light on anything they ask for from a system service. You must, therefore, **be very careful** when talking to system services while running as root. The same applies if you write a command-line utility that mimics the way many of the commands we cover in this section interact with system services.

am

As I mentioned earlier, one of the most important system services is the Activity Manager. It should come as no surprise, therefore, that there's a command that allows us to directly invoke its functionality. Here's its online help in 2.3/Gingerbread:

```
# am
usage: am [subcommand] [options]

       start an Activity: am start [-D] [-W] <INTENT>
           -D: enable debugging
           -W: wait for launch to complete

       start a Service: am startservice <INTENT>

       send a broadcast Intent: am broadcast <INTENT>

       start an Instrumentation: am instrument [flags] <COMPONENT>
           -r: print raw results (otherwise decode REPORT_KEY_STREAMRESULT)
           -e <NAME> <VALUE>: set argument <NAME> to <VALUE>
           -p <FILE>: write profiling data to <FILE>
           -w: wait for instrumentation to finish before returning

       start profiling: am profile <PROCESS> start <FILE>
       stop profiling: am profile <PROCESS> stop

       start monitoring: am monitor [--gdb <port>]
           --gdb: start gdbserv on the given port at crash/ANR

       <INTENT> specifications include these flags:
           [-a <ACTION>] [-d <DATA_URI>] [-t <MIME_TYPE>]
           [-c <CATEGORY> [-c <CATEGORY>] ...]
           [-e|--es <EXTRA_KEY> <EXTRA_STRING_VALUE> ...]
           [--esn <EXTRA_KEY> ...]
           [--ez <EXTRA_KEY> <EXTRA_BOOLEAN_VALUE> ...]
           [-e|--ei <EXTRA_KEY> <EXTRA_INT_VALUE> ...]
           [-n <COMPONENT>] [-f <FLAGS>]
```

```
[--grant-read-uri-permission] [--grant-write-uri-permission]
[--debug-log-resolution]
[--activity-brought-to-front] [--activity-clear-top]
[--activity-clear-when-task-reset] [--activity-exclude-from-recents]
[--activity-launched-from-history] [--activity-multiple-task]
[--activity-no-animation] [--activity-no-history]
[--activity-no-user-action] [--activity-previous-is-top]
[--activity-reorder-to-front] [--activity-reset-task-if-needed]
[--activity-single-top]
[--receiver-registered-only] [--receiver-replace-pending]
[<URI>]
```

In 4.2/Jelly Bean, *am*'s capabilities have expanded, and so, too, has its online help. Since the latter now covers three pages, it's impractical to print it in its entirety in this book. The previous snippet is sufficient for the present discussion; still, I encourage you to read the *am* command's online help in 4.2/Jelly Bean.

As we saw in Chapter 2, there are four types of components available to app developers: activities, services, broadcast receivers, and content providers. The first three types of components are activated through intents, and one of *am*'s major features is its ability to send intents straight from the command line.

Here's how you can use *am* to get the browser to navigate to a given website along with the relevant log excerpts:

```
# am start -a android.intent.action.VIEW -d http://source.android.com
Starting: Intent { act=android.intent.action.VIEW dat=http://source.android.com }

# logcat
...
D/AndroidRuntime(  786):
D/AndroidRuntime(  786): >>>>>> AndroidRuntime START com.android.internal.os.Run
timeInit <<<<<<
D/AndroidRuntime(  786): CheckJNI is ON
D/AndroidRuntime(  786): Calling main entry com.android.commands.am.Am
I/ActivityManager(   62): Starting: Intent { act=android.intent.action.VIEW dat=
http://source.android.com flg=0x10000000 cmp=com.android.browser/.BrowserActivit
y } from pid 786
I/ActivityManager(   62): Start proc com.android.browser for activity com.androi
d.browser/.BrowserActivity: pid=794 uid=10015 gids={3003, 1015}
D/AndroidRuntime(  786): Shutting down VM
D/dalvikvm(  786): GC_CONCURRENT freed 100K, 69% free 317K/1024K, external 0K/0K
, paused 1ms+1ms
D/jdwp    (  786): adbd disconnected
I/ActivityThread(  794): Pub browser: com.android.browser.BrowserProvider
I/BrowserSettings(  794): Selected search engine: ActivitySearchEngine{android.a
pp.SearchableInfo@40593270}
D/dalvikvm(  794): GC_CONCURRENT freed 447K, 51% free 2909K/5831K, external 934K
```

```
    /1038K, paused 5ms+14ms
    I/ActivityManager(    62): Displayed com.android.browser/.BrowserActivity: +1s924
    ms
    D/dalvikvm(  794): GC_EXTERNAL_ALLOC freed 51K, 50% free 2953K/5831K, external 9
    51K/1038K, paused 62ms
    ...
```

That's a rather straightforward example. Let's look at something a little more customized. Here's a broadcast receiver declaration from a custom application:

```
            <receiver android:name="FastBirdApproaching">
                <intent-filter >
                    <action android:name="com.acme.coyotebirdmonitor.FAST_BIRD"/>
                </intent-filter>
            </receiver>
```

And here's the corresponding code:

```
    public class FastBirdApproaching extends BroadcastReceiver {
      private static final String TAG = "FastBirdApproaching";

      @Override
      public void onReceive(Context context, Intent intent) {
      // TODO Auto-generated method stub
      Log.i(TAG, "**********");
      Log.i(TAG, "Meep Meep!");
      Log.i(TAG, "**********");
      }
    }
```

Here's how you can use *am* to trigger this broadcast receiver and the resulting output in the logs:

```
    # am broadcast -a com.acme.coyotebirdmonitor.FAST_BIRD
    Broadcasting: Intent { act=com.acme.coyotebirdmonitor.FAST_BIRD }
    Broadcast completed: result=0

    # logcat
    ...
    I/ActivityManager(    62): Start proc com.acme.coyotebirdmonitor for broadcast co
    m.acme.coyotebirdmonitor/.FastBirdApproaching: pid=466 uid=10029 gids={}
    I/FastBirdApproaching(  466): **********
    I/FastBirdApproaching(  466): Meep Meep!
    I/FastBirdApproaching(  466): **********
    ...
```

As you can see from *am*'s online help, you can specify a lot of details regarding the intent to be sent. Whereas the previous two examples used implicit intents, you can also send explicit intents to activate designated components:

```
    # am start -n com.android.settings/.Settings
```

In this case, this will start the Settings activity of the settings app in the system. Interestingly, *am* can start components in ways you can't replicate using the officially published app development API. That's because it's built as part of the AOSP and has therefore access to hidden calls available only to code building within the AOSP.

am is in fact a shell script, as you can see in *frameworks/based/cmds/am/am/*:

```
# Script to start "am" on the device, which has a very rudimentary
# shell.
#
base=/system
export CLASSPATH=$base/framework/am.jar
exec app_process $base/bin com.android.commands.am.Am "$@"
```

The script uses *app_process* to start Java code that implements *am*'s functionality. All parameters passed on the command line are actually passed on to the Java code as is.

You can also use *am* for instrumentation, profiling, and monitoring. Have a look at the Testing Fundamentals (*http://bit.ly/Z5VAWj*) and Testing from Other IDEs (*http://bit.ly/13JVmJN*) sections of the Android developer manual for more information on Android testing and the use of the *am instrument* command.

The *am profile* commands allow us to generate data that can then be visualized on the host using the *traceview* command. You can find more information about *traceview* in the relevant section of the Android developer manual (*http://bit.ly/ZxmMxc*). Note that the documentation says there are two ways to create trace files, and the use of the *am* command on the command line isn't listed as one of them.

Finally, the *am monitor* command allows us to monitor apps run by the Activity Manager. Here's a session where I start the command and then start several apps:

```
# am monitor
Monitoring activity manager...   available commands:
(q)uit: finish monitoring
** Activity starting: com.android.browser
** Activity resuming: com.android.launcher
** Activity starting: com.android.settings
** Activity resuming: com.android.launcher
** Activity starting: com.android.browser
** Activity starting: com.android.launcher
...
```

Note that when you start an app and click Back, the command reports that the Launcher is resuming, whereas if you click the Home button, the Launcher is reported as starting. This monitoring capability will also allow you to catch ANRs (Application Not Responding) and enable you to attach *gdb* to a crashing process.

Don't let this brief coverage of *am* mislead you: This is an extremely powerful and useful command that you should keep well in mind. If you ever need to script the starting of apps from the command line, you will find it to be very useful.

pm

Another very important system service is the Package Manager and, much like the Activity Manager, it's got its own command-line tool. Here's its online help from 2.3/Gingerbread:

```
# pm
usage: pm [list|path|install|uninstall]
       pm list packages [-f] [-d] [-e] [-u] [FILTER]
       pm list permission-groups
       pm list permissions [-g] [-f] [-d] [-u] [GROUP]
       pm list instrumentation [-f] [TARGET-PACKAGE]
       pm list features
       pm list libraries
       pm path PACKAGE
       pm install [-l] [-r] [-t] [-i INSTALLER_PACKAGE_NAME] [-s] [-f] PATH
       pm uninstall [-k] PACKAGE
       pm clear PACKAGE
       pm enable PACKAGE_OR_COMPONENT
       pm disable PACKAGE_OR_COMPONENT
       pm setInstallLocation [0/auto] [1/internal] [2/external]

The list packages command prints all packages, optionally only
those whose package name contains the text in FILTER.  Options:
  -f: see their associated file.
  -d: filter to include disabled packages.
  -e: filter to include enabled packages.
  -u: also include uninstalled packages.

The list permission-groups command prints all known
permission groups.

The list permissions command prints all known
permissions, optionally only those in GROUP.  Options:
  -g: organize by group.
  -f: print all information.
  -s: short summary.
  -d: only list dangerous permissions.
  -u: list only the permissions users will see.

The list instrumentation command prints all instrumentations,
or only those that target a specified package.  Options:
  -f: see their associated file.

The list features command prints all features of the system.

The path command prints the path to the .apk of a package.

The install command installs a package to the system.  Options:
  -l: install the package with FORWARD_LOCK.
  -r: reinstall an existing app, keeping its data.
  -t: allow test .apks to be installed.
```

```
-i: specify the installer package name.
-s: install package on sdcard.
-f: install package on internal flash.

The uninstall command removes a package from the system. Options:
    -k: keep the data and cache directories around.
after the package removal.

The clear command deletes all data associated with a package.

The enable and disable commands change the enabled state of
a given package or component (written as "package/class").

The getInstallLocation command gets the current install location
0 [auto]: Let system decide the best location
1 [internal]: Install on internal device storage
2 [external]: Install on external media

The setInstallLocation command changes the default install location
0 [auto]: Let system decide the best location
1 [internal]: Install on internal device storage
2 [external]: Install on external media
```

 Much like *am*, *pm*'s capabilities have grown through the versions, and the online help in 4.2/Jelly Bean for this tool is now much larger than can reasonably fit in this book. I still encourage you to take a look at it.

Fortunately, this command is actually pretty well documented, as you can see from the output above. Listing the installed packages, for example, is as simple as:

```
# pm list packages
package:android
package:android.tts
package:com.android.bluetooth
package:com.android.browser
package:com.android.calculator2
package:com.android.calendar
package:com.android.camera
package:com.android.certinstaller
package:com.android.contacts
package:com.android.defcontainer
...
```

Installing an app (the command used by the adb install command covered in the last chapter):

```
# pm install FastBirds.apk
    pkg: FastBirds.apk
Success
```

Note that removing the app requires knowing its package name, not the original *.apk*'s name:

```
# pm uninstall com.acme.fastbirds
Success
```

pm is also a shell script that starts Java code:

```
# Script to start "pm" on the device, which has a very rudimentary
# shell.
#
base=/system
export CLASSPATH=$base/framework/pm.jar
exec app_process $base/bin com.android.commands.pm.Pm "$@"
```

As with *am*, there's much more to *pm* than I can cover in this book. I encourage you to explore its many uses, as it can be very helpful for scripts, either during development and/or in production.

svc

Unlike the two previous commands, *svc* is something of a Swiss Army knife in attempting to provide you with the ability to control several system services. Here's the online help for 2.3/Gingerbread:

```
# svc
Available commands:
       help    Show information about the subcommands
       power   Control the power manager
       data    Control mobile data connectivity
       wifi    Control the Wi-Fi manager
```

The online help for 4.2/Jelly Bean shows that it can now also deal with USB:

```
root@android:/ # svc
Available commands:
       help    Show information about the subcommands
       power   Control the power manager
       data    Control mobile data connectivity
       wifi    Control the Wi-Fi manager
       usb     Control Usb state
```

Note how *svc*'s capabilities are limited to enabling and disabling the behavior of the designated system services:

```
# svc help power
Control the power manager

usage: svc power stayon [true|false|usb|ac]
         Set the 'keep awake while plugged in' setting.
```

```
# svc help data
Control mobile data connectivity

usage: svc data [enable|disable]
        Turn mobile data on or off.

       svc data prefer
            Set mobile as the preferred data network

# svc help wifi
Control the Wi-Fi manager

usage: svc wifi [enable|disable]
        Turn Wi-Fi on or off.

       svc wifi prefer
            Set Wi-Fi as the preferred data network
```

Overall, you should be aware of *svc*, but it's unlikely that you'll make regular use of it. Like *am* and *pm*, *svc* is also a script that uses *app_process* to start Java code.

ime

The *ime* command lets you communicate with the Input Method system service to control the system's use of available input methods, and it's the same in 2.3/Gingerbread and 4.2/Jelly Bean:

```
# ime
usage: ime list [-a] [-s]
       ime enable ID
       ime disable ID
       ime set ID

The list command prints all enabled input methods.  Use
the -a option to see all input methods.  Use
the -s option to see only a single summary line of each.

The enable command allows the given input method ID to be used.

The disable command disallows the given input method ID from use.

The set command switches to the given input method ID.
```

Here's the list of input methods available on the 2.3/Gingerbread emulator, for example:

```
# ime list
com.android.inputmethod.latin/.LatinIME:
  mId=com.android.inputmethod.latin/.LatinIME mSettingsActivityName=com.android.
inputmethod.latin.LatinIMESettings
  mIsDefaultResId=0x7f080001
  Service:
    priority=0 preferredOrder=0 match=0x108000 specificIndex=-1 isDefault=false
    ServiceInfo:
```

```
name=com.android.inputmethod.latin.LatinIME
packageName=com.android.inputmethod.latin
labelRes=0x7f0c001f nonLocalizedLabel=null icon=0x0
enabled=true exported=true processName=com.android.inputmethod.latin
permission=android.permission.BIND_INPUT_METHOD
```

Again, *ime* uses *app_process* from within a script to start Java code. Like *svc*, *ime* is a command worth keeping in mind, but you're unlikely to use it very often.

input

input connects to the Window Manager system service and injects text or key events into the system. Here's how it operates on 2.3/Gingerbread:

```
# input
usage: input [text|keyevent]
       input text <string>
       input keyevent <event_code>
```

Here's how it works on 4.2/Jelly Bean:

```
root@android:/ # input
usage: input ...
       input text <string>
       input keyevent <key code number or name>
       input [touchscreen|touchpad] tap <x> <y>
       input [touchscreen|touchpad] swipe <x1> <y1> <x2> <y2>
       input trackball press
       input trackball roll <dx> <dy>
```

input's functionality is very simple, however. It doesn't, for instance, know anything about what's receiving the events, just that the events are sent to whatever presently has focus. It's therefore up to you to make sure that whatever needs to receive your input actually has focus. Evidently this is difficult when you're not in front of the screen and are, instead, trying to script such behavior. Still, *input* gives you a tool to provide raw input from the command line. And, in some cases, the meaning of the input you send doesn't require focus. Here's how to click the Home button from the command line, for example:

```
# input keyevent 3
```

You're probably wondering how I know that 3 is the Home key. Have a look at *frameworks/base/core/java/android/view/KeyEvent.java* and *frameworks/base/native/include/android/keycodes.h* in 2.3/Gingerbread or *frameworks/native/include/android/keycodes.h* in 4.2/Jelly Bean for the full list of key codes recognized by Android. The former, for example, contains code such as this:

```
...
    public static final int KEYCODE_HOME           = 3;
    /** Key code constant: Back key. */
    public static final int KEYCODE_BACK           = 4;
```

```
            /** Key code constant: Call key. */
            public static final int KEYCODE_CALL           = 5;
            /** Key code constant: End Call key. */
            public static final int KEYCODE_ENDCALL        = 6;
            /** Key code constant: '0' key. */
            public static final int KEYCODE_0              = 7;
        ...
```

Like all other commands, *input* is a script that relies on *app_process*.

monkey

There's another tool that allows you to provide input to Android. It's called *monkey*, and there's an entire section about it in the app developer documentation entitled UI/Application Exerciser Monkey (*https://developer.android.com/tools/help/monkey.html*). As the documentation says, *monkey* can be used to provide random yet repeatable input to your application. This command, for instance, will send 50 pseudo-random inputs to the browser app:

```
# monkey -p com.android.browser -v 50
```

monkey can, however, do much more, as you can see from this output on 2.3/Gingerbread (4.2/Jelly Bean's is fairly similar):

```
# monkey
usage: monkey [-p ALLOWED_PACKAGE [-p ALLOWED_PACKAGE] ...]
              [-c MAIN_CATEGORY [-c MAIN_CATEGORY] ...]
              [--ignore-crashes] [--ignore-timeouts]
              [--ignore-security-exceptions]
              [--monitor-native-crashes] [--ignore-native-crashes]
              [--kill-process-after-error] [--hprof]
              [--pct-touch PERCENT] [--pct-motion PERCENT]
              [--pct-trackball PERCENT] [--pct-syskeys PERCENT]
              [--pct-nav PERCENT] [--pct-majornav PERCENT]
              [--pct-appswitch PERCENT] [--pct-flip PERCENT]
              [--pct-anyevent PERCENT]
              [--pkg-blacklist-file PACKAGE_BLACKLIST_FILE]
              [--pkg-whitelist-file PACKAGE_WHITELIST_FILE]
              [--wait-dbg] [--dbg-no-events]
              [--setup scriptfile] [-f scriptfile [-f scriptfile] ...]
              [--port port]
              [-s SEED] [-v [-v] ...]
              [--throttle MILLISEC] [--randomize-throttle]
              [--profile-wait MILLISEC]
              [--device-sleep-time MILLISEC]
              [--randomize-script]
              [--script-log]
              [--bugreport]
              COUNT
```

Most interestingly, you can provide a script to *monkey* for running a predefined set of input instead of providing random input. This is a very useful feature for development, testing, and in-the-field diagnostics. Unfortunately, there's virtually no documentation whatsoever on this very powerful feature of *monkey*. So, for reference, here's a sample script file:

```
# This is a sample test script
# Lines starting with '#' are comments

# This part is the "header"
# monkey doesn't actually look for 'type', but does require 'count', 'speed' and
# 'start data >>'
type= custom
count= 100
speed= 1.0
start data >>

# These are the actual instructions to carry out
LaunchActivity(com.android.contacts,com.android.contacts.TwelveKeyDialer)
# Use this instead in 4.2./Jelly Bean (line-wrap is for book, remove to run)
#   LaunchActivity(com.android.contacts,com.android.contacts.activities.Dialtact
#   sActivity)
UserWait(2500)
DispatchPress(KEYCODE_1)
UserWait(200)
DispatchPress(KEYCODE_8)
UserWait(200)
DispatchPress(KEYCODE_0)
UserWait(200)
DispatchPress(KEYCODE_0)
UserWait(200)
DispatchPress(KEYCODE_8)
UserWait(200)
DispatchPress(KEYCODE_8)
UserWait(200)
DispatchPress(KEYCODE_9)
UserWait(200)
DispatchPress(KEYCODE_8)
UserWait(200)
DispatchPress(KEYCODE_9)
UserWait(200)
DispatchPress(KEYCODE_6)
UserWait(200)
DispatchPress(KEYCODE_9)
UserWait(200)
DispatchPress(KEYCODE_ENTER)
UserWait(10000)
DispatchPress(KEYCODE_ENDCALL)
UserWait(200)
RunCmd(input keyevent 3)
UserWait(1000)
RunCmd(service call statusbar 1)
UserWait(2000)
RunCmd(service call statusbar 2)
```

To run this script, use this command line:

```
# monkey -f myscript 1
```

This script will essentially start the standard dialer, dial 1-800-889-8969,[3] wait 10 seconds, hang up, return to the home screen, and then expand and collapse the status bar. Notice that the last part uses the RunCmd instruction to make the script run commands straight from the command line; incidentally these are commands we saw earlier. Of course this script is rather short and simple. You can create much longer scripts; you can possibly even integrate the invocation of such scripts into much more complicated shell scripts.

For a detailed understanding of the scripting language understood by *monkey*, along with the parameters each command can take, I invite you to take a look at *monkey*'s script interpreting code in *development/cmds/monkey/src/com/android/commands/monkey/MonkeySourceScript.java* and look for EVENT_KEYWORD_. You should then find event keywords such as DispatchPress, UserWait, and many others.

To do its magic, *monkey* communicates with the Activity Manager, the Window Manager, and the Package Manager. It too is a shell script that relies on *app_process* to start the Java code that implements the utility.

If you look into the tool's sources in *development/cmds/monkey/*, you will find a file called *example_script.txt* that appears to contain some scripted instructions. It's unclear why this file is in the sources, as the semantics in that file do not correspond to the actual semantics expected by the *monkey* utility.

bmgr

Since 2.2/Froyo, Android has included a backup capability, allowing users to have their data backed up into the cloud so it can be restored later should they lose or change their device. Google itself provides some of this capability by acting as one of the possible *transports*,[4] but others could provide alternative transports. The API provided within Android and to app developers is transport-independent. This remains, however, a functionality that is very specific to the use of Android for phones and tablets and may not be required in an embedded environment. There's a tool that allows you to control the behavior of the Backup Manager system service from the command line:[5]

```
# bmgr
usage: bmgr [backup|restore|list|transport|run]
```

---

3. The publisher's phone number, if you're wondering.
4. A "transport" in the context of *bmgr* is the required engine to interface with a given cloud service.
5. This is the output on 2.3/Gingerbread. 4.2/Jelly Bean's is fairly similar.

```
bmgr backup PACKAGE
bmgr enable BOOL
bmgr enabled
bmgr list transports
bmgr list sets
bmgr transport WHICH
bmgr restore TOKEN
bmgr restore PACKAGE
bmgr run
bmgr wipe PACKAGE
```

The 'backup' command schedules a backup pass for the named package.
Note that the backup pass will effectively be a no-op if the package
does not actually have changed data to store.

The 'enable' command enables or disables the entire backup mechanism.
If the argument is 'true' it will be enabled, otherwise it will be
disabled. When disabled, neither backup or restore operations will
be performed.

The 'enabled' command reports the current enabled/disabled state of
the backup mechanism.

The 'list transports' command reports the names of the backup transports
currently available on the device. These names can be passed as arguments
to the 'transport' command. The currently selected transport is indicated
with a '*' character.

The 'list sets' command reports the token and name of each restore set
available to the device via the current transport.

The 'transport' command designates the named transport as the currently
active one. This setting is persistent across reboots.

The 'restore' command when given a restore token initiates a full-system
restore operation from the currently active transport. It will deliver
the restore set designated by the TOKEN argument to each application
that had contributed data to that restore set.

The 'restore' command when given a package name initiates a restore of
just that one package according to the restore set selection algorithm
used by the RestoreSession.restorePackage() method.

The 'run' command causes any scheduled backup operation to be initiated
immediately, without the usual waiting period for batching together
data changes.

The 'wipe' command causes all backed-up data for the given package to be
erased from the current transport's storage. The next backup operation
that the given application performs will rewrite its entire data set.

If this is relevant to your use of Android, have a look at the Data Backup (*https://devel oper.android.com/guide/topics/data/backup.html*) section of the app developer manual, along with the information provided by Google (*https://developers.google.com/android/ backup/*) regarding its own backup transport. Much like many of the other commands we saw, *app_process* is used to start the actual Java code that interfaces with the Backup Manager service.

stagefright

One of Android's key features is its rich media layer, and the AOSP includes tools that enable you to interact with it. More specifically, the *stagefright* command interacts with the Media Player service to allow you to do media playback. Here's its online help in 2.3/Gingerbread (4.2/Jelly Bean's is slightly expanded):

```
# stagefright -h
usage: stagefright
       -h(elp)
       -a(udio)
       -n repetitions
       -l(ist) components
       -m max-number-of-frames-to-decode in each pass
       -b bug to reproduce
       -p(rofiles) dump decoder profiles supported
       -t(humbnail) extract video thumbnail or album art
       -s(oftware) prefer software codec
       -o playback audio
       -w(rite) filename (write to .mp4 file)
       -k seek test
```

Here's how you can play an .mp3 file, for example:

```
# stagefright -a -o /sdcard/trainwhistle.mp3
```

You might also want to investigate the *record* and *audioloop* utilities found alongside *stagefright*'s sources in *frameworks/base/cmds/stagefright/* in 2.3/Gingerbread and *frameworks/av/cmds/stagefright/* in 4.2/Jelly Bean. Their documentation is severely lacking, though, and few examples of their uses can be found online or elsewhere. Interestingly, though, all three utilities are coded in C, unlike the majority of the system service-specific utilities we've seen thus far, which were mostly written in Java and activated through a script using *app_process*. Also, while *stagefright* directly communicates with the Media Player service, the *record* and *audioloop* commands use an OMXClient, which conveniently wraps the communication to the same service.

Dalvik Utilities

We've already seen how we can send intents with the *am* command and therefore trigger the starting of new apps, each of which comes with its own Zygote-forked Dalvik instances. We've also seen how the *app_process* command can be used to start Java-coded command-line tools using the Android Runtime. There are some cases, however, where you may want to forgo all the Android-specific layers and dabble directly with Dalvik. Here are the commands that allow you to do just that.

dalvikvm

If you haven't yet already asked yourself if there's a way to actually start just a Dalvik VM without any Android-specific functionality, here's the command you've been looking for:[6]

```
# dalvikvm -help

dalvikvm: [options] class [argument ...]
dalvikvm: [options] -jar file.jar [argument ...]

The following standard options are recognized:
  -classpath classpath
  -Dproperty=value
  -verbose:tag  ('gc', 'jni', or 'class')
  -ea[:<package name>... |:<class name>]
  -da[:<package name>... |:<class name>]
  (-enableassertions, -disableassertions)
  -esa
  -dsa
  (-enablesystemassertions, -disablesystemassertions)
  -showversion
  -help

The following extended options are recognized:
  -Xrunjdwp:<options>
  -Xbootclasspath:bootclasspath
  -Xcheck:tag  (e.g. 'jni')
  -XmsN  (min heap, must be multiple of 1K, >= 1MB)
  -XmxN  (max heap, must be multiple of 1K, >= 2MB)
  -XssN  (stack size, >= 1KB, <= 256KB)
  -Xverify:{none,remote,all}
  -Xrs
  -Xint  (extended to accept ':portable', ':fast' and ':jit')

These are unique to Dalvik:
  -Xzygote
  -Xdexopt:{none,verified,all}
  -Xnoquithandler
  -Xjnigreflimit:N  (must be multiple of 100, >= 200)
  -Xjniopts:{warnonly,forcecopy}
  -Xjnitrace:substring (eg NativeClass or nativeMethod)
  -Xdeadlockpredict:{off,warn,err,abort}
  -Xstacktracefile:<filename>
```

---

6. This is the output from 2.3/Gingerbread. 4.2/Jelly Bean's output is fairly similar.

```
-Xgc:[no]precise
-Xgc:[no]preverify
-Xgc:[no]postverify
-Xgc:[no]concurrent
-Xgc:[no]verifycardtable
-Xgenregmap
-Xcheckdexsum
-Xincludeselectedop
-Xjitop:hexopvalue[-endvalue][,hexopvalue[-endvalue]]*
-Xincludeselectedmethod
-Xjitthreshold:decimalvalue
-Xjitblocking
-Xjitmethod:signature[,signature]* (eg Ljava/lang/String\;replace)
-Xjitcheckcg
-Xjitverbose
-Xjitprofile
-Xjitdisableopt

Configured with: debugger profiler hprof jit(armv5te) show_exception=1

Dalvik VM init failed (check log file)
```

dalvikvm is actually a raw Dalvik VM without any connection to "Android" whatsoever. It doesn't rely on the Zygote, nor does it include the Android Runtime. It simply starts a VM to run whatever class or JAR file you provide it. It's actually not used very often in the AOSP itself, probably because there isn't much in the AOSP that doesn't run in the context of "Android." The "preload" Java library in 2.3/Gingerbread, for example, uses it in *frameworks/base/tools/preload/MemoryUsage.java* in conjunction with *adb* to check the amount of memory used by a class on the target.

dvz

Yet another way to start a Dalvik VM is the *dvz* command:

```
# dvz --help
Usage: dvz [--help] [-classpath <classpath>]
[additional zygote args] fully.qualified.java.ClassName [args]

Requests a new Dalvik VM instance to be spawned from the zygote
process. stdin, stdout, and stderr are hooked up. This process remains
while the spawned VM instance is alive and forwards some signals.
The exit code of the spawned VM instance is dropped.
```

As the description implies, *dvz* actually acts in a similar fashion to the Activity Manager by requesting the Zygote to fork and start a new process. The only difference here is that the resulting process isn't managed by the Activity Manager. Instead, it's very much standalone.

It's unclear whether this utility is meant to be heavily used, as the only instances of its use within 2.3/Gingerbread are in test code, specifically in *dalvik/tests/etc/push-and-*

run-test-jar, and it's not even included in the default builds in 4.2/Jelly Bean. Nevertheless, there might be instances where having this in your arsenal could be useful.

The Many Ways to Start Dalvik

Up to now, we've seen four different ways to start a Dalvik VM. It's worth taking a moment to put them all in perspective. Table 7-1 describes each way to get a working Dalvik VM, along with what's included in the VM and how it's started.

Table 7-1. Ways to start Dalvik

Command	Dalvik VM	Android Runtime	Zygote	Activity Manager	Mechanism
dalvikvm	X				Uses *libdvm.so*
app_process	X	X			Uses *libandroid_run time.so*
dvz	X	X	X		Uses *libcutils*[a]
am	X	X	X	X	Talks to Activity Manager service

[a] See *system/core/libcutils/zygote.c*, which contains a `zygote_run_wait()` and a `zygote_run_one shot()`.

am is the only command that provides us with a Dalvik VM instance that's actually controlled by the Activity Manager. In all other cases, the VM is independent and does not have its lifecycle managed. *am* is also the only command that allows us to automatically trigger the execution of code contained in an *.apk*. All other commands require us to provide a specific class or JAR file.

---

dexdump

If you'd like to reverse-engineer Android apps or JAR files, you can do so with *dexdump*:

```
# dexdump
dexdump: no file specified
Copyright (C) 2007 The Android Open Source Project

dexdump: [-c] [-d] [-f] [-h] [-i] [-l layout] [-m] [-t tempfile] dexfile...

 -c : verify checksum and exit
 -d : disassemble code sections
 -f : display summary information from file header
 -h : display file header details
 -i : ignore checksum failures
 -l : output layout, either 'plain' or 'xml'
 -m : dump register maps (and nothing else)
 -t : temp file name (defaults to /sdcard/dex-temp-*)
```

Here's how it can be used on a JAR file:

```
# dexdump /system/framework/services.jar
Processing '/system/framework/services.jar'...
Opened '/system/framework/services.jar', DEX version '035'
Class #0            -
  Class descriptor  : 'Lcom/android/server/AccessibilityManagerService$1;'
  Access flags      : 0x0000 ()
  Superclass        : 'Landroid/os/Handler;'
  Interfaces        -
  Static fields     -
  Instance fields   -
    #0              : (in Lcom/android/server/AccessibilityManagerService$1;)
      name          : 'this$0'
      type          : 'Lcom/android/server/AccessibilityManagerService;'
      access        : 0x1010 (FINAL SYNTHETIC)
  Direct methods    -
    #0              : (in Lcom/android/server/AccessibilityManagerService$1;)
      name          : '<init>'
      type          : '(Lcom/android/server/AccessibilityManagerService;)V'
      access        : 0x10000 (CONSTRUCTOR)
      code          -
      registers     : 2
      ins           : 2
      outs          : 1
      insns size    : 6 16-bit code units
      catches       : (none)
      positions     :
        0x0000 line=113
      locals        :
        0x0000 - 0x0006 reg=0 this Lcom/android/server/AccessibilityManagerServi
ce$1;
  Virtual methods   -
    #0              : (in Lcom/android/server/AccessibilityManagerService$1;)
      name          : 'handleMessage'
...
```

You can also ask it to dissassemble code:

```
# dexdump -d /system/app/Launcher2.apk
...
00ea5c:                                         |[00ea5c] com.android.common.Arra
yListCursor.<init>:([Ljava/lang/String;Ljava/util/ArrayList;)V
00ea6c: 1206                                    |0000: const/4 v6, #int 0 // #0
00ea6e: 1a07 e804                               |0001: const-string v7, "_id" //
string@04e8
00ea72: 7010 b400 0800                          |0003: invoke-direct {v8}, Landro
id/database/AbstractCursor;.<init>:()V // method@00b4
00ea78: 2190                                    |0006: array-length v0, v9
00ea7a: 1201                                    |0007: const/4 v1, #int 0 // #0
00ea7c: 1202                                    |0008: const/4 v2, #int 0 // #0
00ea7e: 3502 0f00                               |0009: if-ge v2, v0, 0018 // +000
f
```

```
00ea82: 4604 0902                         |000b: aget-object v4, v9, v2
00ea86: 1a05 e804                         |000d: const-string v5, "_id" //
string@04e8
00ea8a: 6e20 dd07 7400                    |000f: invoke-virtual {v4, v7}, L
java/lang/String;.compareToIgnoreCase:(Ljava/lang/String;)I // method@07dd
00ea90: 0a04                              |0012: move-result v4
00ea92: 3904 3e00                         |0013: if-nez v4, 0051 // +003e
00ea96: 5b89 3600                         |0015: iput-object v9, v8, Lcom/a
ndroid/common/ArrayListCursor;.mColumnNames:[Ljava/lang/String; // field@0036
00ea9a: 1211                              |0017: const/4 v1, #int 1 // #1
00ea9c: 3901 1400                         |0018: if-nez v1, 002c // +0014
00eaa0: d804 0001                         |001a: add-int/lit8 v4, v0, #int
1 // #01
00eaa4: 2344 d901                         |001c: new-array v4, v4, [Ljava/l
ang/String; // class@01d9
00eaa8: 5b84 3600                         |001e: iput-object v4, v8, Lcom/a
ndroid/common/ArrayListCursor;.mColumnNames:[Ljava/lang/String; // field@0036
00eaac: 5484 3600                         |0020: iget-object v4, v8, Lcom/a
ndroid/common/ArrayListCursor;.mColumnNames:[Ljava/lang/String; // field@0036
...
```

Obviously the topic of reverse-engineering Android goes way beyond the scope of this book, but if this topic is of general interest, I recommend taking a look at your favorite online bookstore for books that specialize in Android security and forensics.

Support Daemons

While the bulk of Android's intelligence is implemented in system services, there are a number of cases where a system service acts partly as intermediary to a native daemon that actually does the key operations required. There are likely two main reasons why this approach has been favored instead of conducting the actual operations directly as part of a system server: security and reliability.

As I explained in Chapter 1, Android's permission model requires app developers who need to call on privileged operations to request specific permissions at build time. Typically, these permissions will resemble something like this in an app's manifest file:

```
...
    <uses-permission android:name="android.permission.INTERNET" />
    <uses-permission android:name="android.permission.WAKE_LOCK" />
...
```

In this case, these permissions ask for the ability to open sockets and grab wakelocks. There are obviously a whole lot more permissions than this. Have a look at the app developer documentation on the full list of permissions available (*http://developer.android.com/reference/android/Manifest.permission.html*). Without these permissions, an app can't conduct some of the most critical Android operations. And the main reason is that apps run as unprivileged users that can't, for instance, invoke any system call that requires root privileges or access most of the key devices in */dev*. Instead, apps must ask system services to act on their behalf and, in turn, system services check apps' permissions before following through with any requests they get.

System services don't, however, themselves run as root. Instead, the *system_server* process runs as `system`; the *mediaserver* process runs as `media`; and the Phone app runs as `radio`. And if you check in */dev*, you'll see that some entries belong exclusively to some of these users. You'll also see quite a few entries that belong to the `root` user. Hence, much like apps, system services can't typically use system calls that require root privileges nor access key devices in */dev*.

Instead, many key operations require system services to communicate through Unix domain sockets in */dev/socket/* with native daemons running as either root or as a specific user to conduct privileged operations. Many of those daemons are Android-specific, though some, such as *bluetoothd* prior to 4.2/Jelly Bean, we've already covered in Chapter 6 as being legacy Linux daemons.

In some specific cases, such as *rild*, for example, which takes care of the communication with the Baseband Processor, it seems that the choice to run as a separate process might likely have more to do with reliability. Indeed, the phone functionality of a smartphone is so critical that it's worth ensuring that its operation is independent of any potential issues that could affect the system services housed in the `system_server` process.

Let's take a look at the main support daemons used by system services, their configuration, and related command-line tools. Note that we won't cover the daemons we covered earlier, such as the Zygote; or those that aren't tied to system services, such as *ueventd* and *dumpsys*; or those, such as *bluetoothd* or *wpa_supplicant*, that are not Android specific.

installd

While the Package Manager service's job is to deal with the management of *.apk* files, it doesn't have the proper privileges to carry out many of the manipulations and/or operations required to set up an app to run. Instead, it relies on *installd*, which runs as root in 2.3/Gingerbread and as the `install` user in 4.2/Jelly Bean, for key filesystem operations and commands. Running *dexopt* on an *.apk* to generate JIT-optimized *.dex* files for Dalvik, for instance, is done by *installd* on the Package Manager's behalf at install time.

installd is started by this section of *init.rc* in 2.3/Gingerbread (4.2/Jelly Bean does something fairly similar):

```
service installd /system/bin/installd
    socket installd stream 600 system system
```

It then opens */dev/socket/installd* and listens for a connection, and thereafter listens for commands from the Package Manager. It doesn't have a configuration file, nor does it take any command-line parameters. Neither is there any command-line tool to communicate with it independently of the Package Manager. Hence, the only way to activate *installd* from the command line is to use the *pm* command, which will communicate with the Package Manager, which will, in turn, communicate with *installd* if required.

installd's sources are in *frameworks/base/cmds/installd/*, and you may want to take a look at *install.c* and *commands.c*. The former contains the list of commands recognized by *installd*, and the latter contains the actual implementation of those commands. For reference, here's the snippet from 2.3/Gingerbread's *install.c* that lists the commands recognized by *installd* (4.2/Jelly Bean adds a few more commands to that list):

```
struct cmdinfo cmds[] = {
    { "ping",        0, do_ping },
    { "install",     4, do_install },
    { "dexopt",      3, do_dexopt },
    { "movedex",     2, do_move_dex },
    { "rmdex",       1, do_rm_dex },
    { "remove",      2, do_remove },
    { "rename",      3, do_rename },
    { "freecache",   1, do_free_cache },
    { "rmcache",     2, do_rm_cache },
    { "protect",     2, do_protect },
    { "getsize",     4, do_get_size },
    { "rmuserdata",  2, do_rm_user_data },
    { "movefiles",   0, do_movefiles },
    { "linklib",     2, do_linklib },
    { "unlinklib",   1, do_unlinklib },
};
```

Note that, much like many of the other daemons we'll see below, the wire protocol between *installd* and the Package Manager is string based. Hence, the above snippet contains three entries per command: the command's string as sent "on the wire," the number of parameters expected, and the function within *install.c* to call when the command is received.

vold

vold takes care of many of the key operations required by the Mount Service, such as mounting and formatting volumes. Unlike *installd*, *vold* runs as root in both 2.3/Gingerbread and 4.2/Jelly Bean, while the Mount Service is part of the System Server. *vold* is started by this section of 2.3/Gingerbread's *init.rc* (the snippet in 4.2/Jelly Bean is similar):

```
service vold /system/bin/vold
    socket vold stream 0660 root mount
    ioprio be 2
```

Unlike the rest of the support daemons covered here, *vold* actually has a configuration file, */etc/vold.fstab*. Here's a snippet from the default *vold.fstab* found in *system/core/rootdir/etc/* describing the file's semantics:

```
########################
## Regular device mount
##
## Format: dev_mount <label> <mount_point> <part> <sysfs_path1...>
## label        - Label for the volume
## mount_point  - Where the volume will be mounted
## part         - Partition # (1 based), or 'auto' for first usable partition.
## <sysfs_path> - List of sysfs paths to source devices
########################
```

Here's the section that relates to the SD card in the emulator, for example:

```
dev_mount sdcard /mnt/sdcard auto /devices/platform/goldfish_mmc.0 /devices/plat
form/msm_sdcc.2/mmc_host/mmc1
```

When *vold* starts, it parses this file and then opens */dev/socket/vold* to listen for connections and commands. Unlike *installd*, there's a command-line tool to communicate directly with *vold*:

```
Usage: vdc <monitor>|<cmd> [arg1] [arg2...]
```

The actual parameters expected by *vdc* on the command line are the same as those expected by *vold* from the Mount Service when it connects through the designated socket. There is, unfortunately, no document or online help that describes the complete command set. Instead, you must look at the *CommandListener.cpp* file in *system/vold/* to see the implementation of *vold*'s command set.

You can, for instance, dump *vold*'s internal status:

```
# vdc dump
000 Dumping loop status
000 Dumping DM status
000 Dumping mounted filesystems
000 rootfs / rootfs ro 0 0
000 tmpfs /dev tmpfs rw,mode=755 0 0
000 devpts /dev/pts devpts rw,mode=600 0 0
000 proc /proc proc rw 0 0
000 sysfs /sys sysfs rw 0 0
000 none /acct cgroup rw,cpuacct 0 0
000 tmpfs /mnt/asec tmpfs rw,mode=755,gid=1000 0 0
000 tmpfs /mnt/obb tmpfs rw,mode=755,gid=1000 0 0
000 none /dev/cpuctl cgroup rw,cpu 0 0
000 /dev/block/mtdblock0 /system yaffs2 ro 0 0
000 /dev/block/mtdblock1 /data yaffs2 rw,nosuid,nodev 0 0
000 /dev/block/mtdblock2 /cache yaffs2 rw,nosuid,nodev 0 0
200 dump complete
```

In some cases, *vdc* actually offers online help:

```
# vdc volume format
500 Usage: volume format <path>
```

To customize the list of storage devices for your device in 4.2/Jelly Bean, have a look at *frameworks/base/core/res/res/xml/storage_list.xml*. You may want to create an overlay version of that file in your *device/acme/coyotepad/overlay/* to customize it for your device.

netd

The Network Management Service relies on *netd* for critical network configuration operations such as configuring network interfaces, setting up tethering, and running *pppd*. In this case, too, *netd* runs as root, while the Network Management Service is part of the System Server. *netd* is started by the following section of *init.rc* in 2.3/Gingerbread:

```
service netd /system/bin/netd
    socket netd stream 0660 root system
```

In 4.2/Jelly Bean, however, the declaration has changed:

```
service netd /system/bin/netd
    class main
    socket netd stream 0660 root system
    socket dnsproxyd stream 0660 root inet
    socket mdns stream 0660 root system
```

netd opens */dev/socket/netd* and listens for connections and commands. It doesn't take any command-line parameters, nor does it rely on any configuration file. Like *vold*, however, it has a command-line tool to communicate with it. Here's the online help for that command in 2.3/Gingerbread:

```
# ndc
Usage: ndc <monitor>|<cmd> [arg1] [arg2...]
```

Here's the same help on 4.2/Jelly Bean:

```
root@android:/ # ndc
Usage: ndc [sockname] <monitor>|<cmd> [arg1] [arg2...]
```

Like *vdc*, the command-line parameters expected by *ndc* are the same as those expected by *netd* on its socket. And as with *vold*, you need to look at *netd*'s *CommandListener.cpp* in *system/netd/* to understand its command semantics.

As with *vdc*, you can request *netd* status info with *ndc*:

```
# ndc interface list
110 lo
110 eth0
110 tunl0
110 gre0
200 Interface list completed
```

The Command Sets of vold and netd

Both *vold* and *netd* are constructed using the same C++ mechanism provided by *libsysutils* and rely on a *CommandListener.cpp* to parse and dispatch commands sent to them. To understand the specific commands accepted by each, have a look at the constructors in *CommandListener.cpp*:

```
CommandListener::CommandListener() :
                FrameworkListener("...") {
...
```

Each will contain calls to registerCmd(), which register objects defined farther below in the same file. Here's an excerpt from *vold* for the *dump* command in 2.3/Gingerbread:

```
CommandListener::CommandListener() :
                FrameworkListener("vold") {
    registerCmd(new DumpCmd());
    registerCmd(new VolumeCmd());
...
CommandListener::DumpCmd::DumpCmd() :
                VoldCommand("dump") {
}

int CommandListener::DumpCmd::runCommand(SocketClient *cli,
                                         int argc, char **argv) {
    cli->sendMsg(0, "Dumping loop status", false);
    if (Loop::dumpState(cli)) {
        cli->sendMsg(ResponseCode::CommandOkay, "Loop dump failed", true);
    }
...
```

Every command accepted by *vold* or *netd* has a corresponding runCommand() that parses the parameters passed to that command. By running *vdc dump* on the command line as we did earlier, for instance, we're invoking the runCommand() in the snippet above. Conversely, typing *vdc volume list* will invoke the following function and pass list as one part of the arguments:

```
int CommandListener::VolumeCmd::runCommand(SocketClient *cli,
                                           int argc, char **argv) {
...
```

rild

The Phone system service, which is hosted in the Phone app, uses *rild* to communicate with the Baseband Processor. *rild* itself uses dlopen() to load a baseband-specific *.so* to interface to the actual baseband hardware. As I mentioned before, *rild* likely exists to ensure that the phone side of the system remains active even if a problem occurs with the rest of the stack.

In the case of the emulator, *rild* is started by this portion of the *init.rc* file in 2.3/Gingerbread (4.2/Jelly Bean's version is practically identical):

```
service ril-daemon /system/bin/rild
    socket rild stream 660 root radio
    socket rild-debug stream 660 radio system
    user root
    group radio cache inet misc audio sdcard_rw
```

While it doesn't have a configuration file, *rild* itself can take a few command-line parameters:

```
Usage: rild -l <ril impl library> [-- <args for impl library>]
```

If no RIL implementation library is provided on the command line, *rild* will attempt to locate the library using the `rild.libpath` global property. If that isn't specified either, it'll assume there's no radio on the system loop around calls to `sleep()`. In the case of the emulator, the system relies on */system/lib/libreference-ril.so*, which, as its name implies, is a reference implementation for manufacturers that need to implement real RIL libraries.

There are two Unix domain sockets used by *rild*: */dev/socket/rild*, which is used by the Phone system service, and */dev/socket/rild-debug*, which can be used by the *radiooptions* command to interact. Indeed, the latter is a command-line tool to communicate with *rild*:

```
Usage: radiooptions [option] [extra_socket_args]
            0 - RADIO_RESET,
            1 - RADIO_OFF,
            2 - UNSOL_NETWORK_STATE_CHANGE,
            3 - QXDM_ENABLE,
            4 - QXDM_DISABLE,
            5 - RADIO_ON,
            6 apn- SETUP_PDP apn,
            7 - DEACTIVE_PDP,
            8 number - DIAL_CALL number,
            9 - ANSWER_CALL,
            10 - END_CALL
```

If you'd like to know more about *rild* and *radiooptions*, have a look at their sources in *hardware/ril/rild*. The reference RIL implementation is itself in *hardare/ril/reference-ril/*.

keystore

Unlike the rest of the daemons I've presented thus far, *keystore* doesn't actually service any of the system services. Instead, it's used by a variety of different pieces of the system for the storage and retrieval of key-value pairs. The values it maintains are mainly security keys for connecting to networks or network infrastructure such as access points and VPNs, and the means to secure the values is a user-defined password. Clearly, the goal of having a separate daemon for the storage of this information is to increase the system's overall security.

keystore is started by this portion of the *init.rc* file in 2.3/Gingerbread (4.2/Jelly Bean does substantially the same):

```
service keystore /system/bin/keystore /data/misc/keystore
    user keystore
    group keystore
    socket keystore stream 666
```

keystore doesn't have a configuration file, but it does expect to be provided with a directory to store each key-pair value. Typically, this is */data/misc/keystore*, as you can see before. *keystore* then listens in to */dev/socket/keystore* for connections and commands. Several native daemons connect to *keystore* to retrieve keys, such as *wpa_supplicant*, *mtpd*, and *racoon*. But the Settings app also connects to *keystore* to list and insert new keys.

There's also a command-line utility for communicating with *keystore*:

```
Usage: keystore_cli action [parameter ...]
```

You'll find both the sources of *keystore* and *keystore_cli* in *frameworks/base/cmds/keystore/* in 2.3/Gingerbread and in *system/security/keystore/* in 4.2/Jelly Bean.

Other Support Daemons

There are a few additional daemons that play a more minor role, which we won't cover here, such as *mtpd* and *racoon*. The former is used for VPNs and is found in *external/mtpd/*, and the latter is for IPsec and is found in *external/ipsec-tools/*.

There are possibly, of course, other daemons that may be running on your system for specific purposes, and/or you may want to add your own custom daemons. Have a look back at Chapter 4 for instructions on how to add your own custom binaries to the AOSP's build system. Remember that if you want a daemon to be started at startup by *init*, you need to add a `service` declaration for it in either the main *init.rc* or in the board-specific *init.<device_name>.rc*.

Hardware Abstraction Layer

As I explained in Chapter 2, Android relies on a Hardware Abstraction Layer (HAL) to interface with hardware. Indeed, system services almost never interact with devices through */dev* entries directly. Instead, they go through HAL modules, typically shared libraries, to talk to hardware, as is detailed in Table 2-1.

Android's HAL implementation is found in *hardware/*. Most importantly, you'll find the definitions of the interfaces between the Framework and the HAL modules in header files in *hardware/libhardware/include/hardware/* and *hardware/libhardware_legacy/include/hardware_legacy/*. The header files therein provide the exact API required for each type of hardware to be supported under Android. You'll also find example implementations of some of those HAL modules in the sources for the lead devices in *device/*.

Ideally, you want to avoid having to implement your own HAL modules for existing system services. Instead, you should query your SoC or board vendor for such modules. HAL module writing requires intricate knowledge of the internals of the system server that the module has to interact with and the specific Linux device driver required to interact with the hardware. Learning how to do this right can be a very time-intensive process, especially since the HAL interface tends to evolve with every new version of Android. I therefore strongly recommend that you use components/boards for which most HAL modules have already been made by the manufacturer or the SoC vendor.

Generally, given Android's market success, component and SoC vendors make a big effort to ensure that Android runs well with their products. This means they either provide you with fully functional AOSPs and Android-ready kernels for eval boards, and/or HAL modules and Linux drivers for their components. So, at the risk of sounding redundant, implement your own HAL modules for hardware types already recognized by Android only as a last resort. Instead, talk to your SoC or component vendor to get your hands on the HAL modules and drivers (or kernel) required to run Android on your hardware.

All major SoC vendors provide—in one way or another—access to ready-to-use AOSPs and kernels for running on the eval boards. Such is the case for TI, Qualcomm, Freescale, Samsung, and many others. If you're building your own custom board based on one of their designs, I recommend that you grab those reference AOSP trees and customize them for your own use. Attempting to start from scratch to port Android to your hardware using the AOSP trees provided directly from Google is not likely to be a good use of your time or fit your time-to-market requirements.

If you absolutely must implement your own HAL modules for existing system services, then refer to the header files I alluded to previously, which define the APIs required by each HAL module type, and take as much inspiration as possible from the reference HAL implementations provided for the lead devices in the *device/* directory. For 2.3/Gingerbread, for example, have a look at the various *lib*/* directories in *device/samsung/crespo/*. In the case of 4.2/Jelly Bean, have a look at *device/asus/grouper/* and *device/samsung/tuna/*.

Hardware Abstraction Layer | 305

APPENDIX A
Legacy User-Space

As I explained in Chapter 2, despite being based on the Linux kernel, Android bears little resemblance to any other Linux system out there. Indeed, as you can see in Figure 2-1, Android's user-space, which we explored in Chapters 6 and 7, is a custom creation of Google. Hence, if you're familiar with "legacy" Linux systems or come from an embedded Linux background, you may find yourself reminiscing about classic Linux tools and components you've been using for a long time. This appendix will show you how to get a legacy Linux user-space to coexist side by side with the AOSP on top of the same Linux kernel.

Basics

To start, we need to agree on what exactly a "legacy" Linux user-space is. For the present discussion, we'll assume we're talking about a Filesystem Hierarchy Standard (FHS)-compliant root filesystem. As I mentioned earlier, Android's root filesystem isn't FHS-compliant, and it crucially doesn't use key FHS directories such as */bin* and */lib*, allowing us to superimpose, side by side with it, a root filesystem that does use these directories.

Now, I'm not saying you'll be able to use these instructions to get yourself a root filesystem that houses both the AOSP and, say, a large distribution like Ubuntu. There are a lot more details about Ubuntu as a distribution and the AOSP that you'd need to take into account than resolving how to match a few of the top-level directories of the root filesystem. Nevertheless, if you are familiar with how to create a basic root filesystem for an embedded Linux system, it should become relatively clear how you could get your favorite tools and libraries, such as BusyBox and glibc, loaded on the same root filesystem as the AOSP. And if you're interested in something more ambitious, such as getting Ubuntu or Fedora to sit side by side with the AOSP in the same root filesystem, these explanations offer a good introduction to getting started.

Before starting on this path, though, it's worth answering a general question on this approach: Why bother? Indeed, why take the time to try to get any sort of legacy Linux software package to sit on the same kernel alongside the AOSP? Why not just use the AOSP, since it's already got a C library, command-line tools, a rich user-space, etc.? Can't the AOSP do everything needed? No?

The main reason a developer would want a legacy Linux user-space alongside Android is to be able to port existing Linux applications over to a system that runs Android **without** having to port them over to Android. For instance, if you have legacy code that works just fine on glibc, it might be easier to just get glibc onto your root filesystem than to try to port your legacy code over to Bionic. Indeed, as you can see by reading Bionic's own documentation in *bionic/libc/*, especially those files in the *docs/* directory, Bionic has many limitations and differences when compared with something more mainstream like glibc. It's not Posix-compliant, for example, nor does it expose System V IPC calls. By relying on a well-known C library such as glibc, you avoid any of these portability issues.

Another good reason for reusing components from classic Linux systems is to avoid having to deal with Android's build system. As we saw in Chapter 4, Android's build system is nonrecursive. Therefore, if you would like to reuse large, legacy software packages, you'd typically have to convert their build systems to use Android's build system *.mk* files. As a matter of fact, some of the very well-known packages imported into the AOSP's *external/* directory have had their build files re-created for use within the AOSP. D-Bus, for instance, which is traditionally based on autoconf/automake, has had *Android.mk* files added to its sources in *external/dbus/* so it will build within the AOSP. None of the files originally used for its build, such as the *configure* script, are used when it's built within the AOSP. An easy way out of this is to generate a root filesystem independently of the AOSP for those legacy packages you need and then merge the result with the AOSP.

Put another way, there's benefit to reusing existing legacy build systems. For example, there's no reason not to use something like Yocto or Buildroot to generate a root filesystem that fits your needs and then merge the result with the AOSP. Indeed, there are a lot of existing build systems and packaging systems that can generate very useful output using legacy methods to mix with the AOSP. In some cases, the cost/benefit equation might make it inconceivable to port a package's build system over to the AOSP's simply because of the original project's codebase size.

None of the present explanations should preclude you from trying to build your legacy code against Bionic. There is a slight chance that the changes required are marginal. Also, as I showed in Chapter 4, you can put together *Android.mk* files that call on existing recursive make-based build scripts.

Still, knowing how to circumvent Bionic is a very useful trick. So I encourage you to read on.

Theory of Operation

Once you've decided that you want to get legacy Linux user-space components to work alongside with the AOSP, the next question is how. This is actually a two-part question. First, how do we get the legacy user-space and the AOSP onto the same filesystem images? And second, how does this legacy user-space interact with the AOSP's components? Let's start by addressing the former.

Assuming you're using a method like that covered in *Building Embedded Linux Systems, 2nd ed.* to generate a glibc-based root filesystem, Figure A-1 illustrates the general approach of how this root filesystem can be made to integrate with the AOSP. Essentially, the project environment PRJROOT is made to host the creation of a glibc-based root filesystem. The AOSP build system is then modified to copy the contents of that root filesystem into the images generated by the AOSP. And since the AOSP doesn't originally contain a */bin* and a */lib*, these directories will be created and populated by the contents of the glibc-based root filesystem.

The rest of these explanations assume that you either already have a glibc-based root filesystem that you want to merge with the AOSP or you know how to create one. If you don't have one and don't know how to create one, I recommend you take a look at *Building Embedded Linux Systems, 2nd ed.* (which was originally written by yours truly).

Once the matter of merging the legacy components into the AOSP is solved, the other key issue to discuss is how to use those components and/or interact with them within the AOSP. Put simply, all command-line utilities and binaries can be used as is, straight from Android's command line. For example, if you have */bin/foo* and */bin* is in the Android path, you can just go ahead and type something like *adb shell* and then type *foo* on the command line to run the binary. There's likely more you'll want to do, such as integrating into Android's *init*; we'll discuss this shortly.

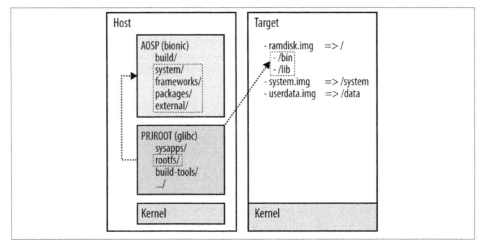

Figure A-1. Merging a legacy Linux user-space with the AOSP

Basic command-line operations and *init* configuration aside, though, a more fundamental discussion point is how to get components running on different C libraries to communicate together. How does a daemon linked against glibc, for instance, sync with a daemon linked against Bionic? Or how does a command-line tool linked against glibc communicate with a Bionic-linked daemon?

Remember that despite being linked against different C libraries, everything is running on the same kernel. Hence, whatever IPC mechanisms exist in the kernel can still be used by whatever binary is running on it. And as you can see in Figure A-2, it's perfectly feasible to have a glibc-based component use regular IPC mechanisms to communicate with a Bionic-based component within the AOSP. Sockets, for instance, are a prime candidate, given that they're implemented in both glibc and Bionic. System V IPC mechanisms, on the other hand, are available only in glibc. You could also look at using Binder, though you'd have to get libbinder to compile against glibc.

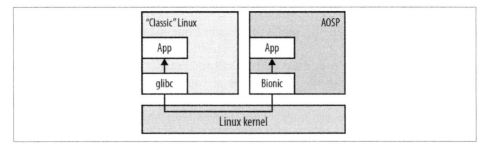

Figure A-2. Communication between a glibc-based stack and the AOSP

Many development teams I work with, for instance, have developed substantial glibc-based stacks over the years that they typically run in embedded Linux systems. And while working on integrating Android in their product lines, they're often confronted with having to make a choice between porting those stacks and their control logic over to Bionic or figuring out a way for those legacy stacks to coexist in a friendly fashion with the AOSP. One potential path for most of these teams is to create a setup like the one I just described and then have the control logic of the legacy stack communicate with newly created Android components using sockets. It's not a silver bullet, but it's a useful trick to master in case it applies to your design, or to part of it.

Merging with the AOSP

Now that we've covered the essentials, let's actually put this method into practice. The first thing you'll need, of course, is a functional legacy filesystem to merge with the AOSP. In this specific case, assume that I followed the instructions described in *Building Embedded Linux Systems, 2nd ed.* to create a root filesystem based on glibc that contains BusyBox. Hence, we have something like this:

```
$ ls -l ${PRJROOT}/rootfs
total 16
drwxr-xr-x 2 karim karim 4096 2012-10-26 23:12 bin
drwxr-xr-x 2 karim karim 4096 2012-10-26 23:12 lib
lrwxrwxrwx 1 karim karim   11 2012-10-26 23:12 linuxrc -> bin/busybox
drwxr-xr-x 2 karim karim 4096 2012-10-26 23:12 sbin
drwxr-xr-x 4 karim karim 4096 2012-10-26 23:12 usr
```

To make things simpler, I'm going to copy that root filesystem into a new directory in my AOSP:

```
$ cp -a ${PRJROOT}/rootfs path_to_my_aosp/rootfs-glibc/
```

I now have a *rootfs-glibc* directory at the top level of my AOSP. This directory won't be of much use, however, given that there's no *Android.mk* that takes it into account, and if you build the AOSP at this point, it'll be completely ignored. To fix this, we can create such an *Android.mk* to force the AOSP's build system to copy the content of our glibc-based root filesystem. Here's my *rootfs-glibc/Android.mk*, as an example of making this work in 2.3/Gingerbread:

```
LOCAL_PATH:= $(call my-dir)
include $(CLEAR_VARS)

# This part is a hack, we're doing "addprefix" because if we don't,
# this dependency will be stripped out by the build system
GLIBC_ROOTFS := $(addprefix $(TARGET_ROOT_OUT)/, rootfs-glibc)

$(GLIBC_ROOTFS):
	mkdir -p $(TARGET_ROOT_OUT)
	cp -af $(TOPDIR)rootfs-glibc/* $(TARGET_ROOT_OUT)
```

```
        rm $(TARGET_ROOT_OUT)/Android.mk
        # The last command just gets rid of this very .mk since it's copied as is

ALL_PREBUILT += $(GLIBC_ROOTFS)
```

This will cause the content of *rootfs-glibc* to be merged into the *ramdisk.img* generated by the AOSP. That, though, is insufficient to make our glibc-based stack function properly on the resulting root filesystem. Indeed, as I explained in Chapter 6, the filesystem permissions of all files in the rootfs are dictated by the *system/core/include/private/android_filesystem_config.h*, and it has to be amended in order to keep the files in the */lib* directory executable. Otherwise, the glibc components are put into the root filesystem's */lib* directory but aren't executable and, therefore, all the binaries linked against glibc will fail to run. Hence, as I did in Chapter 6, you need to find the an droid_files array in *android_filesystem_config.h* and modify it so that it looks something like this in 2.3/Gingerbread:

```
    ...
    { 00750, AID_ROOT,      AID_SHELL,      "sbin/*" },
    { 00755, AID_ROOT,      AID_ROOT,       "bin/*" },
    { 00755, AID_ROOT,      AID_ROOT,       "lib/*" },
    { 00750, AID_ROOT,      AID_SHELL,      "init*" },
    { 00644, AID_ROOT,      AID_ROOT,       0 },
};
```

With these modifications, our glibc-linked binaries will work just fine in the root filesystem generated by the AOSP. Yet this isn't ideal since we're using Android's shell and Toolbox's commands, both of which are severely limited when compared with BusyBox's capabilities. Ideally, we should use BusyBox's shell and its command-line utilities. A few more changes are required to make that a reality. First, we need to modify *init.rc* so that the newly added */bin*, which contains BusyBox's commands, appears in the PATH prior to */system/bin*, which contains Toolbox's commands. Here's the modified *system/core/rootdir/init.rc* from 2.3/Gingerbread:

```
...
# setup the global environment
    export PATH /bin:/sbin:/vendor/bin:/system/sbin:/system/bin:/system/xbin
    export LD_LIBRARY_PATH /vendor/lib:/system/lib
    export ANDROID_BOOTLOGO 1
    export ANDROID_ROOT /system
...
```

Finally, at least in the case of 2.3/Gingerbread, we'll want to use BusyBox's shell instead of the default Android shell. There are two things to change to do that. First, we need to modify *init.rc* so that it uses BusyBox's shell for the console. By default, here's how *init.rc* starts the console:

```
service console /system/bin/sh
...
```

To use BusyBox's shell instead of the default Android shell, all we need to do is make *init.rc* run */bin/sh* instead of */system/bin/sh*:

```
service console /bin/sh
...
```

Also, it would be great if *adb shell* gave us access to BusyBox's shell as well. The shell run by *adbd* on the target is defined in *system/core/adb/services.c*:

```
...
#if ADB_HOST
#define SHELL_COMMAND "/bin/sh"
#else
#define SHELL_COMMAND "/system/bin/sh"
#endif
...
```

All we need to do here is comment out the default and make *adbd* run */bin/sh* instead:

```
...
#if ADB_HOST
#define SHELL_COMMAND "/bin/sh"
#else
//#define SHELL_COMMAND "/system/bin/sh"
#define SHELL_COMMAND "/bin/sh"
#endif
...
```

The sum of these changes will give us a new AOSP root filesystem that contains glibc and BusyBox, and which uses BusyBox's shell as its default shell and BusyBox's commands as its default commands.

If you're using 4.2/Jelly Bean, replacing the default shell or Toolbox's default commands may not be as useful as in 2.3/Gingerbread. The reason is that the AOSP has replaced the old *sh* with *mksh*, which provides many of the features of modern shells, and some of the Toolbox's basic commands, such as *ls*, have been fixed to remove their most obvious annoyances.

Using the Combined Stacks

Once you boot the system with the new root filesystem, you'll get all the benefits of having BusyBox and glibc. Here's a shell session in 2.3/Gingerbread with Android's shell and Toolbox's commands:

```
# ls
config
cache
sdcard
acct
```

```
mnt
vendor
d
etc
...
init
default.prop
data
root
dev
# grep -A 5 -i "\-Xzygote" init.rc
grep: not found
# ls sys[TAB]    [TAB]         [TAB]
```

As you can see, *ls*'s output is not alphabetically ordered, *grep* is an unrecognized command, and tab completion simply doesn't exist. Here are the same commands with BusyBox:

```
/ # ls
acct                init                sdcard
bin                 init.goldfish.rc    sys
cache               init.rc             system
config              lib                 ueventd.goldfish.rc
d                   linuxrc             ueventd.rc
data                mnt                 usr
default.prop        proc                vendor
dev                 root
etc                 sbin
/ # grep -A 5 -i "\-Xzygote" init.rc
service zygote /system/bin/app_process -Xzygote /system/bin --zygote
--start-system-server
    socket zygote stream 666
    onrestart write /sys/android_power/request_state wake
    onrestart write /sys/power/state on
    onrestart restart media
    onrestart restart netd
/ # ls sys[TAB][TAB]
sys/      system/
/ # ls sys
```

Furthermore, while Android's shell doesn't have any sort of color-coding to differentiate file types or files from directories, BusyBox's does, as you can see in Figure A-3.

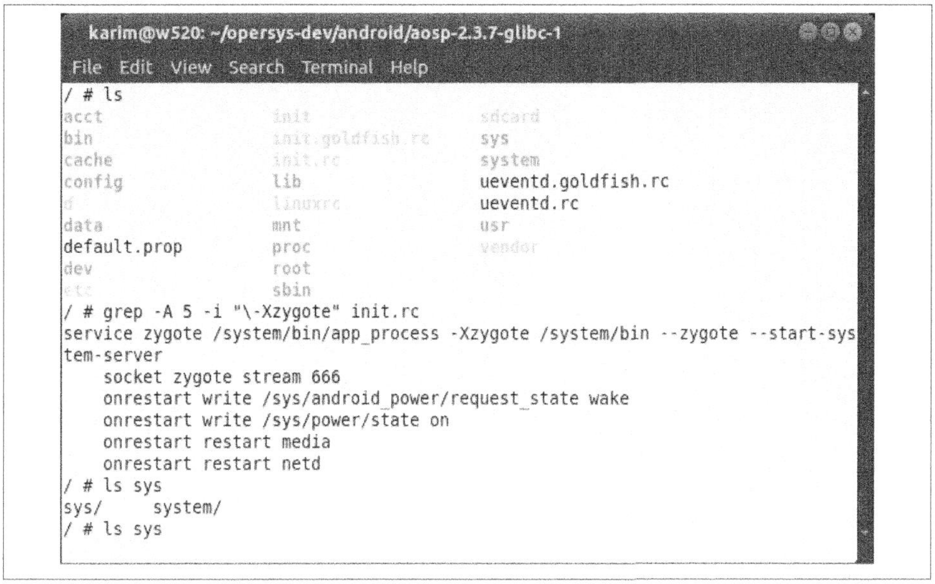

```
karim@w520: ~/opersys-dev/android/aosp-2.3.7-glibc-1
File  Edit  View  Search  Terminal  Help
/ # ls
acct                    init                    sdcard
bin                     init.goldfish.rc        sys
cache                   init.rc                 system
config                  lib                     ueventd.goldfish.rc
d                       linuxrc                 ueventd.rc
data                    mnt                     usr
default.prop            proc                    vendor
dev                     root
etc                     sbin
/ # grep -A 5 -i "\-Xzygote" init.rc
service zygote /system/bin/app_process -Xzygote /system/bin --zygote --start-sys
tem-server
    socket zygote stream 666
    onrestart write /sys/android_power/request_state wake
    onrestart write /sys/power/state on
    onrestart restart media
    onrestart restart netd
/ # ls sys
sys/        system/
/ # ls sys
```

Figure A-3. Sample BusyBox shell session

But BusyBox doesn't stop there. In addition to including commands such as *vi*, thereby allowing you to edit files straight on the target, BusyBox also includes some common daemons like *httpd* and *sendmail*. If you try to connect to port 80 using the regular browser on a typical Android device, you'll get something like Figure A-4.

If BusyBox is available on your target, however, you can add a service declaration for *httpd* in *init.rc*:

```
service httpd /usr/sbin/httpd
    oneshot
```

And then you can actually connect to it as you can see in Figure A-5—the 404 message is in fact the proper message from the web server, indicating that there's no *in dex.html* available.

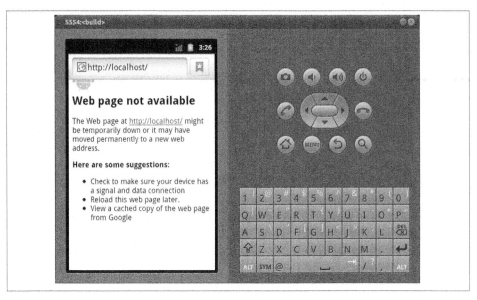

Figure A-4. The browser trying to connect to localhost

Figure A-5. The browser connecting to BusyBox's httpd

As a general rule, BusyBox's command set is far larger than Toolbox's. Here's Toolbox's command set in 2.3/Gingerbread, for instance:

cat, chmod, chown, cmp, date, dd, df, dmesg, getevent, getprop, hd, id, ifconfig, iftop, insmod, ioctl, ionice, kill, ln, log, ls, lsmod, lsof, mkdir, mount, mv, nandread, netstat, newfs_msdos, notify, printenv, ps, reboot, renice, rm, rmdir, rmmod, route, schedtop, sendevent, setconsole, setprop, sleep, smd, start, stop, sync, toolbox, top, umount, uptime, vmstat, watchprops, wipe

4.2/Jelly Bean has about a half-dozen more commands. In contrast, here's BusyBox's command set:

[, [[, acpid, add-shell, addgroup, adduser, adjtimex, arp, arping, ash, awk, base64, basename, beep, blkid, blockdev, bootchartd, brctl, bunzip2, bzcat, bzip2, cal, cat, catv, chat, chattr, chgrp, chmod, chown, chpasswd, chpst, chroot, chrt, chvt, cksum, clear, cmp, comm, cp, cpio, crond, crontab, cryptpw, cttyhack, cut, date, dc, dd, deallocvt, delgroup, deluser, depmod, devmem, df, dhcprelay, diff, dirname, dmesg, dnsd, dnsdomainname, dos2unix, du, dumpkmap, dumpleases, echo, ed, egrep, eject, env, envdir, envuidgid, ether-wake, expand, expr, fakeidentd, false, fbset, fbsplash, fdflush, fdformat, fdisk, fgconsole, fgrep, find, findfs, flock, fold, free, freeramdisk, fsck, fsck.minix, fsync, ftpd, ftpget, ftpput, fuser, getopt, getty, grep, gunzip, gzip, halt, hd, hdparm, head, hexdump, hostid, hostname, httpd, hush, hwclock, id, ifconfig, ifdown, ifenslave, ifplugd, ifup, inetd, init, insmod, install, ionice, iostat, ip, ipaddr, ipcalc, ipcrm, ipcs, iplink, iproute, iprule, iptunnel, kbd_mode, kill, killall, killall5, klogd, last, length, less, linux32, linux64, linuxrc, ln, loadfont, loadkmap, logger, login, logname, logread, losetup, lpd, lpq, lpr, ls, lsattr, lsmod, lspci, lsusb, lzcat, lzma, lzop, lzopcat, makedevs, makemime, man, md5sum, mdev, mesg, microcom, mkdir, mkdosfs, mke2fs, mkfifo, mkfs.ext2, mkfs.minix, mkfs.vfat, mknod, mkpasswd, mkswap, mktemp, modinfo, modprobe, more, mount, mountpoint, mpstat, mt, mv, nameif, nbd-client, nc, netstat, nice, nmeter, nohup, nslookup, ntpd, od, openvt, passwd, patch, pgrep, pidof, ping, ping6, pipe_progress, pivot_root, pkill, pmap, popmaildir, poweroff, powertop, printenv, printf, ps, pscan, pwd, raidautorun, rdate, rdev, readahead, readlink, readprofile, realpath, reboot, reformime, remove-shell, renice, reset, resize, rev, rm, rmdir, rmmod, route, rpm, rpm2cpio, rtcwake, run-parts, runlevel, runsv, runsvdir, rx, script, scriptreplay, sed, sendmail, seq, setarch, setconsole, setfont, setkeycodes, setlogcons, setsid, setuidgid, sh, sha1sum, sha256sum, sha512sum, showkey, slattach, sleep, smemcap, softlimit, sort, split, start-stop-daemon, stat, strings, stty, su, sulogin, sum, sv, svlogd, swapoff, swapon, switch_root, sync, sysctl, syslogd, tac, tail, tar, tcpsvd, tee, telnet, telnetd, test, tftp, tftpd, time, timeout, top, touch, tr, traceroute, traceroute6, true, tty, ttysize, tunctl, udhcpc, udhcpd, udpsvd, umount, uname, unexpand, uniq, unix2dos, unlzma, unlzop, unxz, unzip, uptime, usleep, uudecode, uuencode, vconfig, vi, vlock, volname, wall, watch, watchdog, wc, wget, which, who, whoami, xargs, xz, xzcat, yes, zcat, zcip

Hence, even if you were to include BusyBox during development only and stripped it out for the production images, the benefits are obvious. In fact, if you've been used to BusyBox, being forced to use plain Toolbox is likely akin to torture.

Also, if you look in */lib*, you'll find all the regular glibc components you're used to, whereas none of this exists if you're using the plain AOSP:

```
/ # ls /lib
ld-2.9.so               libm-2.9.so             libnss_nisplus-2.9.so
```

```
ld-linux.so.3              libm.so.6                  libnss_nisplus.so.2
libBrokenLocale-2.9.so     libmemusage.so             libpcprofile.so
libBrokenLocale.so.1       libnsl-2.9.so              libpthread-2.9.so
libSegFault.so             libnsl.so.1                libpthread.so.0
libanl-2.9.so              libnss_compat-2.9.so       libresolv-2.9.so
libanl.so.1                libnss_compat.so.2         libresolv.so.2
libc-2.9.so                libnss_dns-2.9.so          librt-2.9.so
libc.so.6                  libnss_dns.so.2            librt.so.1
libcrypt-2.9.so            libnss_files-2.9.so        libthread_db-1.0.so
libcrypt.so.1              libnss_files.so.2          libthread_db.so.1
libdl-2.9.so               libnss_hesiod-2.9.so       libutil-2.9.so
libdl.so.2                 libnss_hesiod.so.2         libutil.so.1
libgcc_s.so                libnss_nis-2.9.so
libgcc_s.so.1              libnss_nis.so.2
```

Caveats and Pending Issues

Now that you can see what can be done, let's look at what this type of configuration entails. First, the new C library and whatever binaries you're adding are going to make the root filesystem larger. Whereas the default *ramdisk.img* built by a 2.3.x AOSP is about 144KB, the one containing the glibc and BusyBox above is 2.6MB. You can of course trim the glibc-based root filesystem as embedded Linux developers have always done, by removing unnecessary glibc components and using the *strip* command. It may also be that storage is a nonissue in your embedded system. After all, on this same build, *system.img* is 66MB.

 You could, of course, also install glibc libraries in another location from */lib* and avoid using */bin* if you wanted to. For instance, you could create a */legacy* directory and put all your legacy content in that directory and mount it from a separate image to keep the root filesystem RAM disk minimal in size, as it is by default. Still, it's obviously simpler to just use the traditional */bin* and */lib* as spelled out by the FHS.

There's also the fact that you've now got two C libraries that need to be loaded into RAM, Bionic and glibc. Again, this might be a nonissue in your design, but you should be aware of this. One area where adding libraries has no impact, however, is CPU performance. Only the load imposed by the additional binaries you package will actually impact the CPU.

A more subtle problem is what to do with */etc*. Indeed, in Android's root filesystem, */etc* is a symbolic link to */system/etc*. This is a departure from the FHS and works fine for the AOSP. If you've got a legacy embedded Linux filesystem you want to merge with the AOSP's root filesystem, you're going to have to make a choice. Either copy the contents of your */etc* to */system/etc* and keep the symbolic link as is, or copy the contents of */system/etc* to your */etc*. This is an annoyance, but it shouldn't stop you from using the technique explained here.

At runtime you may encounter a few quirks, because Toolbox's tools operate under different assumptions from their regular Linux counterparts. Usually, for instance, *ps* uses */etc/passwd* to match UIDs to user names. In the case of Android, there's no */etc/passwd*. Instead, users and groups are hardcoded into the *android_filesystem_config.h* file we covered earlier. Hence, BusyBox's *ps* is unable to match processes with usernames:

```
/ # ps
PID   USER      TIME   COMMAND
    1 0                0:08 /init
...
   26 0                0:00 /sbin/ueventd
   27 1000             0:00 /system/bin/servicemanager
   28 0                0:00 /system/bin/vold
   29 0                0:00 /system/bin/netd
   30 0                0:00 /system/bin/debuggerd
   31 1001             0:00 /system/bin/rild
   32 0                0:10 zygote /bin/app_process -Xzygote /system/bin --zygote --s
   33 1013             0:00 /system/bin/mediaserver
   34 1002             0:00 /system/bin/dbus-daemon --system --nofork
   35 0                0:00 /system/bin/installd
   36 1017             0:00 /system/bin/keystore /data/misc/keystore
   38 0                0:00 /system/bin/qemud
   40 2000             0:00 /system/bin/sh
   41 0                0:00 /sbin/adbd
   64 1000             0:22 system_server
  116 10018            0:01 com.android.inputmethod.latin
  124 1001             0:03 com.android.phone
  125 1000             0:18 com.android.systemui
...
```

Toolbox's *ps* has no such issues:

```
# ps
USER       PID  PPID  VSIZE  RSS    WCHAN     PC            NAME
root         1     0    268  180    c009b74c  0000875c  S  /init
...
root        26     1    232  136    c009b74c  0000875c  S  /sbin/ueventd
system      27     1    804  188    c01a94a4  afd0b6fc  S  /system/bin/servicemanager
root        28     1   3864  300    ffffffff  afd0bdac  S  /system/bin/vold
root        29     1   3836  316    ffffffff  afd0bdac  S  /system/bin/netd
root        30     1    664  176    c01b52b4  afd0c0cc  S  /system/bin/debuggerd
radio       31     1   5396  432    ffffffff  afd0bdac  S  /system/bin/rild
root        32     1  60876  16396  c009b74c  afd0b844  S  zygote
media       33     1  17976  1000   ffffffff  afd0b6fc  S  /system/bin/mediaserver
bluetooth   34     1   1256  216    c009b74c  afd0c59c  S  /system/bin/dbus-daemon
root        35     1    812  220    c02181f4  afd0b45c  S  /system/bin/installd
keystore    36     1   1744  200    c01b52b4  afd0c0cc  S  /system/bin/keystore
root        38     1    824  260    c00b8fec  afd0c51c  S  /system/bin/qemud
shell       40     1    732  192    c0158eb0  afd0b45c  S  /system/bin/sh
```

```
root        41   1    3364   168   ffffffff 00008294 S /sbin/adbd
system      64   32   119832 26144 ffffffff afd0b6fc S system_server
app_18      116  32   77272  17604 ffffffff afd0c51c S com.android.inputmethod.
                                                        latin
radio       124  32   86120  17996 ffffffff afd0c51c S com.android.phone
system      125  32   73320  19012 ffffffff afd0c51c S com.android.systemui
...
```

Also, Toolbox commands sometimes have different parameters from traditional Linux commands. Toolbox's *ps* for instance, accepts the -t parameter to list the threads in addition to the processes:

```
# ps -t
...
system    64   32   119832 26144 ffffffff afd0b6fc S system_server
system    65   64   119832 26144 c0059e24 afd0c738 S HeapWorker
system    66   64   119832 26144 c0059e24 afd0c738 S GC
system    67   64   119832 26144 c0047be8 afd0bfec S Signal Catcher
system    68   64   119832 26144 c02181f4 afd0c22c S JDWP
system    69   64   119832 26144 c0059e24 afd0c738 S Compiler
system    70   64   119832 26144 c01a94a4 afd0b6fc S Binder Thread #
system    71   64   119832 26144 c01a94a4 afd0b6fc S Binder Thread #
system    72   64   119832 26144 c0059e24 afd0c738 S SurfaceFlinger
system    74   64   119832 26144 c0047be8 afd0bfec S DisplayEventThr
system    75   64   119832 26144 c00b8fec afd0c51c S er.ServerThread
system    77   64   119832 26144 c00b8fec afd0c51c S ActivityManager
system    81   64   119832 26144 c0059f2c afd0c738 S ProcessStats
system    82   64   119832 26144 c00b8fec afd0c51c S PackageManager
system    83   64   119832 26144 c00b7db0 afd0b45c S FileObserver
system    84   64   119832 26144 c00b8fec afd0c51c S AccountManagerS
system    86   64   119832 26144 c00b8fec afd0c51c S SyncHandlerThre
...
```

BusyBox's *ps* expects -T (uppercase T instead of lowercase t) instead and complains:

```
/ # ps -t
ps: invalid option -- 't'
BusyBox v1.18.3 (2011-03-09 09:33:40 PST) multi-call binary.

Usage: ps [-o COL1,COL2=HEADER] [-T]

Show list of processes

Options: -o COL1,COL2=HEADER Select columns for display -T    Show threads
```

In most cases, these incompatibilities cause annoyances, not actual breakage. And, ultimately, we haven't gotten rid of Toolbox or any of the default AOSP commands. So you can still invoke any of Toolbox's commands by providing the full command path:

```
# / /system/bin/ps
USER    PID  PPID VSIZE RSS   WCHAN    PC          NAME
root    1    0    268   180   c009b74c 0000875c S /init
root    2    0    0     0     c004e72c 00000000 S kthreadd
```

320 | Appendix A: Legacy User-Space

```
root         3     2     0     0     c003fdc8 00000000 S ksoftirqd/0
root         4     2     0     0     c004b2c4 00000000 S events/0
root         5     2     0     0     c004b2c4 00000000 S khelper
root         6     2     0     0     c004b2c4 00000000 S suspend
root         7     2     0     0     c004b2c4 00000000 S kblockd/0
...
```

There is at least one case I have noticed where putting BusyBox ahead of Toolbox in the PATH causes breakage. In the case of *dumpstate*, for instance, the default *ps* command from the path is used to retrieve the list of running threads. Yet, since BusyBox's *ps* expects -T instead of -t, the corresponding parts of *dumpstate*'s output are broken.

Another area of substantial difference worth mentioning is name resolution. Indeed, the way Android manages DNSes is very different from the way it's done in glibc and BusyBox. So this may be an issue in your case.

Some people are of the opinion that there's a benefit to Toolbox's very restricted command set: It limits that attack surface that a malicious user or third party could leverage against the system. From that point of view, using BusyBox would lead to an increased security risk. Caveat emptor.

Linking BusyBox Against Bionic

As demonstrated in this section, BusyBox shines when compared with the AOSP's default command-line tools. So much so, in fact, that many people felt the need to get it to work with their AOSP trees. Hence, the default tree from *http://busybox.net* now contains support for Android out of the box. Namely, patches have been added to enable the running of BusyBox against Bionic in addition to the libraries that it already supported, such as glibc. Also, there's an *android-build* script in the *examples/* directory of BusyBox's sources for building it against a given set of AOSP sources.

Whether you link it against Bionic or glibc, however, you still have to find a way to get it to coexist with the rest of the AOSP on the same filesystem. Hence, the above explanations remain relevant regardless of the library you link against.

Moving Forward

There's obviously a lot more you can do with this approach than I've showed you. Even, for instance, if you were to not include BusyBox or if you chose to link it against a library other than glibc, such as uClibc or eglibc, knowing how to get a "classic" C library onto your root filesystem is a useful trick.

I would encourage you to look at projects like Buildroot and Yocto to see how you can leverage their work to gain additional tools and libraries to merge with your AOSP, for an even more versatile end result. Remember that Android's vision and development approach restricts admission to the AOSP to only the packages conforming to Google's plans. Your specific project may, in fact, have nothing in common with any of Google's current market aims, so the plain AOSP may be seriously lacking with regard to your project.

In no way are the explanations given here the only way to achieve the targeted result. There are many ways to skin this cat. Generally speaking, this explanation should allow you to see that you can constructively break from the AOSP's stringent mold and incorporate into your final root filesystem elements that derive from classic embedded Linux work. And this is huge, because it opens the door for leveraging the very large body of work that has been created through the years for Linux in embedded systems. This includes being able to tap into mailing lists, conferences, books, and, most importantly, a very large development community.

APPENDIX B
Adding Support for New Hardware

There are cases where your embedded system includes hardware that isn't already supported in Android. And while some of the work you can do inside the AOSP is modular, adding support for new types of hardware is trickier since it requires knowledge of some of Android's internals. This appendix shows you how to extend Android's various layers to support your own type of hardware.

While you may not be interested in actually adding support for new types of hardware in your system, you might find this appendix instructive if you're trying to understand the intricate details of how the various layers of the Android stack actually come together.

Also, while this appendix demonstrates the modifications using a 2.3/Gingerbread codebase, the mechanisms and Java code being modified are very similar in 4.2/Jelly Bean. Where major differences exist, they will be pointed out in the text.

The Basics

As we discussed in Chapter 2, contrary to standard "vanilla Linux," Android requires more than just proper device drivers to function on hardware. It in fact defines a new Hardware Abstraction Layer (HAL), which defines an API for each type of hardware supported by Android's core. In order for a hardware component to properly interface with Android, it must have a corresponding hardware "module" (unrelated to kernel modules) that conforms to the API specified for that type of hardware.

Generally, each type of hardware supported by Android has a corresponding system service and HAL definition. There's a Lights Service and a lights HAL definition. There's a Wifi Service and a WiFi HAL definition. The same goes for power management, location, sensors, etc. Figure 2-3 illustrates the overall architecture of Android's hardware support. Most of these system services are, of course, typically running within the System Server as we discussed earlier.

There are two general categories of HAL modules: those loaded explicitly (through a runtime call to dlopen()) and those automatically loaded by the dynamic linker (since they're all linked into *libhardware_legacy.so*). The APIs for the former are in *hardware/libhardware/include/hardware/*, and the APIs for the latter are in *hardware/libhardware_legacy/include/hardware_legacy/*. The trend seems to be that Android is moving away from "legacy." The interface between those *.so* files and the actual drivers through */dev* entries or otherwise is up to the manufacturer to specify. Android doesn't care about that. It cares only about finding the appropriate HAL *.so* modules.

One of the questions I often get is, "How do I add support for my own type of hardware in Android?" To illustrate this, I've created an *opersys-hal-hw* type and have posted the code that implements this HAL type on GitHub (*https://github.com/opersys/opersys-hal-hw*), along with a very basic circular buffer driver (*https://github.com/opersys/circular-driver*).

If you copy the content of the opersys-hal-hw project over an existing 2.3.7_r1 release of the AOSP and build it for the emulator, you should get yourself an image that comes up with the opersys service. The latter relies on the circular buffer to implement a very basic new hardware type. Obviously, this is but a skeleton to give you an idea of what it takes to add support for a new hardware type. Your hardware is likely going to have completely different interfaces.

The System Service

To illustrate how a new system service is implemented, I first added a *OpersysService.java* in *frameworks/base/services/java/com/android/server/*. This file implements the OpersysService class, which provides two very basic calls to the outside world:

```
public String read(int maxLength)
{
...
}

public int write(String mString)
{
...
}
```

If you follow the code for the new type of hardware, you will see how I added an implementation corresponding to each of these calls at every layer of Android. So, for example, if you look at the system service's read() function, it does something like this:

```
public String read(int maxLength)
{
```

```
        int length;
        byte[] buffer = new byte[maxLength];

        length = read_native(mNativePointer, buffer);
        return new String(buffer, 0, length);
    }
```

The most important part here being the call to read_native(), which is itself declared as follows in the OpersysService class:

```
    private static native int read_native(int ptr, byte[] buffer);
```

By declaring the method as native, we instruct the compiler not to look for the method in any Java code. Instead, it'll be provided to Dalvik at runtime through JNI. And, indeed, if you look at the *frameworks/base/services/jni/* directory, you'll notice that *Android.mk* and *onload.cpp* have been modified to take into account a new *com_android_server_OpersysService.cpp*. The latter has a register_android_server_Opersys Service() function which is called at the loading of *libandroid_servers.so*, which is itself generated by the *Android.mk* I just mentioned. That registration function tells Dalvik about the native methods implemented in *com_android_server_OpersysService.cpp* for the OpersysService class and how they can be called:

```
static JNINativeMethod method_table[] = {
    { "init_native", "()I", (void*)init_native },
    { "finalize_native", "(I)V", (void*)finalize_native },
    { "read_native", "(I[B)I", (void*)read_native },
    { "write_native", "(I[B)I", (void*)write_native },
    { "test_native", "(II)I", (void*)test_native},
};

int register_android_server_OpersysService(JNIEnv *env)
{
    return jniRegisterNativeMethods(env, "com/android/server/OpersysService",
            method_table, NELEM(method_table));

};
```

The above structure contains three fields per method. The first field is the name of the method as defined in the Java class, while the last field is the corresponding C implementation in the present file. In this case the names match, as they do in most cases in Android, but that doesn't have to be the case. The middle parameter might seem a little bit more mysterious. The content of the parentheses are the parameters passed from Java, and the letter on the right of the parentheses is the return value. init_native() for instance takes no parameters and returns an integer, while read_native() has two parameters, an integer, and a byte array, and returns an integer.

As you start playing around wtih Android's internals, you will often have to deal with JNI-isms such as these. I recommend you take a look at *Java Native Interface: Programmer's Guide and Specificaition* by Sheng Liang (Addison-Wesley) for more information on the use of JNI.

And here's the implementation of read_native():

```
static int read_native(JNIEnv *env, jobject clazz, int ptr, jbyteArray buffer)
{
    opersyshw_device_t* dev = (opersyshw_device_t*)ptr;
    jbyte* real_byte_array;
    int length;

    real_byte_array = env->GetByteArrayElements(buffer, NULL);

    if (dev == NULL) {
        return 0;
    }

    length = dev->read((char*) real_byte_array, env->GetArrayLength(buffer));

    env->ReleaseByteArrayElements(buffer, real_byte_array, 0);

    return length;
}
```

First, notice that there are two more parameters than in the JNI declaration above. All JNI'ed calls start with the same two parameters: a handle to the VM making the call (env), and the this object corresponding to the class making the call (clazz). Also, notice that the byte array isn't used as is. Instead, env->GetByteArrayElements() and env->ReleaseByteArrayElements() are used at the begining and the end to obtain and, later, release a C array that can be used by the present C code. Indeed, don't forget that JNI calls are carrying Java-typed objects into the C world. While some things (such as integers) can be used as is, other objects (such as arrays) need to be converted before and after use.

Most importantly, the operative part of read_native() is the call to dev->read(). But what does this function pointer lead to? To understand that part, you need to look at init_native():

```
static jint init_native(JNIEnv *env, jobject clazz)
{
    int err;
    hw_module_t* module;
    opersyshw_device_t* dev = NULL;

    err = hw_get_module(OPERSYSHW_HARDWARE_MODULE_ID, (hw_module_t const**)
            &module);
    if (err == 0) {
```

```
        if (module->methods->open(module, "", ((hw_device_t**) &dev)) != 0)
            return 0;
    }

    return (jint)dev;
}
```

Two important things are happening in this function. First, the call to `hw_get_module()` which requests that the HAL load the module that implements support for the `OPERSYSHW_HARDWARE_MODULE_ID` type of hardware. Second, there's the call to the loaded module's `open()` function. We'll take a look at both of these below, but, for the moment, note that the former will result in a *.so* being loaded into the system service's address space, and the latter will result in the hardware-specific functions implemented in that library's functions, such as `read()` and `write()`, being callable from *com_android_server_OpersysService.cpp*, which is essentially the C side of the new system service we're adding.

The HAL and Its Extension

The HAL, which is in *hardware/*, provides the `hw_get_module()` call above. And if you follow the code, you'll see that `hw_get_module()` ends up calling the classic `dlopen()`, which enables us to load a shared library into a process's address space.

> Type *man dlopen* on any Linux workstation if you'd like to get more information about `dlopen` and its uses.

The HAL won't, however, just load any shared library. When you request a given hardware type, it'll look in */system/lib/hw* for a filename that matches that given hardware type and the device it's running on. So, for instance, in the case of the present new type of hardware, it'll look for *opersyshw.goldfish.so*, `goldfish` being the code name for the emulator. The actual name of the device used for the middle part of the filename is retrieved from one of the following global properties: `ro.hardware`, `ro.product.board`, `ro.board.platform`, or `ro.arch`. Also, the shared library must have a struct that provides HAL information and that is called `HAL_MODULE_INFO_SYM_AS_STR`. We'll see an example next.

The definition for the new hardware type itself is just another header file, in this case *opersyshw.h*, along with the other hardware definitions in *hardware/libhardware/include/hardware/*:

```
#ifndef ANDROID_OPERSYSHW_INTERFACE_H
#define ANDROID_OPERSYSHW_INTERFACE_H
```

```
#include <stdint.h>
#include <sys/cdefs.h>
#include <sys/types.h>

#include <hardware/hardware.h>

__BEGIN_DECLS

#define OPERSYSHW_HARDWARE_MODULE_ID "opersyshw"

struct opersyshw_device_t {
    struct hw_device_t common;

    int (*read)(char* buffer, int length);
    int (*write)(char* buffer, int length);
    int (*test)(int value);
};

__END_DECLS

#endif // ANDROID_OPERSYSHW_INTERFACE_H
```

In addition to the prototype definitions for read() and write(), note that this is where OPERSYSHW_HARDWARE_MODULE_ID is defined. The latter serves as the basis for the filename looked for on the filesystem that contains the actual HAL module implementation.

The HAL Module

The theory is that each device will require a different HAL module to support a given hardware type for Android. Phones from separate vendors, for instance, will likely use different graphic chips and are therefore likely to have different gralloc modules. Typically, the HAL modules are added to the AOSP sources in the *lib** directory within *device/<vendor>/<product>/*. In the case of the emulator, however, the virtual devices it supports are in *sdk/emulator/*, so this is where the Goldfish implementation for our type of hardware is added.

The opersyshw hardware type isn't really fancy, and therefore the implementation for Goldfish fits in a single file, *opersyshw_qemu.c*. In order for the library resulting from the build of this file to be recognized as a real HAL module, it ends with this snippet:

```
static struct hw_module_methods_t opersyshw_module_methods = {
    .open = open_opersyshw
};

const struct hw_module_t HAL_MODULE_INFO_SYM = {
    .tag = HARDWARE_MODULE_TAG,
    .version_major = 1,
    .version_minor = 0,
    .id = OPERSYSHW_HARDWARE_MODULE_ID,
```

```
        .name = "Opersys HW Module",
        .author = "Opersys inc.",
        .methods = &opersyshw_module_methods,
    };
```

Note the presence of the structure called HAL_MODULE_INFO_SYM. Furthermore, note the opersyshw_module_methods and the open() function pointer it contains. This is the very same open() called by init_native() earlier once the HAL module is loaded. And here's what the corresponding open_opersyshw() does:

```
    static int open_opersyshw(const struct hw_module_t* module, char const* name,
            struct hw_device_t** device)
    {
        struct opersyshw_device_t *dev = malloc(sizeof(struct opersyshw_device_t));
        memset(dev, 0, sizeof(*dev));

        dev->common.tag = HARDWARE_DEVICE_TAG;
        dev->common.version = 0;
        dev->common.module = (struct hw_module_t*)module;
        dev->read = opersyshw_read;
        dev->write = opersyshw_write;
        dev->test = opersyshw_test;

        *device = (struct hw_device_t*) dev;

        fd = open("/dev/circchar", O_RDWR);

        D("OPERSYS HW has been initialized");

        return 0;
    }
```

This function's main purpose is to initialize the dev struct, which is of opersyshw_device_t type, the same type defined by *opersyshw.h*, and open the corresponding device entry in */dev*, thereby connecting to the underlying device driver loaded into the kernel. Obviously some device drivers might require some initialization here, but for our purposes this is sufficient.

Finally, here's what opersyshw_read() does:

```
    int opersyshw_read(char* buffer, int length)
    {
        int retval;

        D("OPERSYS HW - read()for %d bytes called", length);

        retval = read(fd, buffer, length);

        return retval;
    }
```

The HAL Module | 329

We're not doing too much error-checking here, but you should in your case. For instance, we're not even checking that the call to open the device driver succeeded. We usually should. Still, the call path should be clear. The system service's read() call results in a JNI call to read_native() which, by way of the HAL, results in a call to the HAL module's opersyshw_read().

Existing system services and HAL components have similar types of call paths. Most, however, have a much larger number of calls defined in their system services and therefore a lot more happening in between the various layers involved in providing support for their specific type of hardware.

Calling the System Service

Up to this point we've mostly focused on how the new system service interfaces to the layers below. We haven't yet discussed how a system service makes itself available to be called through Binder to other system services and apps. At a bare minimum, there must be an interface definition in order for a system service to be callable through Binder. In the case of the opersys service, we can add a *IOpersysService.aidl* file to *frameworks/base/core/java/android/os/*:

```
package android.os;
/**
 * {@hide}
 */
interface IOpersysService {
    String read(int maxLength);
    int write(String mString);
}
```

This addition makes our system service callable from code that builds within the AOSP. We could, for instance, add an app to *device/acme/coyotepad/* or *packages/apps/* and have its onCreate() callback do something like this:

```
@Override
public void onCreate(Bundle savedInstanceState) {
    super.onCreate(savedInstanceState);
    setContentView(R.layout.main);

    IOpersysService om =
      IOpersysService.Stub.asInterface(ServiceManager.getService("opersys"));
    try {
    Log.d(DTAG, "Going to write to the \"opersys\" service");
    om.write("Hello Opersys");
    Log.d(DTAG, "Service returned: " + om.read(20));
    }
    catch (Exception e) {
    Log.d(DTAG, "FAILED to call service");
    e.printStackTrace();
```

```
        }
    }
```

Notice, however, that we're using ServiceManager.getService() to get a Binder handle to the system service, and then we're using IOpersysService.Stub.asInterface() to convert this to an IOpersysService object that we can call. This works fine if we're building within the AOSP but won't work for a regular app. Namely, ServiceManager.getService() isn't exposed in the SDK. Also, if you're familiar with app development, you'll likely notice that this is different from the regular way that handles to system services are usually obtained—through a call to getSystemService().

To make our system service available through an SDK we build using the AOSP, we need to carry out a few more steps. First, we need to create a *manager* class that acts as a shrink-wrap for our Binder-callable system service. We do this by adding a *Opersys Manager.java* file to *frameworks/base/core/java/android/os/*:

```
package android.os;

import android.os.IOpersysService;

public class OpersysManager
{
    public String read(int maxLength) {
        try {
            return mService.read(maxLength);
        } catch (RemoteException e) {
            return null;
        }
    }

    public int write(String mString) {
        try {
            return mService.write(mString);
        } catch (RemoteException e) {
            return 0;
        }
    }

    public OpersysManager(IOpersysService service) {
        mService = service;
    }

    IOpersysService mService;
}
```

Note how all calls are essentially redirected to the system service through Binder. Most predefined managers have similar semantics, although most will have some additional logic before making the calls, and others will define more calls than those available from the system service. This is similar to what a C library does before it makes calls to the kernel it runs on.

To make that manager available through getSystemService(), there are two more steps
required. First, we'll amend *frameworks/base/core/java/android/content/Context.java*
to recognize a new type of system service:

```
/**
 * Use with {@link #getSystemService} to retrieve a
 * {@link android.os.OpersysManager} for using Opersys Service.
 *
 * @see #getSystemService
 */
public static final String OPERSYS_SERVICE = "opersys";
```

Then, we'll patch *frameworks/base/core/java/android/content/app/ContextImpl.java* to
make getSystemService() recognize our new system service:

```
@Override
public Object getSystemService(String name) {
    if (WINDOW_SERVICE.equals(name)) {
        return WindowManagerImpl.getDefault();
    } else if (LAYOUT_INFLATER_SERVICE.equals(name)) {
        synchronized (mSync) {
...
    } else if (DOWNLOAD_SERVICE.equals(name)) {
        return getDownloadManager();
    } else if (NFC_SERVICE.equals(name)) {
        return getNfcManager();

    } else if (OPERSYS_SERVICE.equals(name)) {
        return getOpersysManager();

...

    private OpersysManager getOpersysManager() {
        synchronized (mSync) {
            if (mOpersysManager == null) {
                IBinder b = ServiceManager.getService(OPERSYS_SERVICE);
                IOpersysService service = IOpersysService.Stub.asInterface(b);
                mOpersysManager = new OpersysManager(service);
            }
        }
        return mOpersysManager;
    }

...
```

In 4.2/Jelly Bean, getSystemService()'s internal implementation is very different from the code shown previously. Have a look at how the registerService() is used in the ContextImpl class in *ContextImpl.java* to declare new managers. Specifically, have a look at the way it's done for POWER_SERVICE. You should be able to easily adapt the above snippet to resemble the one used to register a PowerManager object for use by getSystemService().

And now, after we build an SDK using this AOSP, we can create an app that calls on this new system service like any other predefined service:

```
@Override
public void onCreate(Bundle savedInstanceState) {
    super.onCreate(savedInstanceState);
    setContentView(R.layout.main);

    OpersysManager om = (OpersysManager) getSystemService(OPERSYS_SERVICE);

    Log.d(DTAG, "Going to write to the \"opersys\" service");
    om.write("Hello Opersys");
    Log.d(DTAG, "Service returned: " + om.read(20));
}
```

Starting the System Service

There's one last thing I haven't explained—that's how the system service is started in this case. Generally, as I mentioned in Chapter 7, Java-based system services are started in *SystemServer.java*. Hence, we can patch this file to have it instantiate our system service and register it with the Service Manager:

```
...
            try {
                Slog.i(TAG, "DiskStats Service");
                ServiceManager.addService("diskstats",
                                          new DiskStatsService(context));
            } catch (Throwable e) {
                Slog.e(TAG, "Failure starting DiskStats Service", e);
            }

            try {
                Slog.i(TAG, "Opersys Service");
                ServiceManager.addService(Context.OPERSYS_SERVICE,
                                          new OpersysService(context));
            } catch (Throwable e) {
                Slog.e(TAG, "Failure starting OpersysService Service", e);
            }

        }
...
```

Caveats and Recommendations

The method I just showed you and the code I referred you to works just fine for adding new types of hardware to the AOSP. However, it's very version-specific since you need to patch a few files. In essence, I showed you how to add support for a new type of hardware in the AOSP as if it were meant to be upstreamed. Usually that won't be your case and, therefore, as I suggested in Chapter 4, custom extensions are better added into a *device/<manufacturer/product_name>/* directory, which you can just copy into any new AOSP tree you get.

Despite its shortcoming, the benefit of the method I just showed is that you've got plenty of examples of other system services and HAL modules already in the AOSP from which you can easily copy, since you're adding your code in exactly the same location as the built-in components.

Still, you should know that there are various ways you could add a system service to your product-specific directory in *device/* in order to make a new type of hardware accessible to apps and other system services. The most straightforward one is to create an app that has its `persistent` flag set to `true` in its manifest file. As we discussed earlier, apps are lifecycle-managed by the Activity Manager. Hence, implementing hardware support in a regular app can be an issue because it could be stopped and restarted at any time, and if hardware state must be maintained, such restarting will likely cause issues. By enabling the `persistent` flag, you disable lifecycle management for this app. As I explained in Chapter 7, the Phone app, for instance, uses this trick in order to be able to host the Phone Service.

The downside with this approach is that any failure of the System Server, which houses the Activity Manager, will bring your system service down. Note that the same holds true for the method I showed you above. Another, more substantive, downside is that there are few examples to base your work on. You'll also need to create an SDK add-on instead of using the plain SDK generated by the AOSP that would've been patched by the method shown above. Callers to your system service won't, for instance, be able to use the standard `getSystemService()` to get a handle for an object allowing them to talk to your system service, as is the case for the default set of system services.

You can also probably create a standalone system service in Java that is started in a similar fashion as *am* and *pm*, using *app_process*. This would make your system service immune to any failure of the System Server, but I can't currently point you to any examples of system services implemented this way. And again, even if you followed this path, you'd still have a system service that doesn't appear like the other system services to app developers.

Finally, you could also create a native system service (i.e., in C) that starts the same way as the *mediaserver*. In that case, while you'd benefit from running natively, you wouldn't benefit from the *aidl* tool's capability to generate marshaling and unmarshaling code in

Java for callers and callees. Instead, you'd have to marshal and unmarshal everything sent through Binder manually—a very tedious process. And again, your system service will look different from standard system services.

APPENDIX C
Customizing the Default Lists of Packages

As we saw in Chapter 4, the build system can be modified to add new packages to those it builds by default. What we didn't cover in that chapter is how the build system creates the default list of packages that it uses when creating images or how we can customize it. Obviously, playing around with something as fundamental as the default set of packages required to get a functional AOSP has its risks, as you may end up generating stale images. Still, it's worth taking a look at how this works and what's in there. If nothing else, you'll get a better idea of where to look in case you have to get your hands in there.

Overall Dependencies

In 2.3/Gingerbread, there are two main variables that dictate what gets included in the AOSP: GRANDFATHERED_USER_MODULES and PRODUCT_PACKAGES. The first is generated from a static list found in *build/core/user_tags.mk* and contains the bulk of the "core" packages required for the AOSP, with such things as *adbd*, the system services, and Bionic. This file isn't meant to be edited and starts with a warning to that effect:

```
# This is the list of modules grandfathered to use a user tag

# DO NOT ADD ANY NEW MODULE TO THIS FILE
#
# user modules are hard to control and audit and we don't want
# to add any new such module in the system
```

In effect, the list of packages in GRANDFATHERED_USER_MODULES is more or less fixed in stone—what we want to focus our attention on is the packages added to PRODUCT_PACKAGES. There's in fact a whole series of files that gradually help add more packages to PRODUCT_PACKAGES, as the full list of *.mk* files are included one after the other, per the product description found in the relevant files in *device/<vendor>/<product>/*.

In 4.2/Jelly Bean, neither GRANDFATHERED_USER_MODULES nor *build/core/user_tags.mk* exist. Instead, there's a much-trimmed-down GRANDFATHERED_ALL_PREBUILT and a *build/core/legacy_prebuilts.mk* that carries a warning like the previous one. The bulk of 2.3/Gingerbread's GRANDFATHERED_USER_MODULES are now either in *build/target/product/base.mk* or *build/target/product/core.mk* and are added to PRODUCT_PACKAGES, which is used the same way as in 2.3/Gingerbread.

Assembling the Final PRODUCT_PACKAGES

Generally speaking, products will use the `inherit-product` makefile function, as we did when adding the CoyotePad in Chapter 4, to import other *.mk* files that include previous declarations of the PRODUCT_PACKAGES variable on which they can build.

The core file used for most PRODUCT_PACKAGES sets is *build/target/product/core.mk*. In 2.3/Gingerbread, this file doesn't inherit from any other *.mk* file. In 4.2/Jelly Bean, however, it inherits from *build/target/product/base.mk*. In both versions, *build/target/product/core.mk* includes packages such as the SSL library and the Browser app. Most product descriptions, except the one used for building the SDK, don't actually rely solely on the set of packages defined in this file. Instead, they'll at least rely on *build/target/product/generic.mk* in 2.3/Gingerbread and *build/target/product/generic_no_telephony.mk* in 4.2/Jelly Bean, both of which rely on *core.mk* in addition to including packages for many of the main apps such as Calendar, Launcher2, and Settings. The default emulator build in 2.3/Gingerbread, for instance, relies on *generic.mk*. So does the default tree provided by TI for the BeagleBone, which I used in some parts of this book.

Most products will, however, go a step further. In 2.3/Gingerbread they'll use *build/target/product/full.mk*, which depends on *generic.mk*, to get a few additional input methods, such as PinyinIME (the simplified Chinese keyboard) and some language locales. *full.mk*, for instance, is what's used as the baseline for the *device/samsung/crespo/* (Nexus S). And this is what I used in Chapter 4 for the CoyotePad.

In 4.2/Jelly Bean, most products will use *build/target/product/full_base.mk* instead of *build/target/product/full.mk*. The former depends on *generic_no_telephony.mk* instead of depending on *generic.mk*. You can see example uses of *full_base.mk* in *device/asus/grouper/* and *device/samsung/tuna/*.

Trimming Packages

One request I often get from developers is to explain how to trim the size of the AOSP. To do that, you'd have to go through the list of packages included in GRANDFATHERED_USER_MODULES if you're using 2.3/Gingerbread or GRANDFATHERED_ALL_PREBUILT if you're using 4.2/Jelly Bean and PRODUCT_PACKAGES in either case and remove whatever you think isn't necessary for your system. As I alluded to earlier, this is a tricky proposition because you're likely to generate a nonfunctional AOSP. Indeed, the AOSP's build system doesn't provide any type of dependency checks between packages.

You can, however, proceed with a few basic rules. Generally, I would recommend against trying to play around with the list of grandfathered packages or the packages in *base.mk* in 4.2/Jelly Bean, unless you feel pretty confident that you understand the AOSP's internals and the impact of the changes you're making. Starting with *core.mk*, you're in a little bit safer territory for removing packages. And the further you are down in the dependency chain from *core.mk*, the safer it is to remove modules without causing AOSP breakage. You can, for instance, remove the Launcher2 from *generic.mk* in 2.3/Gingerbread or from *generic_no_telephony.mk* in 4.2/Jelly Bean, and you'll generate a functional AOSP. It won't have the home screen you're used to, but it'll still work. The same goes for many of the apps in those same files.

APPENDIX D
Default init.rc Files

This appendix contains the default *init.rc* files found in 2.3/Gingerbread and 4.2/Jelly Bean.[1] I usually dislike books where files are printed for pages on end, and you won't find much of this in my writings. However, *init.rc* is one case where the best way to explain something is to actually show it to you. To make it easier for you to follow the operations conducted in the file, I've added some callouts throughout to provide insight on key parts of the files. Refer to Chapter 6 for more information regarding the actions, triggers, commands, services, and service options used in *init.rc* files.

2.3/Gingerbread's default init.rc

```
on early-init  ❶
    start ueventd

on init  ❷

sysclktz 0

loglevel 3

# setup the global environment  ❸
    export PATH /sbin:/vendor/bin:/system/sbin:/system/bin:/system/xbin
    export LD_LIBRARY_PATH /vendor/lib:/system/lib
    export ANDROID_BOOTLOGO 1
    export ANDROID_ROOT /system
    export ANDROID_ASSETS /system/app
    export ANDROID_DATA /data
    export EXTERNAL_STORAGE /mnt/sdcard
    export ASEC_MOUNTPOINT /mnt/asec
```

---

1. Both files are configuration files part of the AOSP sources and are therefore assumed to be licensed under the Apache license.

```
    export LOOP_MOUNTPOINT /mnt/obb
    export BOOTCLASSPATH /system/framework/core.jar:/system/framework/bouncycast
le.jar:/system/framework/ext.jar:/system/framework/framework.jar:/system/framewo
rk/android.policy.jar:/system/framework/services.jar:/system/framework/core-juni
t.jar

# Backward compatibility
    symlink /system/etc /etc
    symlink /sys/kernel/debug /d

# Right now vendor lives on the same filesystem as system,
# but someday that may change.
    symlink /system/vendor /vendor

# create mountpoints
    mkdir /mnt 0775 root system
    mkdir /mnt/sdcard 0000 system system

# Create cgroup mount point for cpu accounting
    mkdir /acct
    mount cgroup none /acct cpuacct
    mkdir /acct/uid

# Backwards Compat - XXX: Going away in G*
    symlink /mnt/sdcard /sdcard

    mkdir /system
    mkdir /data 0771 system system
    mkdir /cache 0770 system cache
    mkdir /config 0500 root root

    # Directory for putting things only root should see.
    mkdir /mnt/secure 0700 root root

    # Directory for staging bindmounts
    mkdir /mnt/secure/staging 0700 root root

    # Directory-target for where the secure container
    # imagefile directory will be bind-mounted
    mkdir /mnt/secure/asec  0700 root root

    # Secure container public mount points.
    mkdir /mnt/asec  0700 root system
    mount tmpfs tmpfs /mnt/asec mode=0755,gid=1000

    # Filesystem image public mount points.
    mkdir /mnt/obb 0700 root system
    mount tmpfs tmpfs /mnt/obb mode=0755,gid=1000

    write /proc/sys/kernel/panic_on_oops 1  ❹
    write /proc/sys/kernel/hung_task_timeout_secs 0
    write /proc/cpu/alignment 4
```

```
    write /proc/sys/kernel/sched_latency_ns 10000000
    write /proc/sys/kernel/sched_wakeup_granularity_ns 2000000
    write /proc/sys/kernel/sched_compat_yield 1
    write /proc/sys/kernel/sched_child_runs_first 0

# Create cgroup mount points for process groups
    mkdir /dev/cpuctl
    mount cgroup none /dev/cpuctl cpu
    chown system system /dev/cpuctl
    chown system system /dev/cpuctl/tasks
    chmod 0777 /dev/cpuctl/tasks
    write /dev/cpuctl/cpu.shares 1024

    mkdir /dev/cpuctl/fg_boost
    chown system system /dev/cpuctl/fg_boost/tasks
    chmod 0777 /dev/cpuctl/fg_boost/tasks
    write /dev/cpuctl/fg_boost/cpu.shares 1024

    mkdir /dev/cpuctl/bg_non_interactive
    chown system system /dev/cpuctl/bg_non_interactive/tasks
    chmod 0777 /dev/cpuctl/bg_non_interactive/tasks
    # 5.0 %
    write /dev/cpuctl/bg_non_interactive/cpu.shares 52

on fs  ❺
# mount mtd partitions
    # Mount /system rw first to give the filesystem a chance to save a checkpoint
    mount yaffs2 mtd@system /system
    mount yaffs2 mtd@system /system ro remount
    mount yaffs2 mtd@userdata /data nosuid nodev
    mount yaffs2 mtd@cache /cache nosuid nodev

on post-fs  ❻
    # once everything is setup, no need to modify /
    mount rootfs rootfs / ro remount

    # We chown/chmod /data again so because mount is run as root + defaults
    chown system system /data
    chmod 0771 /data

    # Create dump dir and collect dumps.
    # Do this before we mount cache so eventually we can use cache for
    # storing dumps on platforms which do not have a dedicated dump partition.

    mkdir /data/dontpanic
    chown root log /data/dontpanic
    chmod 0750 /data/dontpanic

    # Collect apanic data, free resources and re-arm trigger
    copy /proc/apanic_console /data/dontpanic/apanic_console
    chown root log /data/dontpanic/apanic_console
    chmod 0640 /data/dontpanic/apanic_console
```

```
        copy /proc/apanic_threads /data/dontpanic/apanic_threads
        chown root log /data/dontpanic/apanic_threads
        chmod 0640 /data/dontpanic/apanic_threads

        write /proc/apanic_console 1

        # Same reason as /data above
        chown system cache /cache
        chmod 0770 /cache

        # This may have been created by the recovery system with odd permissions
        chown system cache /cache/recovery
        chmod 0770 /cache/recovery

        #change permissions on vmallocinfo so we can grab it from bugreports
        chown root log /proc/vmallocinfo
        chmod 0440 /proc/vmallocinfo

        #change permissions on kmsg & sysrq-trigger so bugreports can grab kthread
        stacks
        chown root system /proc/kmsg
        chmod 0440 /proc/kmsg
        chown root system /proc/sysrq-trigger
        chmod 0220 /proc/sysrq-trigger

    # create basic filesystem structure
        mkdir /data/misc 01771 system misc
        mkdir /data/misc/bluetoothd 0770 bluetooth bluetooth
        mkdir /data/misc/bluetooth 0770 system system
        mkdir /data/misc/keystore 0700 keystore keystore
        mkdir /data/misc/vpn 0770 system system
        mkdir /data/misc/systemkeys 0700 system system
        mkdir /data/misc/vpn/profiles 0770 system system
        # give system access to wpa_supplicant.conf for backup and restore
        mkdir /data/misc/wifi 0770 wifi wifi
        chmod 0770 /data/misc/wifi
        chmod 0660 /data/misc/wifi/wpa_supplicant.conf
        mkdir /data/local 0771 shell shell
        mkdir /data/local/tmp 0771 shell shell
        mkdir /data/data 0771 system system
        mkdir /data/app-private 0771 system system
        mkdir /data/app 0771 system system
        mkdir /data/property 0700 root root

        # create dalvik-cache and double-check the perms
        mkdir /data/dalvik-cache 0771 system system
        chown system system /data/dalvik-cache
        chmod 0771 /data/dalvik-cache

        # create the lost+found directories, so as to enforce our permissions
        mkdir /data/lost+found 0770
```

```
    mkdir /cache/lost+found 0770

    # double check the perms, in case lost+found already exists, and set owner
    chown root root /data/lost+found
    chmod 0770 /data/lost+found
    chown root root /cache/lost+found
    chmod 0770 /cache/lost+found

on boot ❼
# basic network init
    ifup lo
    hostname localhost
    domainname localdomain

# set RLIMIT_NICE to allow priorities from 19 to -20
    setrlimit 13 40 40

# Define the oom_adj values for the classes of processes that can be
# killed by the kernel.  These are used in ActivityManagerService.
    setprop ro.FOREGROUND_APP_ADJ 0
    setprop ro.VISIBLE_APP_ADJ 1
    setprop ro.PERCEPTIBLE_APP_ADJ 2
    setprop ro.HEAVY_WEIGHT_APP_ADJ 3
    setprop ro.SECONDARY_SERVER_ADJ 4
    setprop ro.BACKUP_APP_ADJ 5
    setprop ro.HOME_APP_ADJ 6
    setprop ro.HIDDEN_APP_MIN_ADJ 7
    setprop ro.EMPTY_APP_ADJ 15

# Define the memory thresholds at which the above process classes will
# be killed.  These numbers are in pages (4k).
    setprop ro.FOREGROUND_APP_MEM 2048
    setprop ro.VISIBLE_APP_MEM 3072
    setprop ro.PERCEPTIBLE_APP_MEM 4096
    setprop ro.HEAVY_WEIGHT_APP_MEM 4096
    setprop ro.SECONDARY_SERVER_MEM 6144
    setprop ro.BACKUP_APP_MEM 6144
    setprop ro.HOME_APP_MEM 6144
    setprop ro.HIDDEN_APP_MEM 7168
    setprop ro.EMPTY_APP_MEM 8192

# Write value must be consistent with the above properties. ❽
# Note that the driver only supports 6 slots, so we have combined some of
# the classes into the same memory level; the associated processes of higher
# classes will still be killed first.
    write /sys/module/lowmemorykiller/parameters/adj 0,1,2,4,7,15

    write /proc/sys/vm/overcommit_memory 1
    write /proc/sys/vm/min_free_order_shift 4
    write /sys/module/lowmemorykiller/parameters/minfree 2048,3072,4096,6144,
    7168,8192
```

```
            # Set init its forked children's oom_adj.
            write /proc/1/oom_adj -16

            # Tweak background writeout
            write /proc/sys/vm/dirty_expire_centisecs 200
            write /proc/sys/vm/dirty_background_ratio  5

            # Permissions for System Server and daemons.
            chown radio system /sys/android_power/state
            chown radio system /sys/android_power/request_state
            chown radio system /sys/android_power/acquire_full_wake_lock
            chown radio system /sys/android_power/acquire_partial_wake_lock
            chown radio system /sys/android_power/release_wake_lock
            chown radio system /sys/power/state
            chown radio system /sys/power/wake_lock
            chown radio system /sys/power/wake_unlock
            chmod 0660 /sys/power/state
            chmod 0660 /sys/power/wake_lock
            chmod 0660 /sys/power/wake_unlock
            chown system system /sys/class/timed_output/vibrator/enable
            chown system system /sys/class/leds/keyboard-backlight/brightness
            chown system system /sys/class/leds/lcd-backlight/brightness
            chown system system /sys/class/leds/button-backlight/brightness
            chown system system /sys/class/leds/jogball-backlight/brightness
            chown system system /sys/class/leds/red/brightness
            chown system system /sys/class/leds/green/brightness
            chown system system /sys/class/leds/blue/brightness
            chown system system /sys/class/leds/red/device/grpfreq
            chown system system /sys/class/leds/red/device/grppwm
            chown system system /sys/class/leds/red/device/blink
            chown system system /sys/class/leds/red/brightness
            chown system system /sys/class/leds/green/brightness
            chown system system /sys/class/leds/blue/brightness
            chown system system /sys/class/leds/red/device/grpfreq
            chown system system /sys/class/leds/red/device/grppwm
            chown system system /sys/class/leds/red/device/blink
            chown system system /sys/class/timed_output/vibrator/enable
            chown system system /sys/module/sco/parameters/disable_esco
            chown system system /sys/kernel/ipv4/tcp_wmem_min
            chown system system /sys/kernel/ipv4/tcp_wmem_def
            chown system system /sys/kernel/ipv4/tcp_wmem_max
            chown system system /sys/kernel/ipv4/tcp_rmem_min
            chown system system /sys/kernel/ipv4/tcp_rmem_def
            chown system system /sys/kernel/ipv4/tcp_rmem_max
            chown root radio /proc/cmdline

    # Define TCP buffer sizes for various networks
    #   ReadMin, ReadInitial, ReadMax, WriteMin, WriteInitial, WriteMax,
            setprop net.tcp.buffersize.default 4096,87380,110208,4096,16384,110208
            setprop net.tcp.buffersize.wifi    4095,87380,110208,4096,16384,110208
            setprop net.tcp.buffersize.umts    4094,87380,110208,4096,16384,110208
            setprop net.tcp.buffersize.edge    4093,26280,35040,4096,16384,35040
```

```
        setprop net.tcp.buffersize.gprs     4092,8760,11680,4096,8760,11680

        class_start default ❾

## Daemon processes to be run by init. ❿
##
service ueventd /sbin/ueventd
    critical

service console /system/bin/sh
    console
    disabled
    user shell
    group log

on property:ro.secure=0
    start console

# adbd is controlled by the persist.service.adb.enable system property
service adbd /sbin/adbd ⓫
    disabled

# adbd on at boot in emulator
on property:ro.kernel.qemu=1
    start adbd

on property:persist.service.adb.enable=1
    start adbd

on property:persist.service.adb.enable=0
    stop adbd

service servicemanager /system/bin/servicemanager ⓬
    user system
    critical
    onrestart restart zygote
    onrestart restart media

service vold /system/bin/vold
    socket vold stream 0660 root mount
    ioprio be 2

service netd /system/bin/netd
    socket netd stream 0660 root system

service debuggerd /system/bin/debuggerd

service ril-daemon /system/bin/rild
    socket rild stream 660 root radio
    socket rild-debug stream 660 radio system
    user root
    group radio cache inet misc audio sdcard_rw
```

```
service zygote /system/bin/app_process -Xzygote /system/bin --zygote --start-sys
tem-server ⓭
    socket zygote stream 666
    onrestart write /sys/android_power/request_state wake
    onrestart write /sys/power/state on
    onrestart restart media
    onrestart restart netd

service media /system/bin/mediaserver ⓮
    user media
    group system audio camera graphics inet net_bt net_bt_admin net_raw
    ioprio rt 4

service bootanim /system/bin/bootanimation
    user graphics
    group graphics
    disabled
    oneshot

service dbus /system/bin/dbus-daemon --system --nofork
    socket dbus stream 660 bluetooth bluetooth
    user bluetooth
    group bluetooth net_bt_admin

service bluetoothd /system/bin/bluetoothd -n
    socket bluetooth stream 660 bluetooth bluetooth
    socket dbus_bluetooth stream 660 bluetooth bluetooth
    # init.rc does not yet support applying capabilities, so run as root and
    # let bluetoothd drop uid to bluetooth with the right linux capabilities
    group bluetooth net_bt_admin misc
    disabled

service hfag /system/bin/sdptool add --channel=10 HFAG
    user bluetooth
    group bluetooth net_bt_admin
    disabled
    oneshot

service hsag /system/bin/sdptool add --channel=11 HSAG
    user bluetooth
    group bluetooth net_bt_admin
    disabled
    oneshot

service opush /system/bin/sdptool add --channel=12 OPUSH
    user bluetooth
    group bluetooth net_bt_admin
    disabled
    oneshot

service pbap /system/bin/sdptool add --channel=19 PBAP
```

```
    user bluetooth
    group bluetooth net_bt_admin
    disabled
    oneshot

service installd /system/bin/installd
    socket installd stream 600 system system

service flash_recovery /system/etc/install-recovery.sh
    oneshot

service racoon /system/bin/racoon
    socket racoon stream 600 system system
    # racoon will setuid to vpn after getting necessary resources.
    group net_admin
    disabled
    oneshot

service mtpd /system/bin/mtpd
    socket mtpd stream 600 system system
    user vpn
    group vpn net_admin net_raw
    disabled
    oneshot

service keystore /system/bin/keystore /data/misc/keystore
    user keystore
    group keystore
    socket keystore stream 666

service dumpstate /system/bin/dumpstate -s  ❶❺
    socket dumpstate stream 0660 shell log
    disabled
    oneshot
```

❶ The early-init action is the earliest part of the *init.rc* that is executed, per the list of actions and triggers run by *init*, as explained in Chapter 6. As you can see, only *ueventd* is run here. In fact, the next step performed by *init* during its initialization is to check that *ueventd* was properly started as part of early-init.

❷ The init action is the first major chunk of commands that *init* is made to run. It sets the time zone to GMT, sets the log level to 3,[2] exports a core set of environment variables, and proceeds to conduct a number of filesystem operations on the root filesystem.

---

2. See the man page for klogctl() for more details as to the specific effect of this.

❸ This part of the initialization is pretty important. This is where the default PATH for all binaries in the system is set. This is also where the dynamic linker's default search path, LD_LIBRARY_PATH, is set. Note that /bin isn't in PATH and /lib isn't in LD_LIBRARY_PATH.

❹ Here, some of the kernel's parameters are tweaked by way of writing values to /proc entries. This and writing values to /sys entries are common ways of controlling the kernel and/or drivers' behavior.

❺ The fs action is where the /system, /data, and /cache partitions are mounted. Note that by default this config file attempts to mount those from MTD partitions using the YAFFS2 filesystem. Your board may neither have MTD devices nor use YAFFS2. In that case, these commands will fail, and that's fine. Nothing precludes you from having an fs action in your board-specific .rc file that mounts other partitions using other filesystems.

❻ The post-fs action is where all filesystem commands that depend on all filesystems having been mounted to operate properly are executed. Again, a large number of filesystem operations are being conducted here.

❼ The boot action is executed once all filesystems are set up, and by the end of the set of commands in here, the entire set of services will be started. This section starts by setting up the basic network functionality, sets up the OOM adjustments and memory thresholds used by the Activity Manager and the kernel, sets permissions for allowing the system server to access entries in /sys, sets networking properties, and finally starts all default services.

❽ This set of /proc and /sys operations are the way that the low-memory driver, which we discussed in Chapter 2, has its parameters set from user-space.

❾ This seemingly innocuous command is actually one of the most important ones in this file. All the services you see declared later in the file are started by this command. The fact is that any service declared in an .rc file is set to have default as its class, unless a specific class option is used in the service's description. And since none of the services listed in this file contains a specific class option, they're all part of the default class and started by this class_start command.

❿ Now that the majority of actions have been defined, the rest of the file focuses on describing the services to run. Since they're all part of the default class, they are started in the order they are found in the file.

⓫ Notice how *adbd* is set to be disabled at startup unless the persist.service.adb.enable property is set to 1.

⓬ This is the all-important Service Manager, which we covered in Chapter 2. Note how it's marked as critical, and its restarting will cause the System Server and Media Service to restart.

❸ This is the Zygote, also described in Chapter 2. Note how the actual binary being started is *app_process*. The latter is in fact a C-based binary that is made to start a Dalvik VM instance, which the Zygote Java class is started from. From there, the System Server will be started by the Zygote.

❹ This is the Media Server proper. Notice how its I/O nice value is set to mimic the "real time" scheduler and how its priority is set to 4.

❺ This `dumpstate` is necessary for Toolbox's *bugreport* command to operate properly. See the explanation in Chapter 6 about *bugreport* for more information on how it interacts with *dumpstate*.

4.2/Jelly Bean's Default init Files

Unlike 2.3/Gingerbread, 4.2/Jelly Bean has three main .rc files for all builds: *init.rc*, *init.usb.rc*, and *init.trace.rc*. Let's take a look at these.

init.rc

Here's the main *init.rc* from 4.2/Jelly Bean. As you can see by comparing this version with 2.3/Gingerbread's, many of the important parts have remained unchanged. Still, some novelties have appeared in this newer version that are worth highlighting.

Even if you're using 4.2/Jelly Bean, I would recommend reading the previous section about 2.3/Gingerbread's *init.rc* before reading this one, as I'm not repeating explanations I've already made for the latter.

```
# Copyright (C) 2012 The Android Open Source Project
#
# IMPORTANT: Do not create world writable files or directories.
# This is a common source of Android security bugs.
#

import /init.usb.rc ❶
import /init.${ro.hardware}.rc
import /init.trace.rc

on early-init
    # Set init and its forked children's oom_adj.
    write /proc/1/oom_adj -16

    # Set the security context for the init process.
    # This should occur before anything else (e.g. ueventd) is started.
    setcon u:r:init:s0 ❷

    start ueventd
```

```
# create mountpoints
    mkdir /mnt 0775 root system

on init

sysclktz 0

loglevel 3

# setup the global environment
    export PATH /sbin:/vendor/bin:/system/sbin:/system/bin:/system/xbin
    export LD_LIBRARY_PATH /vendor/lib:/system/lib
    export ANDROID_BOOTLOGO 1
    export ANDROID_ROOT /system
    export ANDROID_ASSETS /system/app
    export ANDROID_DATA /data
    export ANDROID_STORAGE /storage
    export ASEC_MOUNTPOINT /mnt/asec
    export LOOP_MOUNTPOINT /mnt/obb
    export BOOTCLASSPATH /system/framework/core.jar:/system/framework/core-junit
.jar:/system/framework/bouncycastle.jar:/system/framework/ext.jar:/system/framew
ork/framework.jar:/system/framework/telephony-common.jar:/system/framework/mms-c
ommon.jar:/system/framework/android.policy.jar:/system/framework/services.jar:/s
ystem/framework/apache-xml.jar

# Backward compatibility
    symlink /system/etc /etc
    symlink /sys/kernel/debug /d

# Right now vendor lives on the same filesystem as system,
# but someday that may change.
    symlink /system/vendor /vendor

# Create cgroup mount point for cpu accounting
    mkdir /acct
    mount cgroup none /acct cpuacct
    mkdir /acct/uid

    mkdir /system
    mkdir /data 0771 system system
    mkdir /cache 0770 system cache
    mkdir /config 0500 root root

    # See storage config details at http://source.android.com/tech/storage/
    mkdir /mnt/shell 0700 shell shell
    mkdir /storage 0050 root sdcard_r

    # Directory for putting things only root should see.
    mkdir /mnt/secure 0700 root root
    # Create private mountpoint so we can MS_MOVE from staging
    mount tmpfs tmpfs /mnt/secure mode=0700,uid=0,gid=0
```

```
    # Directory for staging bindmounts
    mkdir /mnt/secure/staging 0700 root root

    # Directory-target for where the secure container
    # imagefile directory will be bind-mounted
    mkdir /mnt/secure/asec  0700 root root

    # Secure container public mount points.
    mkdir /mnt/asec  0700 root system
    mount tmpfs tmpfs /mnt/asec mode=0755,gid=1000

    # Filesystem image public mount points.
    mkdir /mnt/obb 0700 root system
    mount tmpfs tmpfs /mnt/obb mode=0755,gid=1000

    write /proc/sys/kernel/panic_on_oops 1
    write /proc/sys/kernel/hung_task_timeout_secs 0
    write /proc/cpu/alignment 4
    write /proc/sys/kernel/sched_latency_ns 10000000
    write /proc/sys/kernel/sched_wakeup_granularity_ns 2000000
    write /proc/sys/kernel/sched_compat_yield 1
    write /proc/sys/kernel/sched_child_runs_first 0
    write /proc/sys/kernel/randomize_va_space 2
    write /proc/sys/kernel/kptr_restrict 2
    write /proc/sys/kernel/dmesg_restrict 1
    write /proc/sys/vm/mmap_min_addr 32768
    write /proc/sys/kernel/sched_rt_runtime_us 950000
    write /proc/sys/kernel/sched_rt_period_us 1000000

# Create cgroup mount points for process groups
    mkdir /dev/cpuctl
    mount cgroup none /dev/cpuctl cpu
    chown system system /dev/cpuctl
    chown system system /dev/cpuctl/tasks
    chmod 0660 /dev/cpuctl/tasks
    write /dev/cpuctl/cpu.shares 1024
    write /dev/cpuctl/cpu.rt_runtime_us 950000
    write /dev/cpuctl/cpu.rt_period_us 1000000

    mkdir /dev/cpuctl/apps
    chown system system /dev/cpuctl/apps/tasks
    chmod 0666 /dev/cpuctl/apps/tasks
    write /dev/cpuctl/apps/cpu.shares 1024
    write /dev/cpuctl/apps/cpu.rt_runtime_us 800000
    write /dev/cpuctl/apps/cpu.rt_period_us 1000000

    mkdir /dev/cpuctl/apps/bg_non_interactive
    chown system system /dev/cpuctl/apps/bg_non_interactive/tasks
    chmod 0666 /dev/cpuctl/apps/bg_non_interactive/tasks
    # 5.0 %
    write /dev/cpuctl/apps/bg_non_interactive/cpu.shares 52
```

```
    write /dev/cpuctl/apps/bg_non_interactive/cpu.rt_runtime_us 700000
    write /dev/cpuctl/apps/bg_non_interactive/cpu.rt_period_us 1000000

# Allow everybody to read the xt_qtaguid resource tracking misc dev.
# This is needed by any process that uses socket tagging.
    chmod 0644 /dev/xt_qtaguid

on fs
# mount mtd partitions
    # Mount /system rw first to give the filesystem a chance to save a
    checkpoint
    mount yaffs2 mtd@system /system
    mount yaffs2 mtd@system /system ro remount
    mount yaffs2 mtd@userdata /data nosuid nodev
    mount yaffs2 mtd@cache /cache nosuid nodev

on post-fs
    # once everything is setup, no need to modify /
    mount rootfs rootfs / ro remount
    # mount shared so changes propagate into child namespaces
    mount rootfs rootfs / shared rec
    mount tmpfs tmpfs /mnt/secure private rec

    # We chown/chmod /cache again so because mount is run as root + defaults
    chown system cache /cache
    chmod 0770 /cache
    # We restorecon /cache in case the cache partition has been reset.
    restorecon /cache

    # This may have been created by the recovery system with odd permissions
    chown system cache /cache/recovery
    chmod 0770 /cache/recovery
    # This may have been created by the recovery system with the wrong context.
    restorecon /cache/recovery

    #change permissions on vmallocinfo so we can grab it from bugreports
    chown root log /proc/vmallocinfo
    chmod 0440 /proc/vmallocinfo

    chown root log /proc/slabinfo
    chmod 0440 /proc/slabinfo

    #change permissions on kmsg & sysrq-trigger so bugreports can grab kthread
    stacks
    chown root system /proc/kmsg
    chmod 0440 /proc/kmsg
    chown root system /proc/sysrq-trigger
    chmod 0220 /proc/sysrq-trigger
    chown system log /proc/last_kmsg
    chmod 0440 /proc/last_kmsg

    # create the lost+found directories, so as to enforce our permissions
```

```
    mkdir /cache/lost+found 0770 root root

on post-fs-data
    # We chown/chmod /data again so because mount is run as root + defaults
    chown system system /data
    chmod 0771 /data
    # We restorecon /data in case the userdata partition has been reset.
    restorecon /data

    # Create dump dir and collect dumps.
    # Do this before we mount cache so eventually we can use cache for
    # storing dumps on platforms which do not have a dedicated dump partition.
    mkdir /data/dontpanic 0750 root log

    # Collect apanic data, free resources and re-arm trigger
    copy /proc/apanic_console /data/dontpanic/apanic_console
    chown root log /data/dontpanic/apanic_console
    chmod 0640 /data/dontpanic/apanic_console

    copy /proc/apanic_threads /data/dontpanic/apanic_threads
    chown root log /data/dontpanic/apanic_threads
    chmod 0640 /data/dontpanic/apanic_threads

    write /proc/apanic_console 1

    # create basic filesystem structure
    mkdir /data/misc 01771 system misc
    mkdir /data/misc/adb 02750 system shell
    mkdir /data/misc/bluedroid 0770 bluetooth net_bt_stack
    mkdir /data/misc/bluetooth 0770 system system
    mkdir /data/misc/keystore 0700 keystore keystore
    mkdir /data/misc/keychain 0771 system system
    mkdir /data/misc/sms 0770 system radio
    mkdir /data/misc/vpn 0770 system vpn
    mkdir /data/misc/systemkeys 0700 system system
    # give system access to wpa_supplicant.conf for backup and restore
    mkdir /data/misc/wifi 0770 wifi wifi
    chmod 0660 /data/misc/wifi/wpa_supplicant.conf
    mkdir /data/local 0751 root root

    # For security reasons, /data/local/tmp should always be empty.
    # Do not place files or directories in /data/local/tmp
    mkdir /data/local/tmp 0771 shell shell
    mkdir /data/data 0771 system system
    mkdir /data/app-private 0771 system system
    mkdir /data/app-asec 0700 root root
    mkdir /data/app-lib 0771 system system
    mkdir /data/app 0771 system system
    mkdir /data/property 0700 root root
    mkdir /data/ssh 0750 root shell
    mkdir /data/ssh/empty 0700 root root
```

```
    # create dalvik-cache, so as to enforce our permissions
    mkdir /data/dalvik-cache 0771 system system

    # create resource-cache and double-check the perms
    mkdir /data/resource-cache 0771 system system
    chown system system /data/resource-cache
    chmod 0771 /data/resource-cache

    # create the lost+found directories, so as to enforce our permissions
    mkdir /data/lost+found 0770 root root

    # create directory for DRM plug-ins - give drm the read/write access to
    # the following directory.
    mkdir /data/drm 0770 drm drm

    # If there is no fs-post-data action in the init.<device>.rc file, you
    # must uncomment this line, otherwise encrypted filesystems
    # won't work.
    # Set indication (checked by vold) that we have finished this action
    #setprop vold.post_fs_data_done 1

on boot
# basic network init
    ifup lo
    hostname localhost
    domainname localdomain

# set RLIMIT_NICE to allow priorities from 19 to -20
    setrlimit 13 40 40

# Memory management.  Basic kernel parameters, and allow the high
# level system server to be able to adjust the kernel OOM driver
# parameters to match how it is managing things.
    write /proc/sys/vm/overcommit_memory 1
    write /proc/sys/vm/min_free_order_shift 4
    chown root system /sys/module/lowmemorykiller/parameters/adj
    chmod 0664 /sys/module/lowmemorykiller/parameters/adj
    chown root system /sys/module/lowmemorykiller/parameters/minfree
    chmod 0664 /sys/module/lowmemorykiller/parameters/minfree

    # Tweak background writeout
    write /proc/sys/vm/dirty_expire_centisecs 200
    write /proc/sys/vm/dirty_background_ratio  5

    # Permissions for System Server and daemons.
    chown radio system /sys/android_power/state
    chown radio system /sys/android_power/request_state
    chown radio system /sys/android_power/acquire_full_wake_lock
    chown radio system /sys/android_power/acquire_partial_wake_lock
    chown radio system /sys/android_power/release_wake_lock
    chown system system /sys/power/autosleep
    chown system system /sys/power/state
```

```
chown system system /sys/power/wakeup_count
chown radio system /sys/power/wake_lock
chown radio system /sys/power/wake_unlock
chmod 0660 /sys/power/state
chmod 0660 /sys/power/wake_lock
chmod 0660 /sys/power/wake_unlock

chown system system /sys/devices/system/cpu/cpufreq/interactive/timer_rate
chmod 0660 /sys/devices/system/cpu/cpufreq/interactive/timer_rate
chown system system /sys/devices/system/cpu/cpufreq/interactive/min_sample_
time
chmod 0660 /sys/devices/system/cpu/cpufreq/interactive/min_sample_time
chown system system /sys/devices/system/cpu/cpufreq/interactive/hispeed_freq
chmod 0660 /sys/devices/system/cpu/cpufreq/interactive/hispeed_freq
chown system system /sys/devices/system/cpu/cpufreq/interactive/go_
hispeed_load
chmod 0660 /sys/devices/system/cpu/cpufreq/interactive/go_hispeed_load
chown system system /sys/devices/system/cpu/cpufreq/interactive/above_
hispeed_delay
chmod 0660 /sys/devices/system/cpu/cpufreq/interactive/above_hispeed_delay
chown system system /sys/devices/system/cpu/cpufreq/interactive/boost
chmod 0660 /sys/devices/system/cpu/cpufreq/interactive/boost
chown system system /sys/devices/system/cpu/cpufreq/interactive/boostpulse
chown system system /sys/devices/system/cpu/cpufreq/interactive/input_boost
chmod 0660 /sys/devices/system/cpu/cpufreq/interactive/input_boost

# Assume SMP uses shared cpufreq policy for all CPUs
chown system system /sys/devices/system/cpu/cpu0/cpufreq/scaling_max_freq
chmod 0660 /sys/devices/system/cpu/cpu0/cpufreq/scaling_max_freq

chown system system /sys/class/timed_output/vibrator/enable
chown system system /sys/class/leds/keyboard-backlight/brightness
chown system system /sys/class/leds/lcd-backlight/brightness
chown system system /sys/class/leds/button-backlight/brightness
chown system system /sys/class/leds/jogball-backlight/brightness
chown system system /sys/class/leds/red/brightness
chown system system /sys/class/leds/green/brightness
chown system system /sys/class/leds/blue/brightness
chown system system /sys/class/leds/red/device/grpfreq
chown system system /sys/class/leds/red/device/grppwm
chown system system /sys/class/leds/red/device/blink
chown system system /sys/class/leds/red/brightness
chown system system /sys/class/leds/green/brightness
chown system system /sys/class/leds/blue/brightness
chown system system /sys/class/leds/red/device/grpfreq
chown system system /sys/class/leds/red/device/grppwm
chown system system /sys/class/leds/red/device/blink
chown system system /sys/class/timed_output/vibrator/enable
chown system system /sys/module/sco/parameters/disable_esco
chown system system /sys/kernel/ipv4/tcp_wmem_min
chown system system /sys/kernel/ipv4/tcp_wmem_def
chown system system /sys/kernel/ipv4/tcp_wmem_max
```

```
        chown system system /sys/kernel/ipv4/tcp_rmem_min
        chown system system /sys/kernel/ipv4/tcp_rmem_def
        chown system system /sys/kernel/ipv4/tcp_rmem_max
        chown root radio /proc/cmdline

# Define TCP buffer sizes for various networks
#    ReadMin, ReadInitial, ReadMax, WriteMin, WriteInitial, WriteMax,
        setprop net.tcp.buffersize.default  4096,87380,110208,4096,16384,110208
        setprop net.tcp.buffersize.wifi     524288,1048576,2097152,262144,524288,
                                            1048576
        setprop net.tcp.buffersize.lte      524288,1048576,2097152,262144,524288,
                                            1048576
        setprop net.tcp.buffersize.umts     4094,87380,110208,4096,16384,110208
        setprop net.tcp.buffersize.hspa     4094,87380,262144,4096,16384,262144
        setprop net.tcp.buffersize.hsupa    4094,87380,262144,4096,16384,262144
        setprop net.tcp.buffersize.hsdpa    4094,87380,262144,4096,16384,262144
        setprop net.tcp.buffersize.hspap    4094,87380,1220608,4096,16384,1220608
        setprop net.tcp.buffersize.edge     4093,26280,35040,4096,16384,35040
        setprop net.tcp.buffersize.gprs     4092,8760,11680,4096,8760,11680
        setprop net.tcp.buffersize.evdo     4094,87380,262144,4096,16384,262144

# Set this property so surfaceflinger is not started by system_init
        setprop system_init.startsurfaceflinger 0

        class_start core  ❸
        class_start main

on nonencrypted
    class_start late_start

on charger
    class_start charger

on property:vold.decrypt=trigger_reset_main
    class_reset main

on property:vold.decrypt=trigger_load_persist_props
    load_persist_props

on property:vold.decrypt=trigger_post_fs_data
    trigger post-fs-data

on property:vold.decrypt=trigger_restart_min_framework
    class_start main

on property:vold.decrypt=trigger_restart_framework
    class_start main
    class_start late_start

on property:vold.decrypt=trigger_shutdown_framework
    class_reset late_start
    class_reset main
```

```
## Daemon processes to be run by init.
##
service ueventd /sbin/ueventd
    class core  ❹
    critical
    seclabel u:r:ueventd:s0

on property:selinux.reload_policy=1
    restart ueventd
    restart installd

service console /system/bin/sh
    class core
    console
    disabled
    user shell
    group log

on property:ro.debuggable=1
    start console

# adbd is controlled via property triggers in init.<platform>.usb.rc  ❺
service adbd /sbin/adbd
    class core
    socket adbd stream 660 system system
    disabled
    seclabel u:r:adbd:s0

# adbd on at boot in emulator
on property:ro.kernel.qemu=1
    start adbd

service servicemanager /system/bin/servicemanager
    class core
    user system
    group system
    critical
    onrestart restart zygote
    onrestart restart media
    onrestart restart surfaceflinger
    onrestart restart drm

service vold /system/bin/vold
    class core
    socket vold stream 0660 root mount
    ioprio be 2

service netd /system/bin/netd
    class main  ❻
    socket netd stream 0660 root system
    socket dnsproxyd stream 0660 root inet
```

```
    socket mdns stream 0660 root system

service debuggerd /system/bin/debuggerd
    class main

service ril-daemon /system/bin/rild
    class main
    socket rild stream 660 root radio
    socket rild-debug stream 660 radio system
    user root
    group radio cache inet misc audio log

service surfaceflinger /system/bin/surfaceflinger  ❼
    class main
    user system
    group graphics drmrpc
    onrestart restart zygote

service zygote /system/bin/app_process -Xzygote /system/bin --zygote --start-sys
tem-server
    class main
    socket zygote stream 660 root system
    onrestart write /sys/android_power/request_state wake
    onrestart write /sys/power/state on
    onrestart restart media
    onrestart restart netd

service drm /system/bin/drmserver
    class main
    user drm
    group drm system inet drmrpc

service media /system/bin/mediaserver
    class main
    user media
    group audio camera inet net_bt net_bt_admin net_bw_acct drmrpc
    ioprio rt 4

service bootanim /system/bin/bootanimation
    class main
    user graphics
    group graphics
    disabled
    oneshot

service installd /system/bin/installd
    class main
    socket installd stream 600 system system

service flash_recovery /system/etc/install-recovery.sh
    class main
    oneshot
```

```
service racoon /system/bin/racoon
    class main
    socket racoon stream 600 system system
    # IKE uses UDP port 500. Racoon will setuid to vpn after binding the port.
    group vpn net_admin inet
    disabled
    oneshot

service mtpd /system/bin/mtpd
    class main
    socket mtpd stream 600 system system
    user vpn
    group vpn net_admin inet net_raw
    disabled
    oneshot

service keystore /system/bin/keystore /data/misc/keystore
    class main
    user keystore
    group keystore drmrpc
    socket keystore stream 666

service dumpstate /system/bin/dumpstate -s
    class main
    socket dumpstate stream 0660 shell log
    disabled
    oneshot

service sshd /system/bin/start-ssh
    class main
    disabled

service mdnsd /system/bin/mdnsd
    class main
    user mdnsr
    group inet net_raw
    socket mdnsd stream 0660 mdnsr inet
    disabled
    oneshot
```

❶ 4.2/Jelly Bean uses the import mechanism to bring in other *.rc* files. In this case, three files are imported. *init.usb.rc* and *init.trace.rc* are global to all device builds, and I've included them below for reference. This *init.rc*, however, also imports a board-specific *init.${ro.hardware}.rc*, which will be loaded according to the value of the ro.hardware global property. Have a look at the board-specific *.rc* files in the *device/* directory for examples.

❷ This is new to *init.rc* and is intricately related to the SEAndroid project. Have a look at *http://selinuxproject.org/page/SEAndroid* for more information about SEAndroid.

❸   In the 2.3/Gingerbread *init.rc*, class_start is used only to start the default class of services, which in that version is all services in the default *init.rc*. In 4.2/Jelly Bean, however, two classes are used in this file: core and main. Their names are self-explanatory, and you can see later in the file that the services are marked as either core or main. Generally speaking, the first class is listed first.

❹   Here's the first instance of a service definition where the class property is used to indicate the service's class, which in this case is core.

❺   Unlike in 2.3/Gingerbread, the starting and stopping of *adbd* isn't controlled by the persist.service.adb.enable property. Instead, as the comment suggests, it's controlled in the *init.usb.rc* files. We'll discuss this in more detail below.

❻   *netd* is the first service in the list that's part of the *main* class.

❼   As I mentioned in Chapter 2, the Surface Flinger is no longer part of the System Server. Instead, it's started as a separate process, as we can see here.

init.usb.rc

This *.rc* file is related to all things USB. Specifically, to better understand its operation and the values being set, you need to take a look at the USB system service code in *frameworks/base/services/java/com/android/server/usb/*.

```
# Copyright (C) 2012 The Android Open Source Project
#
# USB configuration common for all android devices
#

on post-fs-data
    chown system system /sys/class/android_usb/android0/f_mass_storage/lun/file
    chmod 0660 /sys/class/android_usb/android0/f_mass_storage/lun/file
    chown system system /sys/class/android_usb/android0/f_rndis/ethaddr
    chmod 0660 /sys/class/android_usb/android0/f_rndis/ethaddr

# Used to disable USB when switching states
on property:sys.usb.config=none  ❶
    stop adbd  ❷
    write /sys/class/android_usb/android0/enable 0
    write /sys/class/android_usb/android0/bDeviceClass 0
    setprop sys.usb.state ${sys.usb.config}

# adb only USB configuration
# This should only be used during device bringup
# and as a fallback if the USB manager fails to set a standard configuration
on property:sys.usb.config=adb
    write /sys/class/android_usb/android0/enable 0
    write /sys/class/android_usb/android0/idVendor 18d1
    write /sys/class/android_usb/android0/idProduct D002
    write /sys/class/android_usb/android0/functions ${sys.usb.config}
    write /sys/class/android_usb/android0/enable 1
```

```
    start adbd ❸
    setprop sys.usb.state ${sys.usb.config}

# USB accessory configuration
on property:sys.usb.config=accessory
    write /sys/class/android_usb/android0/enable 0
    write /sys/class/android_usb/android0/idVendor 18d1
    write /sys/class/android_usb/android0/idProduct 2d00
    write /sys/class/android_usb/android0/functions ${sys.usb.config}
    write /sys/class/android_usb/android0/enable 1
    setprop sys.usb.state ${sys.usb.config}

# USB accessory configuration, with adb
on property:sys.usb.config=accessory,adb
    write /sys/class/android_usb/android0/enable 0
    write /sys/class/android_usb/android0/idVendor 18d1
    write /sys/class/android_usb/android0/idProduct 2d01
    write /sys/class/android_usb/android0/functions ${sys.usb.config}
    write /sys/class/android_usb/android0/enable 1
    start adbd
    setprop sys.usb.state ${sys.usb.config}

# audio accessory configuration
on property:sys.usb.config=audio_source
    write /sys/class/android_usb/android0/enable 0
    write /sys/class/android_usb/android0/idVendor 18d1
    write /sys/class/android_usb/android0/idProduct 2d02
    write /sys/class/android_usb/android0/functions ${sys.usb.config}
    write /sys/class/android_usb/android0/enable 1
    setprop sys.usb.state ${sys.usb.config}

# audio accessory configuration, with adb
on property:sys.usb.config=audio_source,adb
    write /sys/class/android_usb/android0/enable 0
    write /sys/class/android_usb/android0/idVendor 18d1
    write /sys/class/android_usb/android0/idProduct 2d03
    write /sys/class/android_usb/android0/functions ${sys.usb.config}
    write /sys/class/android_usb/android0/enable 1
    start adbd
    setprop sys.usb.state ${sys.usb.config}

# USB and audio accessory configuration
on property:sys.usb.config=accessory,audio_source
    write /sys/class/android_usb/android0/enable 0
    write /sys/class/android_usb/android0/idVendor 18d1
    write /sys/class/android_usb/android0/idProduct 2d04
    write /sys/class/android_usb/android0/functions ${sys.usb.config}
    write /sys/class/android_usb/android0/enable 1
    setprop sys.usb.state ${sys.usb.config}

# USB and audio accessory configuration, with adb
on property:sys.usb.config=accessory,audio_source,adb
```

```
write /sys/class/android_usb/android0/enable 0
write /sys/class/android_usb/android0/idVendor 18d1
write /sys/class/android_usb/android0/idProduct 2d05
write /sys/class/android_usb/android0/functions ${sys.usb.config}
write /sys/class/android_usb/android0/enable 1
start adbd
setprop sys.usb.state ${sys.usb.config}

# Used to set USB configuration at boot and to switch the configuration
# when changing the default configuration
on property:persist.sys.usb.config=*
    setprop sys.usb.config ${persist.sys.usb.config}    ❹
```

❶ The sys.usb.config global property is what controls the state of the USB connection. It's either explicitly set by the code in *frameworks/base/services/java/com/android/server/usb/UsbDeviceManager.java* or updated based on changes to persist.sys.usb.config as is done farther down in the file.

❷ Here's *adbd* being stopped based on a change to sys.usb.config.

❸ This is one of several instances where *adbd* is started based on a change to sys.usb.config.

❹ Whenever persist.sys.usb.config is modified, sys.usb.config is automatically updated here. That, in turn, is likely to trigger other parts of this file based on the above-declared triggers.

init.trace.rc

Since 4.1/Jelly Bean, Android has included a *systrace* command for use by app developers. The *systrace* tool on the host side actually depends on an *atrace* tool on the target, which is invoked via ADB. For its part, *atrace* uses the kernel's ftrace functionality to trace the system. This *init.trace.rc* sets up ftrace for use by Android's tracing tools. A quick search for "ftrace" in your favorite search engine should allow you to easily find more documentation on this mechanism.

```
## Permissions to allow system-wide tracing to the kernel trace buffer.
##
on boot

# Allow writing to the kernel trace log.
    chmod 0222 /sys/kernel/debug/tracing/trace_marker

# Allow the shell group to enable (some) kernel tracing.
    chown root shell /sys/kernel/debug/tracing/trace_clock
    chown root shell /sys/kernel/debug/tracing/buffer_size_kb
    chown root shell /sys/kernel/debug/tracing/options/overwrite
    chown root shell /sys/kernel/debug/tracing/events/sched/sched_switch/enable
    chown root shell /sys/kernel/debug/tracing/events/sched/sched_wakeup/enable
    chown root shell /sys/kernel/debug/tracing/events/power/cpu_frequency/enable
    chown root shell /sys/kernel/debug/tracing/events/power/cpu_idle/enable
```

```
    chown root shell /sys/kernel/debug/tracing/events/power/clock_set_rate/enable
    chown root shell /sys/kernel/debug/tracing/events/cpufreq_interactive/enable
    chown root shell /sys/kernel/debug/tracing/tracing_on

    chmod 0664 /sys/kernel/debug/tracing/trace_clock
    chmod 0664 /sys/kernel/debug/tracing/buffer_size_kb
    chmod 0664 /sys/kernel/debug/tracing/options/overwrite
    chmod 0664 /sys/kernel/debug/tracing/events/sched/sched_switch/enable
    chmod 0664 /sys/kernel/debug/tracing/events/sched/sched_wakeup/enable
    chmod 0664 /sys/kernel/debug/tracing/events/power/cpu_frequency/enable
    chmod 0664 /sys/kernel/debug/tracing/events/power/cpu_idle/enable
    chmod 0664 /sys/kernel/debug/tracing/events/power/clock_set_rate/enable
    chmod 0664 /sys/kernel/debug/tracing/events/cpufreq_interactive/enable
    chmod 0664 /sys/kernel/debug/tracing/tracing_on

# Allow only the shell group to read and truncate the kernel trace.
    chown root shell /sys/kernel/debug/tracing/trace
    chmod 0660 /sys/kernel/debug/tracing/trace
```

APPENDIX E
Resources

There is more to Android than could ever be covered in a single book. For starters, Android has a living ecosystem around it and a lot of community projects. This appendix highlights the major resources you should explore as your work with Android progresses.

Websites and Communities

A vast number of websites and communities are either directly or indirectly related to Android. I've tried to categorize them below as neatly as possible.

Google

Android Open Source Project (http://source.android.com/)
> Google's main site for the Android platform. It historically contained more information about the system, but it has been removed. It still is a very good reference on how to get the sources and how to set up your development system to build the AOSP. It also contains the latest documentation on the Android Compatibility Program, including the Compliance Definition Document.

Android Developer (https://developer.android.com/develop/index.html)
> This is Google's site for app developers. Unlike the platform site, this site is quite rich in documentation. It contains tutorials, an API reference, guidelines for graphic designers, and more. In sum, if you're developing an app, you're in good hands with this site.

Android Tools Project Site (http://tools.android.com/)
> This is the site that contains the information about Android's developer tools. This includes the SDK, the Eclipse plug-in, the NDK, etc.

SoC Vendors

TI Android Development Kit for Sitara (http://www.ti.com/tool/androidsdk-sitara)
This dev kit includes a set of AOSP sources that have been customized to run on boards based on TI's chips such as the BeagleBone. You may also find the porting information available here (*http://processors.wiki.ti.com/index.php/Android*).

Linaro Android (https://wiki.linaro.org/Platform/Android)
Per its website, "Linaro is a not-for-profit engineering organization consolidating and optimizing open source Linux software and tools for the ARM architecture." Effectively, it's an organization serving several SoC vendors, helping them with platform enablement. They maintain an Android tree for their members that is freely available to download.

CodeAurora (https://www.codeaurora.org/)
This is part of Linux Foundation Labs and provides enablement for various open source projects for Qualcomm chips. As such, it maintains an Android tree.

Forks

Apart from the information provided on their sites, many of these forks have public mailing lists that you may find useful.

CyanogenMod (http://www.cyanogenmod.org/)
This is probably the most popular Android fork. It's essentially an aftermarket AOSP distribution aimed at techies and power users, with additional features and enhancements. Most interestingly, all the development is done in the open.

Android-x86 (http://www.android-x86.org/)
This is a separate project from the work done by Intel to get x86 support merged into the main AOSP tree. Instead, this is geared to porting Android to PCs, netbooks, and laptops.

RowBoat (https://code.google.com/p/rowboat/)
This is the community project maintained by TI from which the TI Android Development Kit is derived.

Replicant (http://replicant.us/)
This project aims to replace as many Android components with free software as possible. For instance, it includes F-Droid (*http://f-droid.org/*), a free software application catalog (essentially a free software version of Google Play).

Apart from the above list, there's also a large and growing number of closed-source forks of the AOSP. Remember that Android's licensing is very permissive.

Documentation and Forums

Linux Weekly News (http://lwn.net/)
 The primary news site for all things relating to the kernel's development. Android is covered when relevant, but the focus is certainly on classic Linux distributions and the Linux kernel.

Embedded Linux Wiki (http://www.elinux.org/)
 A wiki site that has a large collection of information related to embedded Linux. For some time now, it's also had an Android section (http://www.elinux.org/Android_Portal).

OMAPpedia (http://omappedia.org/wiki/Main_Page)
 This wiki contains information about the use of Linux and Android on TI's OMAP processors. Some of the articles include a lot of detailed instructions.

xdadevelopers (https://www.xda-developers.com/)
 While this site is traditionally frequented by modders, it sometimes contains information that is very difficult to obtain otherwise. Have a look at the Android section (https://www.xda-developers.com/tag/all-android/). Most of the valuable information found here is in the site's forums.

Slideshare (http://www.slideshare.net/)
 This is a general-purpose site for sharing slides. It contains a large number of Android-related slides, including many about its internals or various internal components.

Vogella (http://www.vogella.com/android.html)
 This site is maintained by Lars Vogel and provides various tutorials about Android app development. It's a very good complement to the official Android app development information distributed by Google.

Embedded Linux Build Tools

BuildRoot (http://buildroot.uclibc.org/)
 This project has been around for over a decade now, and allows you to build a target embedded Linux root filesystem and tools based on a configuration fed to it using a menu-based system.

Yocto Project (https://www.yoctoproject.org/)
 Similar to BuildRoot but much more ambitious in its goals. It contains a framework and tools for generating entire embedded Linux distributions.

Open Hardware Projects

BeagleBoard and BeagleBone (http://beagleboard.org)
There are many inexpensive evaluation boards on the market. However, the BeagleBoard and BeagleBone have accrued a very active community. Schematics provided.

Books

Building Embedded Linux Systems, 2nd ed., by Karim Yaghmour, Jon Masters, Gilad Ben-Yossef, and Philippe Gerum (O'Reilly, 2008)
The classic book on the topic of embedded Linux, originally written by yours truly and since updated under Jon Masters' lead.

Embedded Linux Primer, 2nd ed., by Christopher Hallinan (Prentice Hall, 2010)
Another good embedded Linux book.

Linux Device Drivers, 3rd ed., by Jonathan Corbet, Alessandro Rubini, and Greg Kroah-Hartman (O'Reilly, 2005)
Despite its age, this remains the reference for Linux device driver authors.

Linux Kernel Development, 3rd ed., by Robert Love (Addison-Wesley, 2010)
One of the kernel internals books that has withstood the test of time.

Linux Kernel Architecture, by Wolfgang Mauerer (Wrox, 2008)
Another internals title.

Programming Android, 2nd ed., by Zigurd Mednieks, Laird Dornin, Blake Meike, and Masumi Nakamura (O'Reilly, 2012)
An in-depth book on app development.

Learning Android, by Marko Gargenta (O'Reilly, 2011)
An introductory book on app development.

Professional Android 4 Application Development, by Reto Meier (Wrox, 2012)
An app development book by the tech lead for the Android Developer Relations team at Google.

Conferences and Events

Android Builders Summit (https://events.linuxfoundation.org/events/android-builders-summit)
The primary event for developers doing work inside the AOSP stack.

Embedded Linux Conference (https://events.linuxfoundation.org/events/embedded-linux-conference)
   The main event for all things related to embedded Linux.

Embedded Linux Conference Europe (https://events.linuxfoundation.org/events/embedded-linux-conference-europe)
   The European run of the ELC.

Linaro Connect (http://www.linaro.org/connect)
   The event Linaro uses to bring together its members and developers.

AnDevCon (http://www.andevcon.com/)
   The main app developer conference. Also has some platform talks.

Index

Symbols
2.3/Gingerbread (see Gingerbread, Android 2.3/)
3G, support for, 3
4.0/Ice-Cream Sandwich (see Ice-Cream Sandwich, Android 4.0/)
4.2/Jelly Bean (see see JellyBean, Android 4.2/)

A
accelerometer, support for, 4
access enforcement, using URIs, 31
ACP (Android Compatibility Program), 14, 17–21
acquire() method, 68
activities, as Android component, 26
Activity Manager, 70, 223, 264, 279
adb (Android debug bridge)
    command-line tool, 45
    connecting to USB target using commands, 169
    device connection and status, 195–196
    filesystem commands, 202–204
    local commands, 194–195
    main flags, parameters, and environment variables, 193–194
    setting up, 171
    state-altering commands, 204
    theory of operation, 191–193
    tunneling PPP, 207
    using in AOSP, 101–105
address bus, 165, 167
add_lunch_combo() function, 120
ADT (Android Development Tools) plugin, 31
Affero-licensed FDroid Repository, 15
.aidl files, IDL stored in, 30
aidl tool, 30, 39
alarm driver, 41–42
AlarmManager class, 41
ALSA drivers, 47
am command, 279–282
"Anatomy of contemporary GSM cellphone hardware" (Welte), 158
Android
    about developers, xi–xii
    architecture vs. Linux, 34
    characteristics, 4–5
    daemons, 59–60
    development model, 5–7
    development setup and tools, 22
    ecosystem, 8
    features, 2–4
    finding drivers, 38
    getting to work on embedded system, 9
    hacking and customizing, 10
    hardware
        compliance requirements and, 17–21
        support for, 323–335

We'd like to hear your suggestions for improving our indexes. Send email to index@oreilly.com.

373

support of, 47–49
history of, 1–2
legal framework, 10–17
libraries, 54–57
resources for information about, 367–371
Android 3.x/Honeycomb, 7, 10
Android Compatibility Program (ACP), 14, 17–21
Android debug bridge (adb)
command-line tool, 45
connecting to USB target using commands, 169
device connection and status, 195–196
filesystem commands, 202–204
local commands, 194–195
main flags, parameters, and environment variables, 193–194
remote commands, 197–202
state-altering commands, 204–207
theory of operation, 191–193
tunneling PPP, 207
using in AOSP, 101–105
Android Developers Guide (Google), 105, 109
Android Developers website, xiii
Android Development Tool (ADT) plug-in, 19, 31
Android Inc., 1
Android Market (see Google Play)
Android Open Source Project (AOSP) (see AOSP (Android Open Source Project))
Android Platform
AOSP and, 10
requirement, 9
Android Runtime, 251–253
"Android Simulator Environment" (website post), 117
Android Software Development Kit (SDK)
accessing, 31
building for Mac OS, 135–136
building for Windows, 136
Android.mk files, 112, 128–131
android:persistent, 264
Androidisms, merging into mainline, 36–37
Androidized kernels, 34–35
ANDROID_LOG_TAGS variable, 199, 221
anonymous shared memory (ashmem), 40
ANR (Application Not Responding) dialog box, 70

AOSP (Android Open Source Project)
Android Platform and, 10
basic hacks, 143–152
build environment for, 22
build system setup, 91–94
building Android, 94–99
building without framework, 250
coexisting with legacy Linux user-space, 307–322
communication between glibc-based stack and, 310
development host setup, 79
device support, 51–52
generated libraries, 54–56
getting, 80–86
GPL requirements in, 11
hardware requirements for running, 17
inside, 86–90
legacy Linux user-space merging with, 309
logging within, 43
mastering emulator, 105–109
modifying, 5
packages, 71–73
running Android, 99–101
submitting fixes to code, 6
trimming size of, 338–339
using adb, 101–105
AP (Application Processor), in system architecture, 157
Apache Harmony project, IBM and, 16
Apache License 2.0 (ASL) licensing, 11, 14
API
in application framework, 4
updating, 138–139
Apkudo, testing apps on devices at, 31
app developers, view of Android, 26–33
app development tools, 31–32
app overlay, adding, 149–150
app, adding, 148–149
Apple, mobile patent issues, 17
application components, 26–27
application framework, Android
APIs in, 4
SUB1, 2
Application Not Responding (ANR) dialog box, 70
Application Processor (AP), in system architecture, 157
apps startup, 263–265

374 | Index

APPWIDGET_UPDATE intent, 265
architecture
    Binder as cornerstone of, 39
    build system
        about, 113–115
        cleaning, 127
        configuration of, 115–118
        envsetup.sh, 118–124
        function definitions, 124–125
        main make recipes, 125–127
        module build templates, 127–132
        output, 132–134
    overview of, 33–34, 87
    system, 155–160
ARCH_ARM_HAVE_* variables, 118
ashmem (anonymous shared memory), 40
asInterface(), 331
ASL (Apache License 2.0) licensing, 11–14
Audio Flinger, reliance on ashmem, 40
autosleep mechanisms, 36, 37

B

backing up data, 290–292
Backup Manager service, 292
Baseband Processor (BP), in system architecture, 157–158
battery-powered device, managing, 159
BeagleBoard, 103, 160
Bhoj, Vishal, 35
Binder
    about, 39–40
    against glibc based stacks, 310
    as RPC/IPC mechanism, 39
    calling system services through, 330–331
    developers using aidl tool to, 39
    interaction and Service Manager, 68–70
    OpenBinder Documentation, 39
    using through dev/binder, 30
Binder driver, merging into staging tree, 40
Bionic
    building legacy code against, 309, 310
    BusyBox linking against, 321
    dynamic linker, 225
Bird, Tim, 105
Bitbar's Testdroid products, testing apps on devices at, 31
Bluetooth, support for, 3
bmgr, 290–292
BoardConfig.mk file, 118

boot animation, 245, 257–260
boot logo, 245–247
bootanimation.zip, 259
BOOTCLASS PATH variable, 252
BOOT_COMPLETED intent, 265
Bornstein, Dan, 62
BP (Baseband Processor), in system architecture, 157–158
branding elements, 13
Brin, Sergey, 1
Brisset, Fabien, 161
broadcast receivers, as Android component, 27
Brown, Martin "Improve collaborative build times with ccache", 124
browsers
    connecting to port 80 on Android device using, 314–317
    WebKit-based, 3
BSD license, 11, 13
bug report, adb, 199–200
build commands, seeing, 134–135
build environment, Google supported, 22
build recipes, 134–143
build system
    about, 111
    AOSP hacks in, 143–152
    architecture
        about, 113–115
        cleaning, 127
        configuration of, 115–118
        envsetup.sh, 118–124
        function definitions, 124–125
        main make recipes, 125–127
        module build templates, 127–132
        output, 132–134
    build recipes, 134–143
    comparison with other build systems, 112–113
    configuring, 94–99
    creating and customizing default list of packages, 337–339
    design background of, 113
    filesystem and, 185–190
    reuse large legacy software packages in, 308
    setup, 91–94
Building Embedded Linux Systems (Yaghmour), xii, 96, 140, 309
Buildroot, 308, 322
BUILD_* macros, 132

Index | 375

BUILD_ENV_SEQUENCE_NUMBER variable, 116
BusyBox
    connecting to port 80 on Android device, 314–317
    linking against Bionic, 321
    providing init, 57
    shell session with Android's shell and Toolbox's commands in, 313–314
    using instead of Android shell, 312–313
    vs. Toolbox, 58, 319–321

C

C/C++
    interacting with HAL modules, 49
    vs. Java, 61
camera
    component in SoC, 164
    support for, 4
Canonical, 92
ccache (Compiler Cache), 123
CDD (Compliance Definition Document), 17–21, 159
check_prereq, 225
clean, 127
CLEAR_VARS, 130
Code Licenses, 11–13
command line
    adb tool, 45
    Android, 208–228
    utilities, 60
commands and utilities
    command line, 219–228
    framework
        bmgr, 290–292
        Dalvik Utilities, 293–297
        dumpstate, 270–276
        dumpsys, 268–270
        ime command, 286–287
        input command, 287–288
        monkey, 288–290
        pm command, 282–285
        rawbu, 276–277
        service, 266–268
        stagefright command, 292
        svc command, 285–286
compass, support for, 4
Compliance Definition Document (CDD), 17–21, 159
Compliance Test Suite (CTS), 17–18, 21, 127
component lifecycles, 28
components, 26–27
concepts, Android, 26–30
connectivity, in system architecture, 160
console_init_action(), 245
content providers, as Android component, 27
Copy-on-Write (COW), 75
core components, in system architecture, 158–159
COW (Copy-on-Write), 75
CPUs
    address bus in, 165, 167
    handling SoCs, 161–164
crespo, 118
croot command, 119
CTS (Compliance Test Suite), 17–18, 21, 127, 136–137
CyanogenMod project, 8

D

D-Bus method, 50
daemons, 59–60, 150, 219–228, 297–304
Dalvik Debug Monitor Server (ddms) libraries, 193, 199
Dalvik Virtual Machine
    about, 3
    Android's Java and, 60
    global properties, 252
    in framework, 251–253
    JIT code cache, 40, 100
    starting up, 295
    utilities, 293–297
    vs. JVM, 62–63
dalvikvm command, 293–294
Danger Inc., development of Sidekick phone, 1
/data directory, 53, 134, 182–185
data storage options, 30
data, backing up, 290–292
ddms (Dalvik Debug Monitor Server) libraries, 193, 199
debuggerd, 223
debugging components
    host-target debug setup, 169
    in SoC, 164
    in system architecture, 160–161
debugging, Dalvik, 202
default droid build, 134
default properties, vet, 148

/dev nodes method, 50
dev struct, initializing, 330
dev/binder, 30, 40
development application tools, 31
development components, in system architecture, 160–161
development environment, Android, 4
development setup
  hardware components in, 169–170
development setup and tools, 22, 79
device support details, 51–52
device, adding custom, 144–148
DEVICE_PACKAGE_OVERLAYS variable, 144, 149
Dex Optimization, 260–262
dexdump command, 295–297
display, in SoC, 164
dlopen() method
  hardcoded, 50–52
  loading through HAL, 49, 51–52, 324, 327
  rild using, 302
DMA, in SoC, 164
driver operation, ioctl() as, 217
drivers, finding Android, 38
droid, 134
DSP, in SoC, 164
dump() function, 70
dump() method, 268
dumpstate, 270–276
dumpsys, 268–270
dvz command, 294

E

Eclipse, 31, 68, 148
EDGE, support for, 3
embedded Multi- MediaCard (eMMC) chips, 158, 177
embedding Android, about, xi–xii
eMMC (embedded Multi- MediaCard) chips, 158, 177
emulator, 96
  (see also QEMU-based emulator)
  adb
    controlling, 206–207
    interacting with, 193
  mastering, 105–109
  starting, 99
  vs. QEMU-based emulator, 107
  vs. simulator, 117

enhancements, submitting, 6
envsetup.sh, 118–124
EPOLLWAKEUP, 36
Ethernet connections, 160–161, 169
evaluation boards, 171–173
EventLog class, 45
expand(), 268
expansion components, in system architecture, 160–161
expansion headers, 160
explicit intents, 27–28
external directory, 87

F

Federal Communications Commission (FCC), certification of SDR devices, 157
filesystem
  adb commands, 202–204
  build system and, 185–190
  native user space, 175–185
Filesystem Hierarchy Standard (FHS), 53, 307
filesystem layout, 53–54
FIRST_CALL_TRANSACTION variable, 268
forward, connection types of adb, 200–202
framework
  about, 30–31, 249
  apps startup, 263–265
  boot animation, 257–260
  building AOSP without, 250
  building blocks of, 251–253
  daemons, 297–304
  Dex Optimization, 260–262
  Hardware Abstraction Layer, 304–305
  system services in, 254–257
  utilities and commands, 266–297
    am command, 279–282
    bmgr, 290–292
    Dalvik Utilities, 293–297
    dumpstate, 270–276
    dumpsys, 268–270
    ime command, 286–287
    input command, 287–288
    monkey, 288–290
    pm command, 282–285
    rawbu, 276–277
    service, 266–268
    stagefright command, 292
    svc command, 285–286
full-eng combo, 95, 144

Index | 377

function definitions, build system architecture, 124–125
FUSE (Filesystem in User SpacE), 158

G

game developers, NDK for, 33
generic-eng combo, 95, 122
GetByteArrayElements(), 326
getprop command, 58
getProperty(), 252
getService(), 331
getSystemService(), 70, 331, 333, 334
GID (Group Identifier), 30
Gingerbread, Android 2.3/
    Android code in, 87–91
    availability, 5
    build environment for, 22
    default init.rc files, 341–351
    function definitions in, 124
    generic-eng combo in, 95
    shell session with Android's shell and Toolbox's commands in, 313
    simulator, 117
    Status bar in, 25
    stock apps in, 71, 73
    time to build, 98
    Toolbox commands in, 316
    variables set by lunch in, 122
git rebase command, Androidized kernel using, 35
git web interface, 80
GitHub website, 161
glibc library
    about, 308
    BusyBox linking against, 321
    installing, 317
glibc-based stacks, communication between AOSP and, 310
global properties, 57, 213, 238–243
GNU autotools kernel style, 112
GNU GPLv2 license, 11
GNU make, 113
godir command, 119
Goldfish, 101
Google
    Android Developers Guide, 105, 109
    apps owned by, 15
    aquiring Android Inc., 1
    architecture overview from, 34

build forms, 98
developing in, xi–xii, 5
Initializing a Build Environment, 79, 91
online documentation to set up application development environment, 22
right to decline participation in Android ecosystem, 18
vs. Oracle, 15–16
Google Play
    apps available through, 4, 7
    marketing apps outside of, 15
Gosling, James, 15–16, 60
GPL-licensed components, 11, 13
GPS, support for, 4
Graphics Processing Units (GPUs), 162, 164
Groklaw website, 16
Group Identifier (GID), 30
grouper, 118
GSM telelphony, support for, 3
GStreamer, 4

H

Hackborn, Dianne, 39
hacks, basic AOSP, 143–152
HAL (Hardware Abstraction Layer)
    audio support for, 47
    C/C++ interacting with, 47
    definitions with hardware, 324
    device manufacturers providing, 3
    dlopen() method-loading through, 49, 51–52, 324, 327
    extension, 327–328
    framework, 304–305
    vs. loadable kernel modules, 49
HAL modules
    to support hardware types, 328–330
hardware
    compliance requirements and, 17–21
    support, 46–52, 323–335
Hardware Abstraction Layer (HAL) (see HAL (Hardware Abstraction Layer))
hardware components
    development setup, 169
    evaluation boards, 171–173
    for memory layout and mapping, 165–169
    inside SoC, 161–165
    system architecture, 155–160
High-Resolution Timers (HRT), alarm driver and, 41

Hjønnevåg, Arve, 39
hmm command, 119
home screen, 245, 264–265
Honeycomb, Android 3.x/, 7, 10
host-target debug setup, 169
HRT (High-Resolution Timers), alarm driver and, 41
httpd daemon, 314
hw_get_module(), 49, 327

I

"I, Robot: The Man Behind the Google Phone" (Markoff), 1
IBM, Apache Harmony project and, 16
Ice-Cream Sandwich, Android 4.0/
    library prelinking in, 151
    support for Ethernet, 161
IDL (Interface Definition Language), 30
ime command, 286–287
IMemory interface, 40
IMEs (Input Method Editors), 71
implicit intents, 27
"Improve collaborative build times with ccache" (Brown), 124
in-tree, building recursively, 142–143
include directive, 128
inherit-product function, 144
inherit-product makefile function, 338
init, 341
    (see also .rc files, init)
    about, 228
    boot logo, 245–247
    configuration files, 230–238
    configuration of, 57
    global properties, 238–243
    normal vs. Android, 230–238
    shell scripts and, 238
    theory of operation, 228–230
    ueventd, 243–245
init boot logo, 245
/init directory, 134
Initializing a Build Environment (Google), 79, 91
init_native(), 325, 326
input command, 287–288
input events, 215
input method, 264
Input Method Editors (IMEs), 71
installd, 298–299

intents
    about, 27–28, 280
    APPWIDGET_UPDATE, 265
    BOOT_COMPLETED, 265
    globally defined, 26
Inter-Process Communication (IPC) mechanism, communicating with Binder, 30
Interface Definition Language (IDL), 30
interfacing methods, 49–50
intermediates, 132
Internals, Android
    alarm driver, 41–42
    anonymous shared memory, 40
    AOSP packages, 71
    app developers view of, 26–33
    architecture overview, 33–34
    Binder interaction and Service Manager, 68–70
    Binder mechanism, 39–40
    Dalvik Virtual Machine vs. JVM, 62–63
    hardware support, 46–52, 323–335
    Linux kernel, 34–35
    logging, 42–45
    low-memory killer, 37–38
    native-user space environment, 52–60
    paranoid networking, 45
    physical memory driver, 45
    RAM console, 45
    system services, 63–70
    system startup, 73–77
    WakeLock mechanism, 36–37
ioctl(), 40, 41, 47, 216–217
ioprio option, 256
IPC (Inter-Process Communication) mechanism
    RPC/, 30
    shared memory as, 40
IStatusBarService interface, 267
ITIMER_REAL, 41

J

Java
    Dalvik Virtual Machine and Android's, 60–63
    rights to, 15–16
    terminology, 61
    vs. C/C++, 61
Java ARchives (JAR), 62
Java Debug Wire Protocol (JDWP), 202

Java Development Kit (JDK), 61, 92–94
Java Native Interface (JNI), 63, 326, 330
Java Native Interface (Liang), 326
Java System Properties, 252
Java Virtual Machine (JVM)
    building Android on, 98
Java Virtual Machine (JVM) vs. Dalvic, 62
JDK (Java Development Kit), 61, 92–94
JellyBean, Android 4.1/
    build environment for, 22
    libraries, 57
JellyBean, Android 4.2/
    Android code in, 91
    build environment for, 22
    BusyBox commands in, 316–317
    configuring build system results, 95–96
    default init.rc files, 351–364
    full-eng combo in, 95
    function definitions in, 124
    getSystemService() in, 333
    hmm command, 119
    prebuilts in, 87
    support for Ethernet, 161
    time to build, 98
JIT code cache, Dalvik Virtual Machine, 40, 100
JNI (Java Native Interface), 63, 326, 330
JTAG, 160
JVM (Java Virtual Machine)
    building Android on, 98

K

kernel, 29
    (see also Linux kernel)
    Androidized, 34–35
    boot process of, 57
    code license for, 11
    features of, 9
    images, 127
    loadable modules, vs. HAL modules, 49
    styles, 112
kernel boot screen, 245
kernel.org, 34, 38
keystore, 304
Kroah-Hartman, Greg, The Linux Staging Tree (blog post), 38

L

LCD displays, display bridge for, 159

Learning Android (O'Reilly), xi–xii, 22, 32
legacy Linux user-space
    communication between glibc-based stack and AOSP, 310
    merging with AOSP, 309
legal framework, Android, 10–17
LessPainful, testing apps on devices at, 31
LGPL licensed components, 11
Liang, Sheng, Java Native Interface, 326
liblog functions, 43
libraries, 54–57, 167
library
    adding native, 151–152
    ddms, 193, 199
    prelinking, 151
    RIL, 303
Linaro
    about, 8
    Androidized kernel, 35
    patches for adding Ethernet functionality, 161
linker, 225
Linker method, loaded .so files, 50, 51–52
Linux
    architecture vs. Android, 34
    building SDK for, 135–136
    commands from Toolbox, 212–213
    hardware support of, 46–47
    logging systems vs. Android, 42
    MTD layer, 178
    staging tree, 38
Linux kernel
    Androidization of, 34–35
    code license for, 11
    handling multicore SoC, 164
    hardware running Android and, 17
    Out-of-Memory killing mechanisms, 29
    requirement, 9
Linux Kernel Development, 3rd ed. (Love), 35
LInux Kernel Mailing List (LKML), Userspace low memory killer daemon posted at, 38
The Linux Staging Tree (blog post), 38
Linux user-space, legacy
    coexisting with AOSP, 307–322
Linux user-space, legacy, coexisting with AOSP, 307–322
LKML (LInux Kernel Mailing List), Userspace low memory killer daemon posted at, 38
loading methods, 49–50

LOCAL_, prefix, 127, 130, 131
LOCAL_MODULE variable, 131
LOCAL_MODULE_PATH variable, 131, 185
LOCAL_MODULE_TAGS variable, 131
LOCAL_PACKAGE_NAME variable, 131
LOCAL_PATH variable, 131
LOCAL_PRELINK_MODULE variable, 151
LOCAL_SHARED_LIBRARIES variable, 131
LOCAL_SHARED_LIBRARIES, variable, 151
LOCAL_SRC_FILES variable, 131
Log class, 43
logcat command, 45, 199, 221–223
logging
    about, 42–45
    Android framework for, 43
    using Toolbox, 216
logs, adb, dumping, 197–199
logwrapper command, 226
Love, Robert, Linux Kernel Development, 3rd ed., 35
low-memory killer, 37–38
Low-Voltage Differential Signaling (LVDS), 159
lunch command, 120, 122–124
LVDS (Low-Voltage Differential Signaling), 159

M

m and mm commands, 120
Mac OS X Lion, building Gingerbread on, 98
Mac OS, building SDK for, 135–136
MAIN, 227, 301
MAIN HEADING, 266–268
main() method, 75, 252
make
    clean, 127
    recipes, 125–127
makefile, for building out of tree, 140–142
man dlopen, 327
man page, sh's, 209–210
manifest file, 29
    repo's "manifest" file and, 81
marketing apps, 15
Markoff, John, "I Robot: The Man Behind the Google Phone", 1
McFadden, Andrew, response to post "Android Simulator Environment", 117
Media Service, 64
mediaserver, 256
memory layout and mapping, hardware components for, 165–169

Memory Management Unit (MMU), 167
menuconfig kernel style, 112
methods, loading and interfacing, 49–50
Microsoft, mobile patent issues, 17
Miller, Peter "Recursive Make Considered Harmful", 113
MirBSD Korn Shell, 210
mmap(), 47, 167
MMU (Memory Management Unit), 167
Mobile Network Operator (MNO), 157
mobile patent issues, 16–17
module
    build templates, 127–132
    build templates list, 128–129
    building single, 139–140
    definition of, 112
monkey, 288–290
MTD layer, Linux, 178

N

NAND flash
    embedded systems equipped with, 158
    vs. eMMC, 178
nandread utility, Toolbox, 218
Native Development Kit (NDK), 33, 127, 137
native library, adding, 151–152
native tool, adding, 150
native user-space
    adb (see adb (Android debug bridge))
    Android command line tools, 208–228
    filesystem, 175–190
    init
        about, 228
        boot logo, 245–247
        global properties, 238–243
        theory of operation, 228–230
        ueventd, 243–245
native-user space
    about, 52–60
NDK (Native Development Kit), 33, 127, 137
NetBSD sh utility, 209
netcfg utility, 223
netd, 301
newfs_msdos command, Toolbox, 218
NFC app, 257
non-Linux systems, building Android on, 98
NOR flash
    embedded systems equipped with, 158
    vs. eMMC, 178

notify command, Toolbox, 219

O

obj/ directories, 132
OHA (Open Handset Alliance), 2, 8
On the Go (OTG) connector, 159
onCreate() callback, 330
OOM (Out-of-Memory)
    adjustments, 58
    killing mechanism, 29, 37, 38
Open Binder project, 39
open source projects (classic) vs. Android development model, 5–7
open source software movement, xii
open() function, 327
OpenBinder Documentation, 39
OpenGL ES, 3
OpenJDK, 92, 94
OPERSYSHW_HARDWARE_MODULE_ID type of hardware, 327, 330
OpersysService class, 325
Oracle
    dispute with Canonical, 92
Oracle vs. Google, 15–16
OS X Lion, building Gingerbread on, 98
OTG (On the Go) connector, 159
Out-of-Memory (OOM)
    adjustments, 58
    killing mechanism, 29, 37, 38
output, build, 132–134
OUT_DIR variable, 116
overlays, adding app, 149–150

P

Package Manager Service, 252, 260–262, 282, 298
packages
    trimming, 338–339
packages, AOSP, 71–73
PacketVideo's OpenCore framework, 3
Page, Larry, 1
PandaBoard, 103, 160
paranoid networking, 45
PCB (Printed Circuit Board), address bus on, 165
performance compatibility requirements, 21
permission system, circumventing, 278
permissions and security, 30–31

persistent apps, 264
persistent flag, enabling, 334
physical memory (pmem) driver, 45
physical memory vs. virtual memory, 165–167
PID (Process Identifier), 77
pm command, 282–285
PMIC (Power Management IC), 159
port 80, connecting Android device using browser to, 314–317
port forwarding, adb, 200–202
POSIX SHM vs. Ashmem, 40
Power Management IC (PMIC), 159
PPP connection, using adb for, 207
prebuilt directory, 87
Printed Circuit Board (PCB), address bus on, 165
printk(), 42
Process Identifier (PID), 77
processes and threads, 29
PRODUCT_BRAND variable, 145
PRODUCT_COPY_FILES variable, 145, 186
PRODUCT_DEVICE variable, 118, 145
PRODUCT_MODEL variable, 145
PRODUCT_NAME variable, 145
PRODUCT_PACKAGES variable, 144–145, 149, 150
PRODUCT_PACKAGES, assembling, 338
ps command, Toolbox vs. Busybox, 319–321
Pundir, Amit, 140–142
push functionality, 202–203

Q

QEMU-based emulator, 31, 96, 107
    (see also emulator)
Queru, Jean-Baptiste, 92, 94

R

RAM
    console, 45
    controller in SoC, 164
    location in physical address, 167
rawbu, 276–277
.rc files, init
    about, 230, 236–238
    default
        2.3/Gingerbread, 341–351
        4.2/Jelly Bean, 351–364
read() function, 327, 330

read_native(), 325–326
Real-Time Clock (RTC), 41, 159
Real-Time OS (RTOS), 157
real-world interaction, in system architecture, 159
reboot command, adb, 204
"Recursive Make Considered Harmful" (Miller), 113
registerService(), 333
ReleaseByteArrayElements(), 326
Remote Procedure Calls (RPCs), 30
remount, 202
repo tool, 80–84
Repo, the Android Source Management Tools (blog post), 81
RF transceiver, connection to BP, 157
RIL implementations, 51–52
RIL libraries, 303
RIL, Android, 157
rild, 302–303
root access to devices, 102
root command, adb, 204–206
root directory, 179–180
root filesystem layout, 53–54, 175–176
rootfs-glibc directory, using, 311
Rosenkränzer, Bernhard
　building AOSP with OpenJDK, 94
　building recursively, in-tree, 142
RPCs (Remote Procedure Calls) mechanism, IPC/, 30
RTC (Real-Time Clock), 41, 159
RTOS (Real-Time OS), 157
Rubin, Andy, 1, 5
Runtime, Android, 251–253

S

Safari Books Online, xvii
Samba Project, 123
Samsung, mobile patent issues, 17
SD card, appearance in filesystem, 178
/sdcard directory, 185
SDK (Software Development Kit), Android
　accessing, 31
　building for Linux, 135–136
　building for Mac OS, 135–136
　building for Windows, 136
SDR (Software Defined Radio) devices, 157
security and permissions, 30–31
security model compatibility requirements, 21

sendmail daemon, 314
serial (RS-232), 160
Service Manager, Binder interaction and, 68–70
service-specific utilities, 278–292
　am, 279–282
　bmgr, 290–292
　ime command, 286–287
　input command, 287–288
　monkey, 288–290
　pm, 282–285
　stagefright command, 292
　svc, 285–286
servicemanager, as building block of framework, 251
services
　controlling, 215–216
services, as Android component, 27
　vs.System Server, 40
setconsole command, Toolbox, 220
setitimer(), 41
setProperty(), 252
shared memory, as IPC mechanism, 40
shell
　adb, 197, 204
　init and, 238
　MirBSD Korn, 210
　running from Toolbox, 211–220
　sh's man page, 209–210
show commands target, adding, 134
Sidekick phone, 1
SIM card, connection to BP, 157
simulator, 117, 121
single module, building, 139
sleep(), 303
Slog class, 43
smd command, Toolbox, 220
Sockets method, 50, 51–52
SoCs (System-on-Chips)
　about, 161–165
　connection to PMIC, 159
　in system architecture, 155–156, 160
　vendors for, 304–305
software compatibility testing requirements, 21
Software Defined Radio (SDR) devices, 157
Software Development Kit (SDK), Android
　building for Linux, 135–136
　building for Mac OS, 135–136
　building for Windows, 136
SQLite database, 3

StageFright
    GStreamer replacing, 4
    support for media formats through, 3
stagefright command, 292
staging tree, Linux
    Android and, 38
    Binder driver merged into, 40
startActivity() method, 70
startViaZygote() method, 70
start_kernel() function, 74
state-altering adb commands, 204–207
storage, component in SoC, 164
Sun Java Virtual Machine (VM)
    about, 3
    building Android on, 98
Sun Microsystems, acquisiton by Oracle, 15–16
Surface Flinger
    as first system service, 254
    reliance on ashmem, 40
svc command, 285–286
Swetland, Brian, 11, 178
switch-case, 58
switching connection type, adb, 205–206
Sysfs entries method, 50
syslog, 43
system architecture, 155–160
/system directory, 53, 134, 180–182
System Server
    about, 64
    in framework, 254–257
    Java code in, 68
    system services running within, 323
    vs. services running in services component, 27, 40
system services
    about, 63–70
    calling, 330–333
    implementing new, 324–327
    in framework, 254–257
    starting, 333–334
    to support hardware types, 323
system startup, 73–77
System V IPC mechanisms
    ashmem code and, 40
    available in glibc, 310
System-on-Chips (SoCs)
    about, 161–165
    connection to PMIC, 159
    in system architecture, 155–156, 160

vendors for, 304–305
System.getProperty(), 252
System.setProperty(), 252
/system/bin/system_server, 255
SystemClock class, 41

T

TARGET_ARCH_VARIANT variable, 118
TARGET_BUILD_TYPE variable, 116
TARGET_BUILD_VARIANT variable, 116, 205
TARGET_DEVICE variable, 118
TARGET_PRODUCT variable, 115–116, 118
TARGET_SHELL variable, 211
TARGET_TOOLS_PREFIX variable, 116
telephony support
    about, 158
    GSM, 3
telnet, using emulator console to connect to, 206
templates, module build, 127–132
testing apps on devices, websites for, 31
threads and processes, 29
Toolbox
    about, 58
    commands in Jelly Bean, 316–317
    running shell from, 211–220
    vs. Busybox, 58, 319–321
Torvalds, Linus, 11, 35
tree
    building out of, 140–142
    building recursively in-, 142–143

U

Ubuntu 10.04, 64-bit, as Google supported build environment, 22
udev events, 58, 79
ueventd, 243–245
UI/Application Exerciser Monkey (website), 288
UID (Unique Identifier), 30
Unix domain sockets, 223, 303
URIs (Universal Resource Identifiers), access enforcement using, 31
USB
    connecting to target, 169–170
    controller in SoC, 164
    host, 160
User Interface (UI), 30
user-facing system, Android as, 159
user-space environment, native, 52–60

Userspace low memory killer daemon, 38
USE_CCACHE variable, 123
Using the Android Emulator (posting), 193
utilities and commands
    command line, 219–228
    framework, 266–297
        am command, 279–282
        bmgr, 290–292
        Dalvik Utilities, 293–297
        dumpstate, 270–276
        dumpsys, 268–270
        ime command, 286–287
        input command, 287–288
        monkey, 288–290
        pm command, 282–285
        rawbu, 276–277
        service, 266–268
        stagefright command, 292
        svc command, 285–286

V

vet default properties, 148
vi command, 314
virtual filesystems, 178
virtual machine, 61–63
"Virtual Machine Showdown" (Shi et al.), 62
virtual machines, 98
    (see also Dalvik Virtual Machine)
    building Android on, 98
virtual memory vs. physical memory, 165–167
VM (Sun Java Virtual Machine)
    about, 3
    building Android on, 98
vold, 299–301, 302

W

WakeLock mechanism, 36–37, 68
web browsers
    connecting to port 80 on Android device using, 314–317
    WebKit-based, 3
WebKit-based browser, 3
WebView class, using WebKit engine, 3
Welte, Harald, "Anatomy of contemporary GSM cellphone hardware", 158
WiFi, support for, 3
Windows, building SDK for, 136
wipe command, Toolbox, 217
wireless connection technologies, support for, 3
wireless radio technologies, 158
write() function, 327, 330

X

X Window System, 47

Y

YAFFS2-formatted NAND flash partitions, 177
Yaghmour, Karim, Building Embedded Linux Systems, xii, 96, 140, 309
Yocto, 308, 322

Z

ZIP files, uncompressed, 259
Zores, Benjamin, 161
Zygote daemon, 74–77, 251–254

About the Author

Karim J. Yaghmour is part serial entrepreneur, part unrepentant geek. He is the CEO of Opersys Inc., a company providing development and training services on embedded Android and embedded Linux, and is most widely known for having authored O'Reilly's *Building Embedded Linux Systems*—which sold tens of thousands of copies worldwide and has been translated into several different languages.

Karim pioneered in the world of Linux tracing by introducing the Linux Trace Toolkit (LTT) in the late '90s. He continued maintaining LTT through 2005 and was joined in this effort by developers from several companies, including IBM, HP, and Intel. LTT users have included Google, IBM, HP, Oracle, Alcatel, Nortel, Ericsson, Qualcomm, NASA, Boeing, Airbus, Sony, Samsung, NEC, Fujitsu, SGI, RedHat, Thales, Oerlikon, Bull, Motorola, ARM, and ST Micro. Other contributions include relayfs and Adeos.

Karim has presented and published with a number of peer-reviewed scientific and industry conferences, magazines, and online publications—including Usenix, the Linux Kernel Summit, the Embedded Linux Conference, the Android Builders Summit, AnDevCon, the Embedded Systems Conference, the Ottawa Linux Symposium, Linux Journal, the O'Reilly Network, and the Real-Time Linux Workshop.

Colophon

The animal on the cover of *Embedded Android* is a Moorish wall gecko (*Tarentola mauritanica*), which is a species of gecko native to the Western Mediterranean region of Europe and North Africa and also found in North America and Asia. It is commonly observed on walls in urban environments, mainly in warm coastal areas, though it can spread inland, especially in Spain. The adoption of this species as a pet has led to populations becoming established in Florida and elsewhere.

The Moorish wall gecko is mainly nocturnal or crepuscular, but it is also active during the day, especially on sunny days at the end of the winter. It lays two almost-spherical eggs twice a year around April and June. After 4 months, little *salamanquesas* of less than 5 centimeters in length are born. They are slow to mature, taking 4 to 5 years in captivity.

Adults can measure up to 15 centimeters, including the tail. They have a robust body and flat head and their tubercules are enlarged, which give the species a spiny, armored appearance. They are brownish gray or brown with darker or lighter spots; these colors change in intensity according to the light.

The cover image is from Heck's Nature & Science. The cover font is Adobe ITC Garamond. The text font is Adobe Minion Pro; the heading font is Adobe Myriad Condensed; and the code font is Dalton Maag's Ubuntu Mono.

Get even more for your money.

Join the O'Reilly Community, and register the O'Reilly books you own. It's free, and you'll get:

- $4.99 ebook upgrade offer
- 40% upgrade offer on O'Reilly print books
- Membership discounts on books and events
- Free lifetime updates to ebooks and videos
- Multiple ebook formats, DRM FREE
- Participation in the O'Reilly community
- Newsletters
- Account management
- 100% Satisfaction Guarantee

Signing up is easy:

1. Go to: oreilly.com/go/register
2. Create an O'Reilly login.
3. Provide your address.
4. Register your books.

Note: English-language books only

To order books online:
oreilly.com/store

For questions about products or an order:
orders@oreilly.com

To sign up to get topic-specific email announcements and/or news about upcoming books, conferences, special offers, and new technologies:
elists@oreilly.com

For technical questions about book content:
booktech@oreilly.com

To submit new book proposals to our editors:
proposals@oreilly.com

O'Reilly books are available in multiple DRM-free ebook formats. For more information:
oreilly.com/ebooks

Spreading the knowledge of innovators · oreilly.com